The Geometrical Optics Workbook

The Geometrical Optics Workbook

David S. Loshin, OD, PhD

University of Houston
College of Optometry

Butterworth-Heinemann
An Imprint of Elsevier
Boston London Oxford Singapore Sydney Toronto Wellington

Recognizing the importance of preserving what has been written, it is the policy of Butterworth-Heinemann to have the books it publishes printed on acid-free paper, and we exert our best efforts to that end.

Library of Congress Cataloging-in-Publication Data

Loshin, David S.
The geometrical optics workbook / David S. Loshin.
p. cm.
Includes index,
ISBN-13: 978-0-7506-9052-2 ISBN-10: 0-7506-9052-6 (pbk.)
1. Optics, Geometrical--Problems, exercises, etc. 2. Optometry--Problems, exercises, etc. I. Title
QC382.2.L67 1991
535'.32--dc20 91-3279
CIP

British Library Cataloguing in Publication Data
Loshin, David S.
The geometrical optics workbook.
1. Geometry. Optics
I. Title
535.32

Butterworth–Heinemann
313 Washington Street
Newton, MA 02158–1626

ISBN-13: 978-0-7506-9052-2
ISBN-10: 0-7506-9052-6

15 14

Printed in the United States of America

This book is dedicated to my wife

Mary

and my children

Aaron and Ryan.

Without their understanding and help,

this book would never have been completed.

I thank them.

Contents

Preface

After teaching optics to optometry students for twelve years, I felt as though I had developed some insight into the difficulties students face when attempting to solve optics problems. They often don't know what is being asked or even how to start the problem-solving process. In this workbook, I have attempted to simplify the process by systematically listing what is known, what is unknown, and the equations and theory necessary for solving each example. The problems are then solved step-by-step to guide you through the maze. The best method for improving your ability to solve problems is to practice the procedures. For additional practice, supplemental problems and answers are provided at the end of most chapters.

In each section, a brief explanation of a topic is followed by sample problems and solutions. It is suggested that you attempt these problems before reading the entire solution. Then your answers and procedures can easily be checked. For more details on the content of a topic, read pertinent sections in one or more of these suggested optics texts:

Freeman MH. *Optics*, 10th edition. London: Butterworth-Heinemann, 1990.

Fry GA. *Geometrical Optics*, Philadelphia: Chilton, 1969.

Jenkins FA, and HE White. *Fundamentals of Optics*, 4th edition. New York: McGraw-Hill, 1976.

Keating MP. *Geometric, Physical and Visual Optics*, Boston: Butterworths, 1988.

Meyer-Arendt JR. *Introduction to Classical and Modern Optics*, 2nd edition. Englewood Cliffs, NJ: Prentice Hall, 1984.

Although I have attempted to make this workbook generic, symbols in every text vary slightly. If you find this confusing, simply cross out the symbols in the workbook and write in the ones used in your text. Although every effort was made to incorporate reviewers' suggestions, I am sure that some instructors may be unhappy with the simplification of the subject. I do, however, feel that this type of book is long overdue as it is consistently requested by my students. The workbook does not take the place of a textbook, but rather acts as a supplement. I hope it will help many students who think passing optics is an impossible task.

Acknowledgments

I would like to thank the students who have helped me with this workbook: Laura Branstetter, Elizabeth Stuedle, Mary Nguyen, and especially Teresa Erickson, who was dedicated to the completion. I would also like to thank the 1300 or so students in my classes who have taught me a great deal about teaching optics.

My mentor, Dr. Glenn Fry, had a significant influence on my life as a student and as a person. I thank him for his guidance and advice. I have, with his permission, included some modified diagrams and information from his textbook on optics.

I am grateful to Barbara Murphy at Butterworth-Heinemann, who was patient and helpful in completing this workbook.

Last, I would like to thank my mother and father, Nathan and Jessie Loshin, who instilled in me high standards and the desire to learn.

Introduction

This workbook is intended for students of optometry, physiological optics, and ophthalmology and for individuals studying for the optics portion of state or national board examinations. It should not take the place of a good textbook but rather should supplement it with sample problems and examples. Each chapter has a brief explanation of terms, equations, and methods followed by problems and solutions that demonstrate these concepts. These examples are designed to help you solve problems by showing step-by-step solutions. It is suggested that you work along with the example rather than simply read it and think you understand. Cover the solution and see if you can reach the same answer through a similar process. Although your answer may be the same, you must be careful you used the correct procedure. Your procedure may not follow the example exactly, but should be similar in concept. Remember that you must learn the problem-solving process by actively doing and not by passively observing. At the end of most chapters are unsolved problems with solutions for additional practice.

The major problem many individuals have with optics stems from a weak background in mathematics, especially algebra and trigonometry. A review of general math concepts may be in order before begining the optics course. This text does not cover these concepts; it is recommended that you obtain the necessary references and review deficient areas.

Some equipment can help make problem solving easier. A calculator with trigonometric and exponental functions is a necessity. Many state and national board examinations do not permit the use of calculators with a continuous memory. I suggest you purchase a calculator that is acceptable to both your instructor and the board examiners. Use the same calculator to solve homework problems and examination questions. Become familiar with the keys and functions. This will make you more efficient with your time, and fewer calculator errors should occur. A ruler, pencil, protractor, compass, and eraser may come in handy for certain problems.

In problem solving, some tricks may assist you in finding a valid solution. One scheme is to be sure to identify what is being asked by clearly laying out what is known, what is to be found, and what equations and concepts are necessary to calculate the solution. This can be demonstrated with a simple example:

A corvette driven by a student late for an optics lecture, cruises at 80 miles per hour. If the college is 20 miles from his current position, how long will it take to reach his destination? Assume he was not stopped by traffic lights or the police.

Known	***Unknown***	***Equations/Concepts***
Velocity of car: $v = 80$ mph	Time to reach school	Distance = rate x time or $d = rt$
Distance traveled (20 miles)		Change form of equation to $t = d/r$

Substitute rate and distance traveled in the equation:

$$t = \frac{d}{r} = \frac{20 \text{ miles}}{80 \text{ mph}} = 0.25 \text{ hour}$$

Convert to minutes using 60 minutes/hour:

t (minutes) = (0.25 hour) (60 minutes/hour) = 15 minutes

There are four fundamental concepts in problem solving. The first is dimensional analysis. The units for each value must be carefully noted. In this book, the units are written in the known and unknown columns. If care is taken and the mathematical operations are followed, the final answer will have the correct units. For the previous example, the dimensional analysis is:

When dividing by a fraction, remember to you invert the denominator and multiply, as shown below:

$$\frac{(\text{miles})}{(\text{miles / hour})} = (\text{miles})(\text{hour / miles}) = \text{hour}$$

Converting units also uses dimensional analysis as illustrated in the example:

(hour) (60 minutes/hour) = minutes The hour units cancel.

Remember: For a correct solution, the units are as important as the magnitude.

The second important concept is the number of significant figures in the solution. The answer should have the same number of significant figures as the numbers used to generate the solution. The final solutions in this book will round up answers when the next digit is a 5 or greater and truncate when less than 5.

Ray tracing is a graphical procedure for locating images and objects. Chapter 9 should be consulted for every curved refracting or reflecting element or system. I have included many examples of image-object relationships. The ray tracing process will help you develop an understanding of the location and type of images that may be formed with an imaging element.

The final concept is use of a sign convention. Different values may have a positive or negative sign associated with the magnitude. You must adhere to the sign convention. Different sign conventions will result in different final answers, which could be incorrect. The sign convention used here is one that is commonly used in optics texts.

The Geometrical Optics Workbook

Chapter 1

Introduction to Geometrical Optics

Key Definitions and Concepts

In this chapter, general concepts about **optics** and **light** are discussed. Without light, there would be no optics or vision. Light or, more properly stated, **visible light** usually refers to the small portion of the **electromagnetic spectrum** (Figure 1-1) that stimulates the human visual system. Many sources of light should be familiar: natural (sunlight, stars, static electricity sparks, fire) and artificial (tungsten light bulbs, gas-discharge street lights, explosions). The spectral output, size of the source, or both should be specified (or at least assumed) for most optical applications and problems. Terms that may be used to describe a source are defined here.

ELECTRO-MAGNETIC ENERGY	WAVELENGTH (METERS)	FREQUENCY (HERTZ)
Cosmic rays	10^{-15}	10^{23}
Gamma rays	10^{-13}	10^{21}
X-Rays	10^{-10}	10^{18}
Ultraviolet	10^{-9}	10^{17}
Visible – blue →	3.5×10^{-7}	
Visible – yellow		10^{14}
Visible – red →	7.5×10^{-7}	
Infrared	10^{-5}	10^{13}
Microwave	10^{-3}	10^{11}
FM radio, Television →	10^{0}	10^{8}
AM radio	10^{2}	10^{6}

Figure 1-1. The electromagnetic spectrum.

Source - any object that emits electromagnetic radiation, specifically for our purposes, in the visible region.

Point source - a source that is either infinitely small or sufficiently far away that it acts infinitely small (a star, for example).

Extended source - a source that has a measurable area. Most artificial sources (bulbs, luminous panels, television screens) are extended sources. At times, an extended source may be considered to be a point source if the distance from the source is large compared to the area of the source.

Monochromatic source - a source that emits electromagnetic radiation of a single wavelength or frequency. A laser is a unique source in that it emits true monochromatic light. Other sources, such as a sodium gas source, emit radiation in a narrow region of the EM spectrum and are often considered to be monochromatic. A filter may be used to limit the spectral output of a source to approximate monochromatic light.

Polychromatic source - a source that emits radiation of several wavelengths or frequencies. White light is considered to be a polychromatic source because it emits many of the wavelengths in the visible region of the EM spectrum.

Wave Motion

The propagation of light in space may be represented by two different concepts: **wave motion** and **wavefronts**. Wave motion is used to explain many aspects of **physical optics**, such as **interference** and **diffraction**. As seen in Figure 1-2, wave propagation can be represented as a sine wave that varies in time and space. Electromagnetic energy travels as **transverse waves**, in which components of the wave vibrate perpendicular to the direction of propagation. If you have ever been canoeing on a quiet lake when a powerboat cruised by, you have experienced somewhat similar wave motion. Your canoe moved up and down with the wake, and it also moved in the direction the wave was moving: away from the powerboat. This simple explanation of wave motion is expanded in the discussion of waves in physical optics in other texts. For now, terms that define the electromagnetic wave are presented.

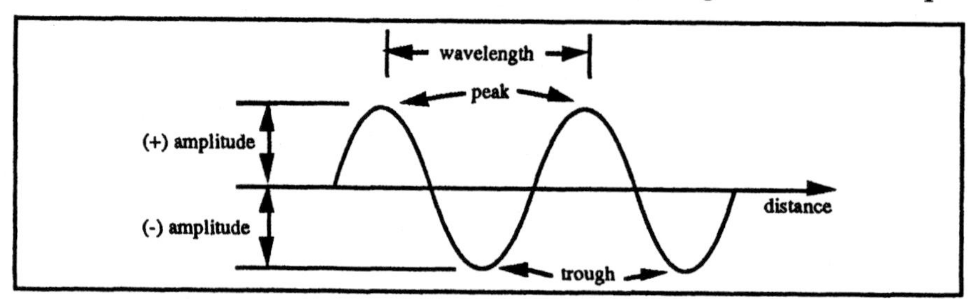

Figure 1-2. Propogation of electromagnetic radiation represented as a sine wave.

Wavelength (λ, lambda; meters, m; nanometers, nm) - the distance from peak to peak, trough to trough, or from any other repeatable position to the same position. (See Figure 1-2.) The visible region of the electromagnetic spectrum consists of waves with wavelengths ranging from 3.50×10^{-7} m (blue) to 7.50×10^{-7} m (red) or 350 nm to 750 nm (1 nanometer (nm) = 10^{-9} m).

Cycle (c) - the completion of a regular periodic event. A wave undergoes a complete cycle before it repeats itself. For example in Figure 1-2, from one peak to the next represents one cycle.

Frequency (f, N; c/s, hertz, Hz) - the number of cycles of a wave that pass a reference position in a certain period of time. The frequency of a wave is constant for all media.

Period (T, P; s/c) -the inverse of the frequency or $T = 1/f$.

Velocity (v; m/s) - the rate of travel or the distance the wave will travel in a certain period of time. In a vacuum, electromagnetic waves travel at 3×10^8 meters per second (m/s). For all practical purposes, light has the same velocity in air as in a vacuum. In most other media, wave velocity is slower than in a vacuum.

Amplitude (A, a) - the maximum displacement of the vibrations of the wave. As shown in Figure 1-2, the amplitude may be positive (peak) or negative (trough), although in most cases the amplitude is considered by magnitude alone.

Velocity, frequency, and wavelength are related by this equation:

$$v = f\lambda \qquad (1\text{-}1)$$

Components of the EM spectrum can be classified by either the frequency or the wavelength (Figure 1-1). Wavelength is usually used to describe the visible spectrum. For example, blue light has a wavelength of 350 nm; yellow-green light 550 nm; and red light 700 nm. The frequency is less commonly used; however, in certain applications such as the calculations of scatter or energy, it becomes an important measure of the EM spectrum.

Example 1-a

The wavelength of a sodium (Na) source in a vacuum is 589 nm. What is the frequency of the Na source in a vacuum?

Known	***Unknown***	***Equations/Concepts***
Wavelength: $\lambda = 589$ nm	Frequency	(1-1) $v = f\lambda$ or $f = v/\lambda$
Velocity of light in a vacuum: 3×10^8 m/s		

Substituting the known values into Equation 1-1 yields

$$f = \frac{3 \times 10^8 \text{ m/s}}{589 \text{ nm}}$$

WAIT A MINUTE -- THE UNITS ARE NOT COMPATIBLE.

The values must be converted to the same unit of distance, either nm or m. This should be done in the ***Known*** column before the problem is attempted. Here are the conversions to remember:

1 cm $= 10^{-2}$ m
1 mm $= 10^{-3}$ m
1 micron $= 10^{-6}$ m
1 nanometer = 1 millimicron $= 10^{-9}$ m
1 Angstrom $= 10^{-10}$ m

Convert the nm units to m by multiplying the wavelength by 10^{-9} m, and the wavelength becomes:

$$589 \times 10^{-9} \text{ m}$$

Substituting into the Equation 1-1 yields

$$f = \frac{3 \times 10^8 \text{ m/s}}{589 \times 10^{-9} \text{ m}} = 5.09 \times 10^{14} \text{ c/s or Hz}$$

Dimensionally the resulting units are actually reciprocal seconds (1/sec), but one must know that the units of frequency are given in cycles per second or hertz.

Example 1-b

Light from the same Na source as in *Example 1-a* enters a material where it travels at 2×10^8 m/s. What is the wavelength of the light in the material?

Known	***Unknown***	***Equations/Concepts***
Wavelength in a vacuum: $\lambda = 589$ nm	Wavelength in media	(1-1) $v = f\lambda$ or $\lambda = v/f$
Velocity $v = 2 \times 10^8$ m/s		Wavelength decreases in media other than a vacuum or air
Frequency from *Example 1-a*: $f = 5.09 \times 10^{14}$ Hz		Frequency is constant in all media

Using the frequency and new velocity in the medium, solve for the wavelength:

$$\lambda = \frac{2\text{x}10^8 \text{ m/s}}{5.09\text{x}10^{14} \text{ c/s}} = 392\text{x}10^{-9} \text{ m} = 392 \text{ nm}$$

The wavelength decreased from 589 nm to 392 nm.

Example 1-c

A profile of a sine wave is diagrammed in Figure 1-3.

a. Determine the wavelength and amplitude of the wave.

b. How many cycles are displayed in the diagram?

c. What is the velocity if the wave travels the distance shown in 2×10^{-15} sec.

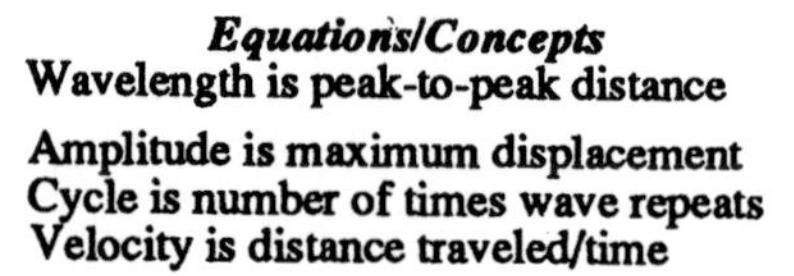

Known	***Unknown***	***Equations/Concepts***
See Figure 1-3 for distance traveled	Wavelength	Wavelength is peak-to-peak distance
Time to travel distance: $t = 2 \times 10^{-15}$ sec	Amplitude	Amplitude is maximum displacement
	Velocity	Cycle is number of times wave repeats
		Velocity is distance traveled/time

a. In Figure 1-3, the x-axis is in units of nanometers; the first peak is located at 25 nm and the next peak is at 125 nm. The distance between the peaks is the wavelength (λ) and thus

$$\lambda = 125 \text{ nm} - 25 \text{ nm} = 100 \text{ nm}$$

The wavelength could be determined with any repeatable portion of the wave (i.e., start at zero on the x-axis and find the position on the x-axis that represents the same position on the wave). In this case, the position would be at 100 nm.

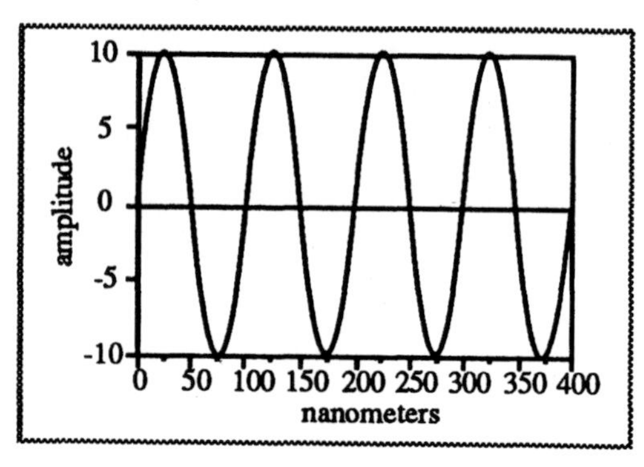

Figure 1-3. *Example 1-c*

The amplitude (a) or maximum displacement of the wave is seen in the figure; the wave goes from 0 to +10 units and from 0 to –10 units. The maximum displacement of the wave is therefore 10 units. Note that amplitude is usually expressed as a positive number.

b. The number of cycles is found by counting how many times the wave repeats itself: four cycles are represented.

c. Velocity is the distance traveled per unit of time. From the diagram four wavelengths are represented in 400 nm. The wave thus travels 400 nm in 2×10^{-15} sec, and by definition:

$$v = \frac{\text{distance}}{\text{time}} = \frac{400 \text{ nm}}{2 \times 10^{-15} \text{ sec}} = 2 \times 10^{17} \text{ nm/sec}$$

Convert to units of m/s:

$$(2 \times 10^{17} \text{ nm/sec})(10^{-9} \text{m/nm}) = 2 \times 10^{8} \text{ m/s}$$

Example 1-d

Red light, with a wavelength of 600 nm in a vacuum, enters a block of glass whereupon the wavelength reduces to 380 nm. What is the velocity and period of the light in the block of glass?

Known	***Unknown***	***Equations/Concepts***
Wavelength in vacuum: $\lambda = 600$ nm	Velocity in glass	Frequency is constant in all media
Wavelength in glass: $\lambda = 380$ nm	Frequency	(1-1) $v = f\lambda$
Velocity in vacuum: $v = 3 \times 10^{8}$ m/s	Period in glass	Period is recriprocal of frequency

Think about the relationship between the frequency, wavelength, and velocity in a vacuum. The velocity and wavelength are known, so solve for the frequency in the vacuum:

$$f = \frac{v}{\lambda} = \frac{3 \times 10^{8} \text{ m/s}}{600 \times 10^{-9} \text{ m}} = 5.00 \times 10^{14} \text{ c/s or Hz}$$

OPTICS WORKBOOK : Ch 1

Period : $T = 1/f$

$v = f\lambda$

ex 1-a $3\times10^{8} = f\,(589)$ $= 589\times10^{-9}$ m , $f = 5.09\times10^{14}$ Hz

ex 2-a $2\times10^{8} = 5.09\times10^{14}\,(\lambda)$, $\lambda = 392$ nm

ex 1-c $\lambda = 100$, $A = 10$, 4 cycles, $\frac{400}{2\times10^{-15}} = 2\times10^{8}$ m/s

ex 1-d $f = \frac{v}{\lambda}$, $\frac{3\times10^{8}}{600\times10^{-9}} = 5\times10^{14}$

$v = f\lambda = 5\times10^{14}$ Hz $(380\times10^{-9}$ m/s$) = 1.9\times10^{8}$ m/s

$T = \frac{1}{5\times10^{14}\text{ Hz}} = 2\times10^{-15}$ sec

Aug 30, 2010

G, P, V Optics - Ch 1

OPTICS, LIGHT & VISION

geometric optics: image forming properties of lenses, mirrors, and prisms.

physical optics: physical character and behavior of light and its interaction with matter (wave, quantum and Fourier optics)

* ability of waves to navigate corners: diffraction, depends on wavelength, longer λ = more bend

* Color is distinguished by 3 types of cones

λ 530 = green
λ 650 = red
λ 400 = blue
} mono chromatic

long → longest → short

* strong absorbers (metals) are strong reflectors

* weak " (water, glass) weak "

$\theta_i = \theta_s$ specular reflection: occurs $\bar{c}$ smooth surface.

diffuse reflection: rough surface

$n = c/v$ $v_m = c/n$

Law of refraction: $n_1 \sin\theta_1 = n_2 \sin\theta_2$ n = index refraction

PROBLEMS

1. $n = 3\times10^8 / 1.8\times10^8 = 1.67$
2. $1.33 = 3\times10^8 / x = 2.25\times10^8$
3. $1 \sin(52°) = 1.523 \sin(\theta)$
   $0.788 = \quad = 31.1°$, toward
4. $1(3.5) = 1.66(\theta)$, 2.108
   $1\sin(35) = 1.66 \sin(\theta)$, 2.103
5. $1.33(4) = 1(\theta)$, 5.32° away
6. $1.586(\sin 37° = 1 \sin(\theta)$, 72.6°
7. yes
8.
9.
10. long, blue 11. short, red

Now, solve for the velocity in the block of glass using Equation 1-1. The frequency and wavelength are known.

$$v = f\lambda = (5 \times 10^{14}\ \mathrm{Hz})(380 \times 10^{-9}\ \mathrm{m/s}) = 1.9 \times 10^{8}\ \mathrm{m/s}$$

The period in the glass is then the recriprocal of the above frequency:

$$T = \frac{1}{5\mathrm{x}10^{14}\,\mathrm{Hz}} = 2\mathrm{x}10^{-15}\,\mathrm{s/c} \text{ or } 2\mathrm{x}10^{-15}\,\mathrm{sec}$$

Wavefronts

In geometrical optics, knowledge about the exact nature of wave propagation is not required to explain imaging phenomena; however, there is a definite relationship between **waves, wavefronts,** and **rays.** Figure 1-4 shows a **point source** emitting waves in all directions. Consider only the waves traveling to the right. If a line is drawn connecting the same position (first peak, for example) on all the waves, the result would be an arc with its center at the source. This arc represents one **wavefront.**

Another wavefront may be constructed in many ways. One may repeat the described process by simply connecting another similar position on each wave. *What if the waves were not provided?* Because the wavefronts represent equal distances from the source (as indicated by the arc), it would be a simple matter to draw with a compass any arc with its center at the source. An infinite number of wavefronts could be drawn this way.

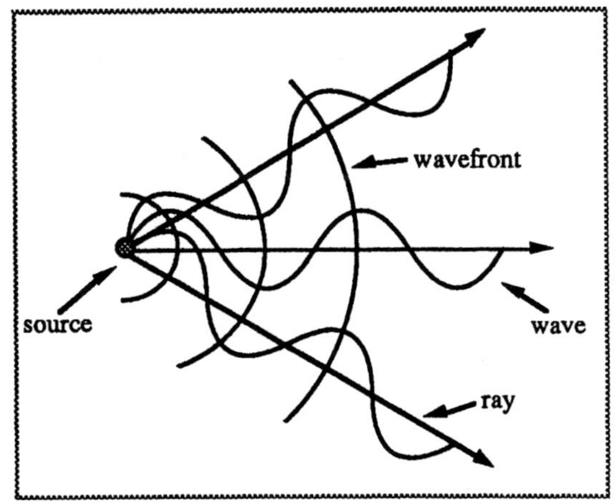

Figure 1-4. The relationship between waves, wavefronts, and rays.

What if the wavefront position were known but no source location was given? **Huygen's method for constructing a new wavefront** could then be used. This method assumes that each point on the original wavefront acts as a point source. Because, in general, point sources emit spherical wavefronts, each point on the original wavefront emits secondary wavefronts called a **wavelets.** These wavelets are arcs of the same radius. The arc that is tangent to these wavelets represents the new wavefront. (See *Example 1-e.*)

Example 1-e

a. Use Huygen's method to locate a new wavefront 1 cm from the wavefront [a] in Figure 1-5.

b. Verify the answer by locating the source and drawing the wavefront with a compass.

a. This problem illustrates Huygen's method of locating a new wavefront. Draw wavelets 1 cm in radius on wavefront [a]. The arc tangent to these wavelets is wavefront [b]. The diagram provided (Figure 1-5) shows this procedure. Try drawing your own diagram, and locate several wavefronts.

b. The easiest way to locate the source is to match the arc of wavefront [a] with a compass. The location of the pointed end of the compass will coincide with the source. Increase the radius of the compass by 1 cm to draw wavefront [b].

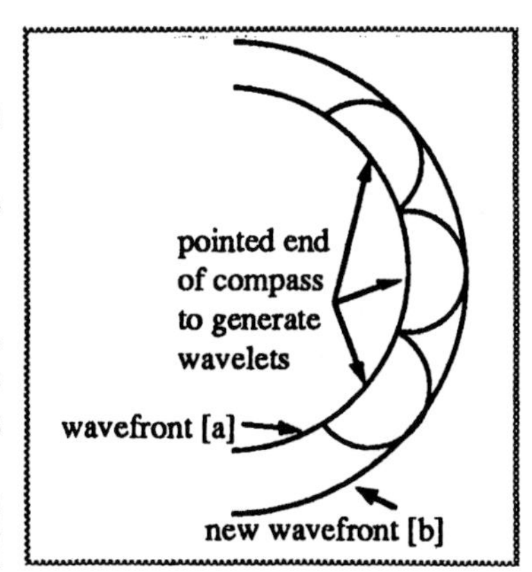

Figure 1-5. *Example 1-e*

Rays

Rays are used to represent the propagation of light in geometrical optics. Figure 1-4 illustrates the relationship between waves, wavefronts, and rays. The **Law of Malus** defines the relationship between rays and wavefronts: *Rays are perpendicular to wavefronts.* You should remember from geometry that a line passing through the center of a circle is perpendicular to any point on the circumference of the circle. Because in this context the source (object) or image represents the center of the wavefront, rays must pass through (or act as though they came from) a point on the source or image.

All rays emanating from a point source or a single point on an extended source may be grouped to form a **pencil** of rays. Depending upon the direction the rays travel within the pencil, pencils may be defined as **convergent, divergent**, or **parallel.**

Axis - in this context, a horizontal line that passes directly through the center of the source. All elements in our systems are centered about the axis. The relationship to the optical axis will be defined in Chapters 4 and 5.

Divergent pencil - rays leaving a point on a source (or an image) that travel away from each other and do not cross at any position. Thus the wavefronts expand and the radius of the wavefront increases with the distance from the source. As seen in Figure 1-6, this pencil may be generated from an on-axis or off-axis source. Remember that a divergent pencil is generated from a point .

Convergent pencil - rays that are aimed toward a single point on an image or object. This means that wavefronts contract (i.e., the radius decreases) as the distance to the point is reduced. Figure 1-6 illustrates on-axis and off-axis convergent pencils.

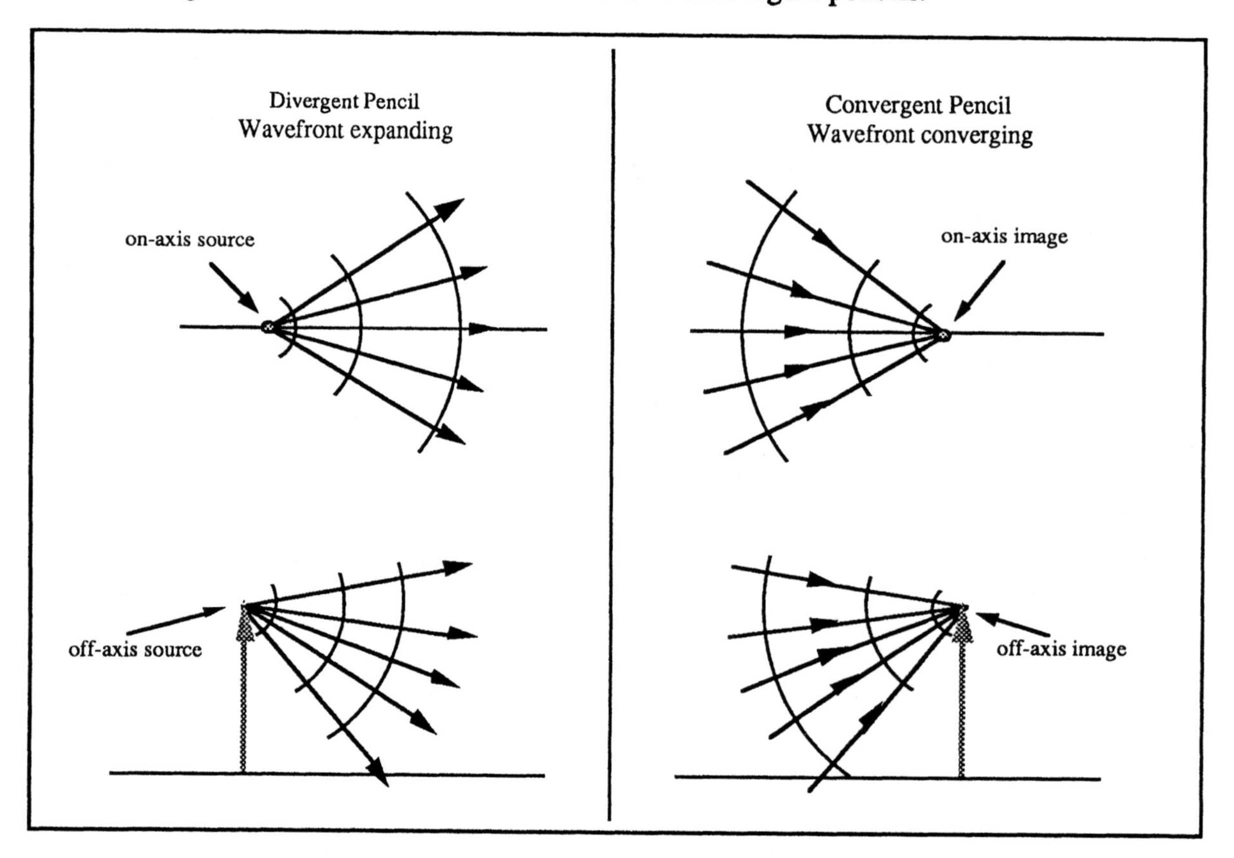

Figure 1-6. On-axis and off-axis divergent and convergent pencils. Note that the divergent rays are traveling away from each other and the radius of the wavefront is increasing. The convergent rays are aimed toward an image point, and the wavefront radius is decreasing.

Parallel pencil - rays emitted by a source at an infinite position or aimed at an image at an infinite position. The resulting wavefronts have extremely large radii that are approximated by straight parallel lines (Figure 1-7). The rays perpendicular to the wavefronts are also parallel to each other. Note that for infinite on-axis sources, the rays that make up the pencil are parallel to the axis, and for infinite off-axis sources, the rays are parallel to each other and oblique to the axis.

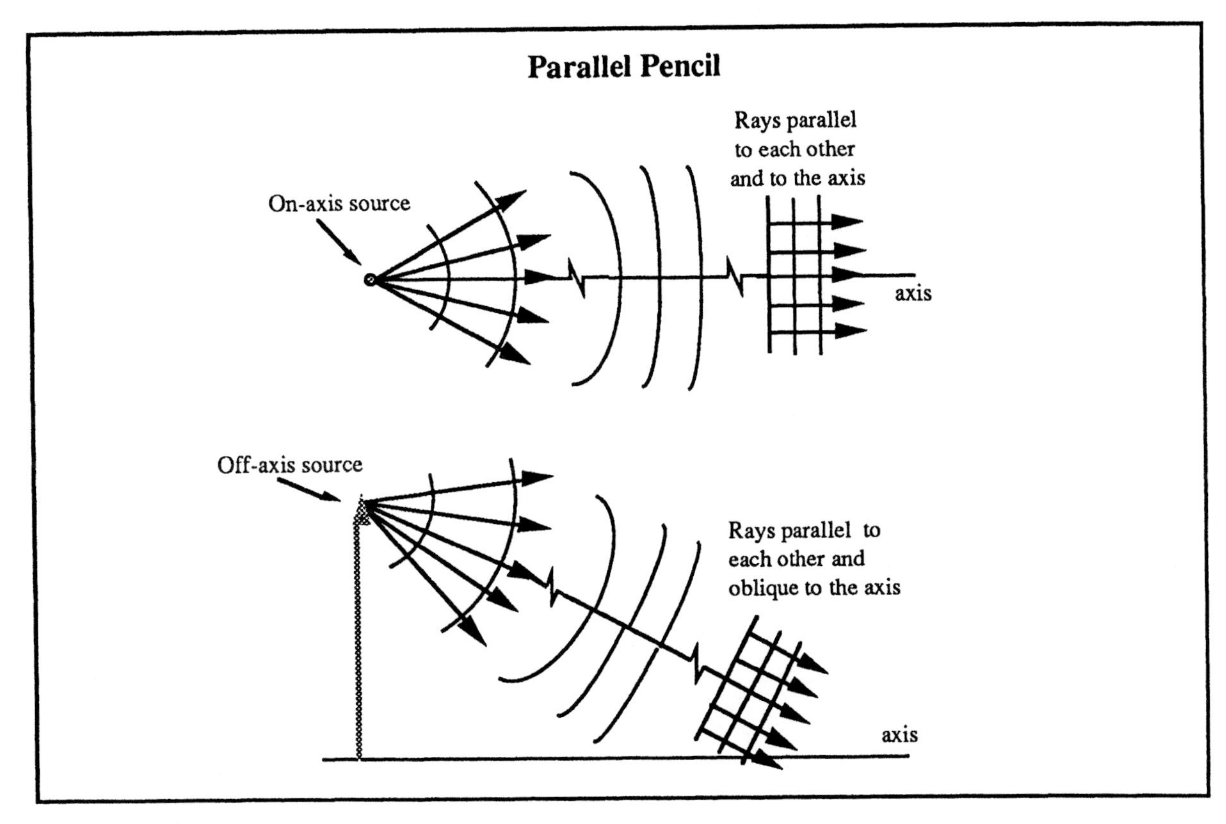

Figure 1-7. Parallel pencils from an infinite on-axis and off-axis sources. The wavefronts have an infinite radius, and the rays are parallel to each other.

Chief ray - the ray that passes through the center of a pencil. For an axial point source, the chief ray coincides with the axis. The chief ray is defined in more detail in other texts when apertures and stops are discussed.

Beam - the sum of all the pencils emanating from a source or aimed toward an image. In the case of a point source, there is only one pencil, and therefore the beam is made up of a single pencil. With extended sources, the beam consists of all pencils emanating from every point on the source. Divergent, convergent, and parallel refer only to pencils, not beams.

Real image point - an image point toward which all wavefronts and rays converge in a convergent pencil. The wavefronts decrease in radius as they approach this point. (See the convergent pencil in Figure 1-6 for an example of a real image point.)

Introduction to Sign Convention Rules

A sign convention is used in all optics texts in measuring distances in an optical system. The convention is essential for solving problems correctly. More details of the most commonly used sign convention (and the one used in this book) are discussed in Chapters 5 and 6. For now, some important rules are stated here and illustrated in Figure 1-8.

1. Assume that *light travels from left to right.* This means that sources must be to the left of the wavefronts, and divergent wavefronts expand toward the right.

2. Distances measured in the *same direction* (left to right) that light travels are *positive*. Distances measured in the *opposite direction* (right to left) of light travel are *negative.*

3. Radii of wavefronts are *measured from the wavefront to the source or the image.* The direction of measurement should be indicated on all diagrams by an arrow, as shown in Figure 1-8.

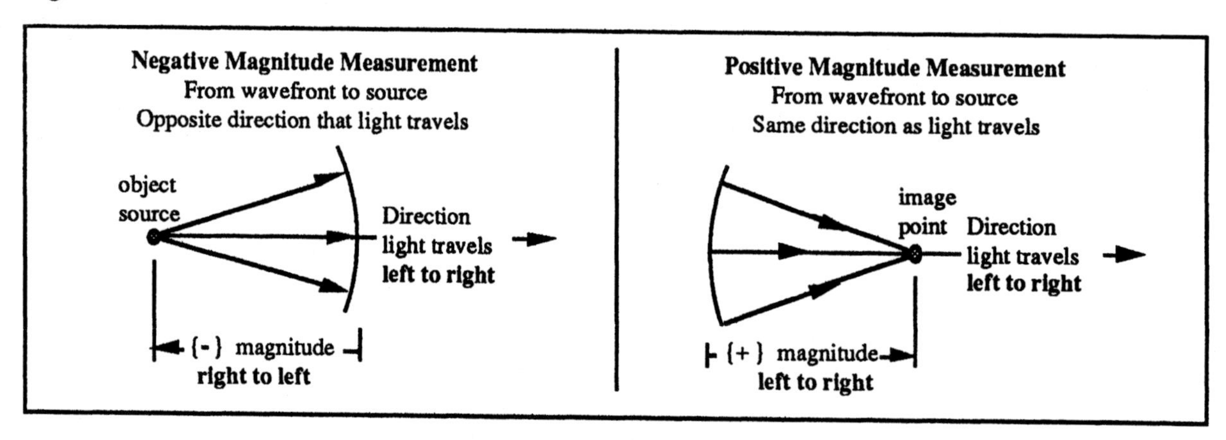

Figure 1-8. Illustration of the sign convention used in this workbook.

Example 1-f
In Figure 1-9, identify the type of wavefront and pencil for positions a, b, and c. Identify the rays in the system. Draw two more rays through the system. Measure and record (in meters) the radius of each wavefront. Be sure to use the proper sign.

This problem should be simple if you remember the definitions of wavefronts, pencils, and rays. First, look at wavefront (a). It originates from the object source. If you put the pointed end of a compass at the source, an infinite number of wavefronts could be generated between the source and the wavefront; the radii of the wavefronts increase as they travel away from the source. This makes the wavefront divergent. Since the rays travel away from the source and each other, as shown by the arrows, they would be defined as components of a divergent pencil.

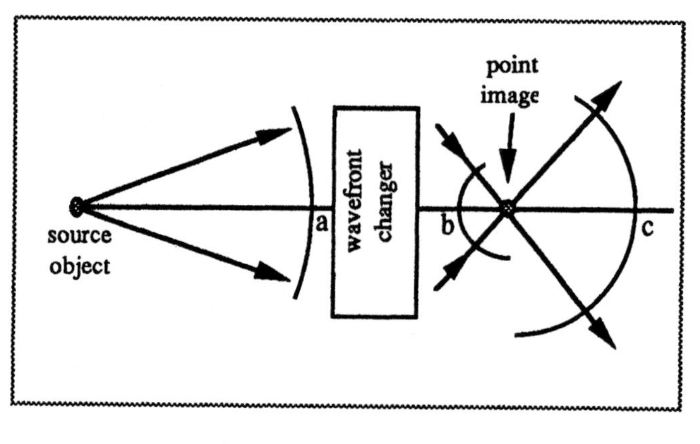

Figure 1-9. *Example 1-f.*

The wavefront changer appears to change the wavefront drastically. You'll find out very soon that a lens is the primary wavefront changer. The emerging wavefront is travels toward a point image. If the pointed end of a compass is placed on the image, radii of the wavefronts between (b) and the image point would be shown to decrease; thus (b) is a convergent wavefront, and the rays drawn with the arrows represent a convergent pencil (rays are headed toward each other). The radius of the wavefront becomes zero at the point image; then it starts to expand again. The point image acts as a source that generated the wavefronts that continue through the system. Try to use the explanation for wavefront (a) to show that the wavefront and pencil at (c) are divergent.

Because rays are perpendicular to wavefronts and must pass through the center of the generating arc, to draw two more rays, simply draw any line from the source or image to the respective wavefront.

Radii measurements must include the proper sign. Measurement of the magnitude of the radius of each wavefront should be straightforward. A ruler calibrated in centimeters show that for position (a), $r_a = 3$ cm = 0.03 m. Since the distance from wavefront (a) to the source is measured from right to left (opposite the direction of light), it has a negative value, or $r_a = -0.03$ m. The distance from the wavefront (b) to the image point is measured from left to right (the same direction that light travels) and has a positive value, or $r_b = +0.60$ cm = +0.006 m. The distance from the wavefront (c) to the image point (which now acts as a source) is measured from right to left, and therefore $r_c = -1.60$ cm = −0.016 m.

Example 1-g

In Figure 1-10a, rays are incident upon a wavefront changer, undergo a change, and then emerge traveling in a different direction. Assume a scale of 1 cm = 1 m for the diagram.

a. Draw the radius of the incident wavefront at the wavefront changer. Of what type are the incident wavefront and pencil?

b. Draw the emerging wavefront. Of what type are the emerging wavefront and pencil?

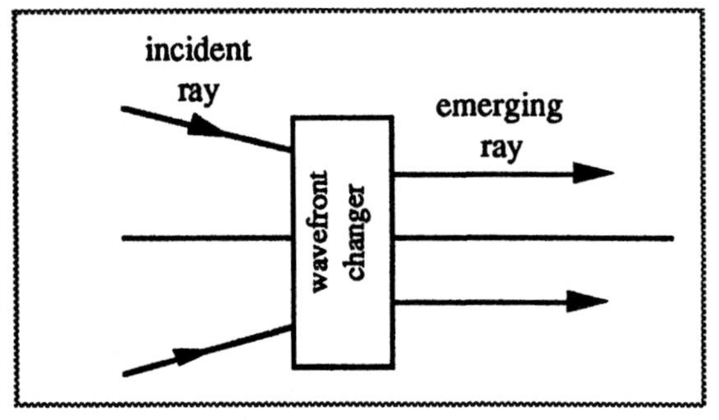

Figure 1-10a. *Example 1-g.*

a. To determine the radius of the incident wavefront, the location of the origin or source of the wave must be determined. Because the incident rays are directed toward and cross on the axis (and rays are perpendicular to wavefront), the origin of the wave is located where the rays cross. This is determined by extending the rays toward the axis (Figure 1-10b.) The distance from the front of the wavefront changer to the center of the wavefront (from left to right and positive) is the radius of the wavefront (measuring + 3.1 cm on the diagram, which, according to the scale, represents + 3.1 m). The wavefront radius decreases as it approaches the axial position and is therefore constricting. The rays are headed toward each other, so the pencil is convergent.

b. The emerging rays are parallel to each other and parallel to the axis. The emerging wavefront is perpendicular to the rays. The wavefront has an infinite radius, and the pencil is parallel.

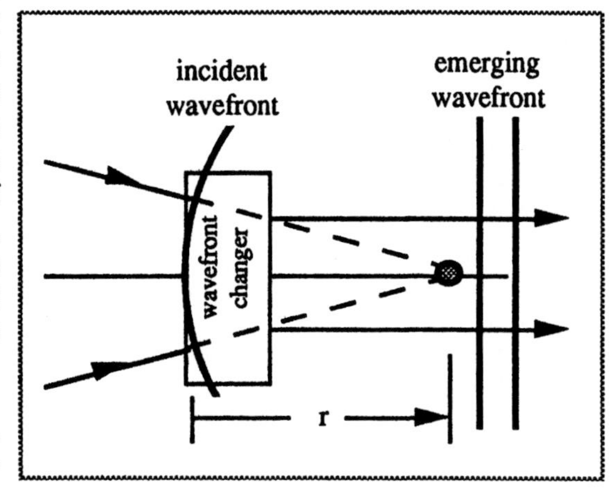

Figure 1-10b. Solution to *Example 1-g.*

It should be clear from *Example 1-g* that it is possible to determine the radius of the wavefront and the type of pencil simply by examining the incident and emerging rays. The wavefront is rarely drawn on the diagrams, and it should be noted that it is unnecessary except for clarification of a problem.

Vergence

Vergence is a very important concept that is used extensively in solving the image - object relationships in geometrical optics. Vergence (V) is defined as the *reciprocal of the radius of the wavefront.* When the radius is measured in meters, the vergence has a special unit called the **diopter (D)**. The reciprocal of the radius of the wavefront may also be called the *curvature* (R) and therefore when the curvature is expressed in diopters, it has the same value as the vergence, as shown in this equation:

$$V_D = R_D = \frac{1}{r_m} \tag{1-2}$$

It is interesting to note that lens power, including ophthalmic prescriptions, is also measured in diopters. Units of diopters are additive.

It is very important to remember the sign convention when calculating the vergence. *Divergent pencils have negative vergence, convergent pencils have positive vergence, and parallel pencils have zero vergence.*

Example 1-h

Determine the vergence for the wavefronts in *Examples 1-f* and *1-g*.

The vergence for each wavefront in *Example 1-d* is easily calculated because the radius of each wavefront is known:

$$V_a = \frac{1}{r_a} = \frac{1}{(-\ 0.03\ \text{m})} = -\ 33.33\ \text{D}$$

$$V_b = \frac{1}{r_b} = \frac{1}{(+\ 0.006\ \text{m})} = +\ 166.67\ \text{D}$$

$$V_c = \frac{1}{r_c} = \frac{1}{(-\ 0.016\ \text{m})} = -\ 62.50\ \text{D}$$

The vergence for the incident wavefront in *Example 1-g* may be calculated using the radius:

$$V = \frac{1}{r} = \frac{1}{(+\ 3.10\ \text{m})} = +\ 0.32\ \text{D}$$

For the emerging vergence, the radius is infinite, and the vergence is therefore calculated as a limit:

$$V = \lim_{r \to \infty} \frac{1}{r} = 0.00\ \text{D}$$

It is a good idea to remember that when the wavefront radius is infinite, the vergence is zero. This concept is extremely important when solving many optics problems.

Example 1-i

In Figure 1-11, rays emitted from an axial object hit a wavefront changer (lens) and are directed toward an axial image. The distances from the lens to the object and the lens to the image are labeled. Find the radius of the incident and emerging wavefronts at the lens, and calculate the vergence for the respective pencils. How much did the lens change the vergence?

Known	***Unknown***	***Equations/Concepts***
Distances from figure:	Vergence	(1-2) Vergence
Lens to object: – 20 cm		Sign convention must be followed
Lens to image: + 40 cm		Distances from lens to object and lens to image represent respective radii of wavefront

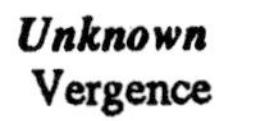

A divergent pencil is leaving the object because the rays are traveling away from each other and do not cross. The radius of the wavefront at the lens (not shown) is measured from the wavefront to the object; in this case, it is a negative number because it is opposite the direction of light, as indicated by the arrow on the diagram. Convert the radius to meters, and check for the proper sign:

$$r = -20 \text{ cm} = -0.20 \text{ m}$$

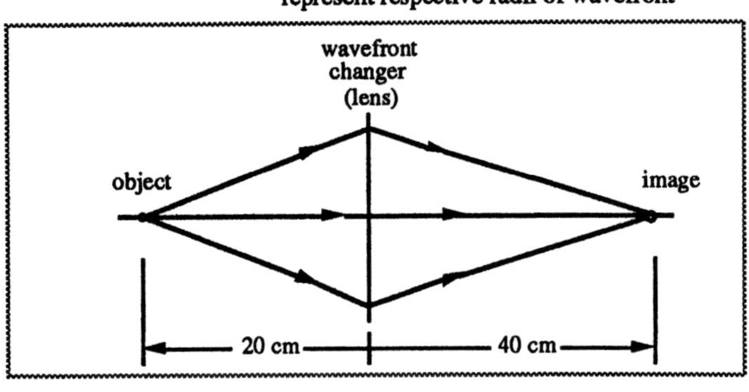

Figure 1-11. *Example 1-i*

The object vergence is calculated by

$$V = \frac{1}{r_m} = \frac{1}{-0.20 \text{ m}} = -5.00 \text{ D}$$

A convergent pencil is emerging from the lens toward an on-axis image point . The radius of this wavefront (not shown) is measured from the wavefront at the lens to the image point (which is in the same direction that light travels) and has a positive value.

$$r = +40 \text{ cm} = +0.40 \text{ m}$$

The image vergence is calculated by

$$V = \frac{1}{r_m} = \frac{1}{+0.40 \text{ m}} = +2.50 \text{ D}$$

An incident vergence of – 5.00 D changes to a vergence of + 2.50 D after passing through the lens. Mathematically, this can be written as

Incident (object) vergence + Lens power = Emergent (image) vergence

Solving for lens power this becomes

$$\text{Lens power} = \text{Image vergence} - \text{Object vergence} \tag{1-3}$$

Substituting the vergence into the formula gives

$$\text{Lens power} = +2.50 \text{ D} - (-5.00 \text{ D}) = +7.50 \text{ D}$$

This means a + 7.50 D lens is required to change the incident divergence of – 5.00 D to the emergent convergence of + 2.50 D. Later you will see how important this concept is for solving optics problems.

Example 1-j

A pencil of rays leaves a lens and converges towards a point 140 cm from the lens as shown in Figure 1-12. What is the vergence at a point 60 cm to the right of the lens?

Known	***Unknown***	***Equations/Concepts***
Pencil converges to focus 140cm to right of lens	Vergence 60 cm to right of lens?	(1-2) Vergence

Rays emerging from the lens are directed toward a position 140 cm to the right of the lens, and thus this position represents the origin of the wavefront. The distance from the point at which the vergence is to be determined to the center of the wavefront is given by

$$140\text{ cm} - 60\text{ cm} = 80\text{ cm}$$

The radius of the wavefront is + 80 cm because it is measured from left to right (in the direction light travels) and the vergence is

$$V = \frac{1}{r} = \frac{1}{+0.80} = +1.25\text{ D}$$

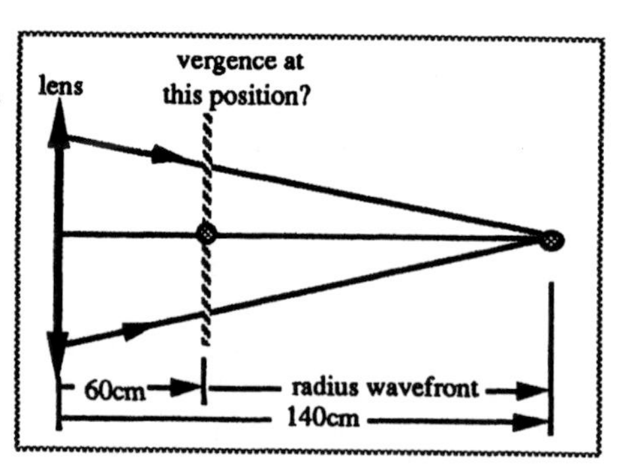

Figure 1-12. *Example 1-j.*

Example 1-k

Light is incident onto a wavefront changer with a vergence of + 3.00 D. The light emerges from the wavefront changer with a vergence of – 4.00 D. Where are the object and image located? What is the power of the wavefront changer?

Known	***Unknown***	***Equations/Concepts***
Incident pencil: +3.00D Emerging pencil: –4.00D	Object location Image location Power of wavefront changer	(1-2) Vergence Object and image located at origin of incident and emerging wavefronts

The incident light is a convergent pencil with a vergence of + 3.00 D. The origin of the wavefront at the wavefront changer may be calculated by solving Equation 1-2 for the radius:

$$r = \frac{1}{V} = \frac{1}{+3.00} = +0.33\text{m} = +33.33\text{cm}$$

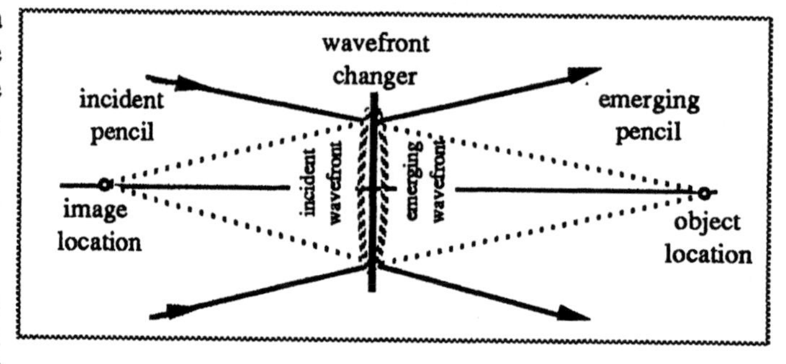

Figure 1-13. *Example 1-k.*

The object location is therefore 33.33 cm to the right of the wavefront changer because the radius is positive. This will be referred to as a **virtual object** because it is formed with converging light. The emerging pencil is divergent (as indicated by the negative vergence), and the origin of the wavefront can be located by solving for the radius:

$$r = \frac{1}{V} = \frac{1}{-4.00\text{ D}} = -0.25\text{ m} = -25.00\text{ cm}$$

The origin of the emerging wavefront is the image; the negative value indicates that its position is to the left of the wavefront change.

The power of the wavefront changer is calculated with the incident and emerging vergence and solving Equation 1-3 for the power of the wavefront changer (lens) yields

$$\text{Lens power} = V_{out} - V_{in} = -4.00\text{ D} - (+3.00\text{ D}) = -7.00\text{ D}$$

Rectilinear Propagation

It has been established that in geometrical optics the propagation of light is described in terms of rays and pencils. This may be restated as the **Law of Rectilinear Propagation of Light**: *Light travels in straight lines.* Although rectilinear propagation is only an approximation, with this law the behavior of light can be predicted for many cases. Light can also interact with obstacles in a way that cannot be predicted by rays. This interaction is called **diffraction**. Diffraction is defined in physical optics (consult your textbook) but for now a simplified explanation is given.

In Figure 1-14, an infinite point source illuminates a pupil (also called an *aperture* or *stop*) with parallel wavefronts (and rays). If the stop is large, the majority of the rays pass through, and only the size of the pencil is reduced. If, however, the stop is sufficiently small to allow only one ray to pass through, a single wavelet will pass through the aperture, and the incident parallel wavefront will emerge as a spherical wavefront with rays diverging from the stop. This is diffraction. For the purposes of this book, pupils are considered to be sufficiently large so that the diffraction effects can be ignored and rectilinear propagation can be assumed.

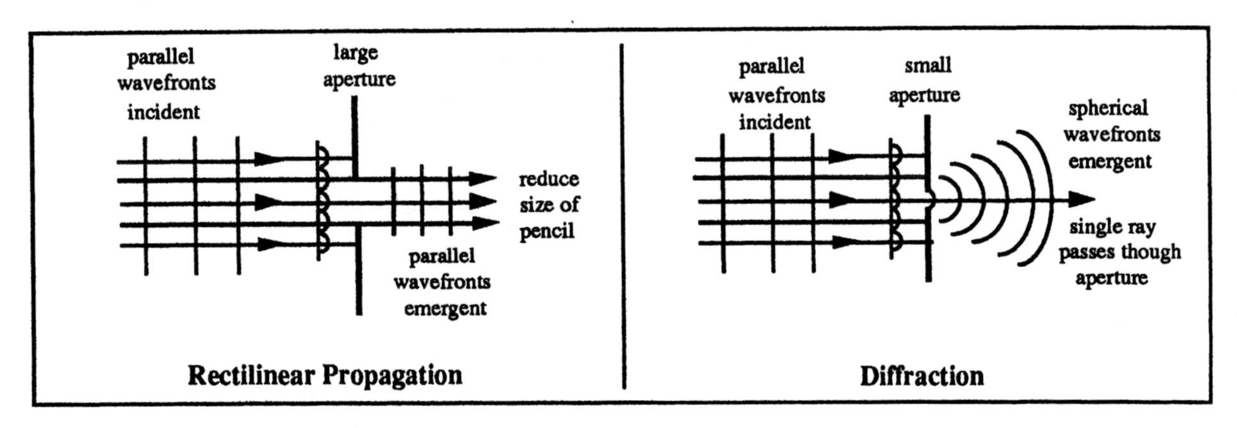

Figure 1-14. Rectilinear propagation of light occurs with large pupils (*left*) where incident parallel rays emerge parallel. With small pupils (*right*), diffraction occurs, and incident parallel light emerges as a spherical wave.

There are several examples of rectilinear propagation of light, including the pinhole camera, shadows, and full and partial illumination.

Pinhole Camera

Most photographic cameras consist of a light-tight body that houses the film, a shutter, and a lens. In the pinhole camera, a pinhole takes the place of the lens. In theory, if the pinhole is small enough to allow only one ray from each point on the object to pass through, each point on the image will be formed by a single ray. Thus each image point is unique. If the pinhole is large, the images formed by the rays overlap and thus create a blurry shadow. However if the pinhole is small, an inverted image is formed, as illustrated in Figure 1-15.

The geometric concept of similar triangles makes solving problems involving rectilinear propagation simple. In Figure 1-15, the two similar triangles used in solving pinhole camera problems are shaded. Similar triangles have equal ratios for equivalent sides. For this example, these ratios can be written as

$$\frac{h}{h'} = \frac{a}{a'} \qquad (1\text{-}4)$$

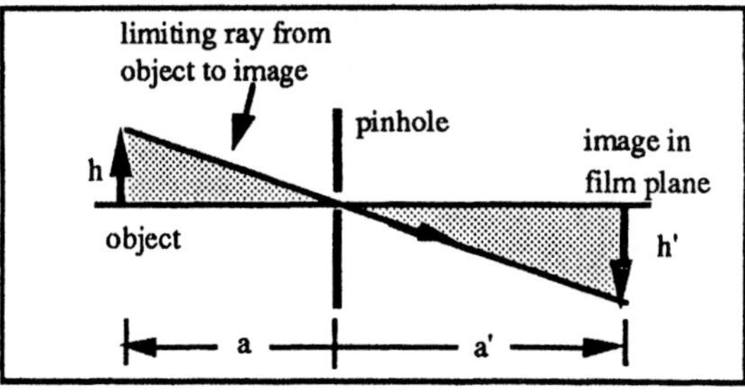

Figure 1-15. The pinhole camera concept.

Do not memorize this formula, but rather, learn how to identify similar triangles and set up ratios for each problem. With practice, this becomes fairly simple.

It should be clear that for the pinhole camera, focusing is not necessary. No matter where the object is placed, the pinhole will limit the rays (Figure 1-16, top) and an image will be formed in the film plane. The size of the image will change, however. A similar effect would be seen if the image plane were to be moved and the object position held constant (i.e., the image would be in focus no matter where the image screen is placed) (Figure 1-16, bottom).

Depth of field - the distance over which an object can be moved without affecting the sharpness of an image in a specific fixed position. For a pinhole, in theory, this range is infinite. Depth of field is important when prescribing bifocal glasses because this is the range over which the reading material can be placed and still form an acceptable image.

Depth of focus - the distance over which an image screen can be moved while maintaining a sharp image for a fixed object position. In theory, this range also is infinite for a pinhole.

Field of view - the maximum angular size of an object imaged by a system. This will depend upon many factors. For a pinhole camera the field of view is almost 180°.

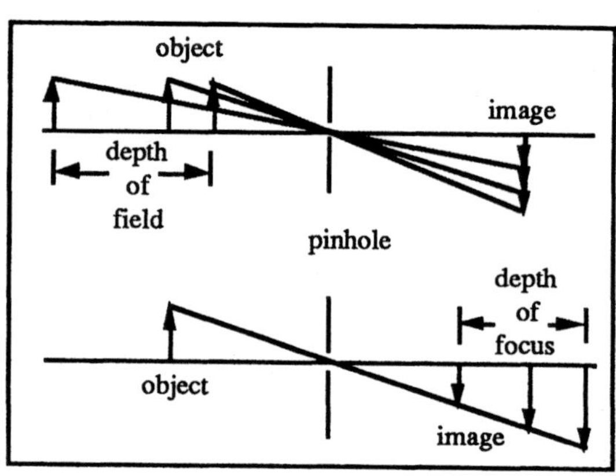

Figure 1-16. Pinhole camera. *Top*: Depth of field. *Bottom*: Depth of focus.

Disadvantages of pinhole cameras include low illumination and poor sharpness, while advantages are no image distortion, a large depth of field, and a large field of view.

Since a pinhole has an infinite depth of field, viewing through a pinhole should produce a clear retinal image for all distances, even without a refractive correction. This may be easily demonstrated by removing your glasses and viewing an object through a small aperture formed with your fingers. A pinhole may also be used to determine if reduced visual acuity is due to optical factors (i.e., if viewing through a pinhole increases acuity). For a corrected individual, a pinhole may actually reduce visual acuity due to diffraction.

Example 1-1

At what distance from a pinhole camera must a 3 cm object be placed so that the image will be three times the size of the object and be located 1 meter behind the pinhole?

Known	***Unknown***	***Equations/Concepts***
Object size: $h = 3$ cm	Object position (a)	(1-4) Solve for object distance (a)
Image size: $h' = 3 \times h = 9$ cm		Must diagram to determine similar triangles
Image position: $a' = 1$ m		

Draw and label a diagram of the problem (Figure 1-17). Changing all units to centimeters is not necessary if the units cancel, but to avoid confusion, it's best to use the same units. Use shaded similar triangles to set up ratios and solve for a:

$$\frac{3\text{ cm}}{a} = \frac{9\text{ cm}}{100\text{ cm}}$$

$$a = \frac{(3\text{ cm})(100\text{ cm})}{9\text{ cm}} = 33.33\text{ cm}$$

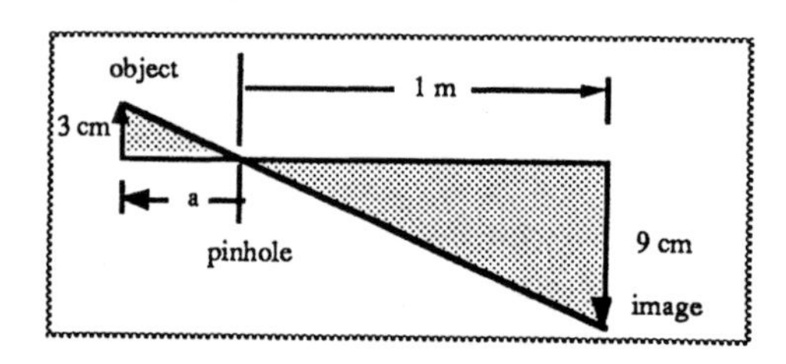

Figure 1-17. *Example 1-1.*

Shadows and Full and Partial Illumination

Shadows and full and partial illumination can also be used to illustrate rectilinear propagation of light. Diagrams are a necessity for solving these problems because similar triangles must be identified. The required similar triangles will vary with the size of the source and obstacle or aperture. Try working at least one of each type of problem.

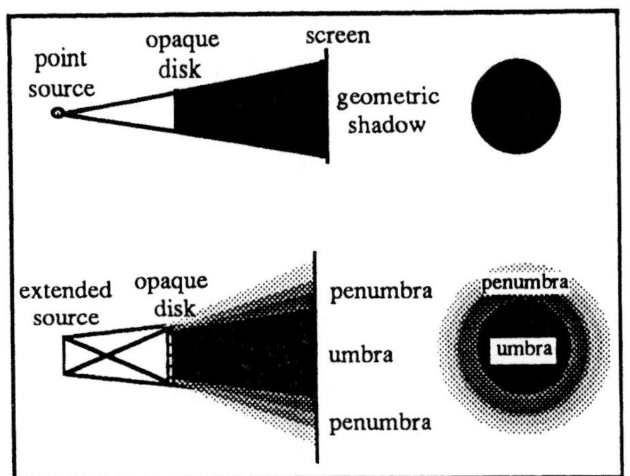

Figure 1-18. Shadows formed by a point source and an extended source.

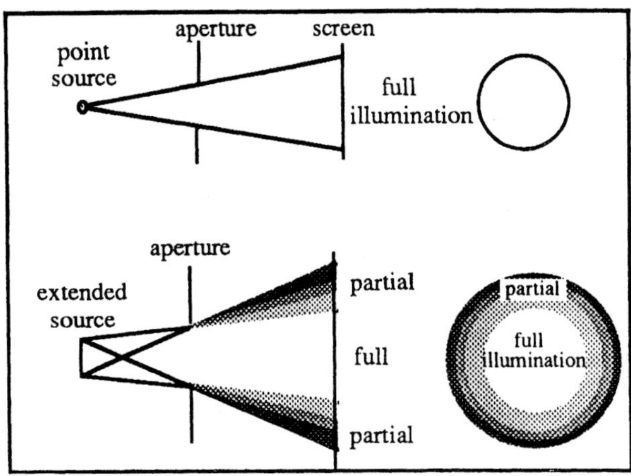

Figure 1-19. Illumination patterns formed by a point source and an extended source.

Shadow - light restricted by an opaque object. If the opaque obstacle is illuminated by a point source (Figure 1-18, top), the shadow formed on a screen will have sharp edges. This type of shadow is called a **geometric shadow**. If the opaque object is illuminated by an extended source (Figure 1-18, bottom), the shadow will have two regions: a central uniformly dark region called the **umbra** and an outer annular region called the **penumbra**. The penumbra varies from dark to light from the edge of the umbra outward.

Region of Illumination - the pattern of light passing through an aperture or stop and falling onto a screen (Figure 1-19). If the aperture is illuminated by a point source, the light on the screen will be uniform and is called **full illumination**. If the aperture is illuminated by an extended source, the pattern formed on the screen will have a uniform central region of full illumination and an outer annulus of **partial illumination**. This annulus will vary from bright at the edge of the full illumination region to dark at the outer edge annulus.

In constructing diagrams for solving shadow and illumination problems, draw two rays from the extreme top and bottom of the source to the extreme top and bottom of the obstacle or aperture. This is illustrated in Figure 1-20 for an extended source that is larger than an opaque disk. Rays from the extreme top and bottom of the source to the opposite sides of the obstacle define the outer diameter of the penumbra. Rays from the source to the same side of the aperture (top to top, bottom to bottom) define the umbra. In cases where the source is larger than the obstacle, the rays that define the umbra cross, and beyond this point the entire region is a partial shadow. To form an umbra, the screen must be placed inside this transition.

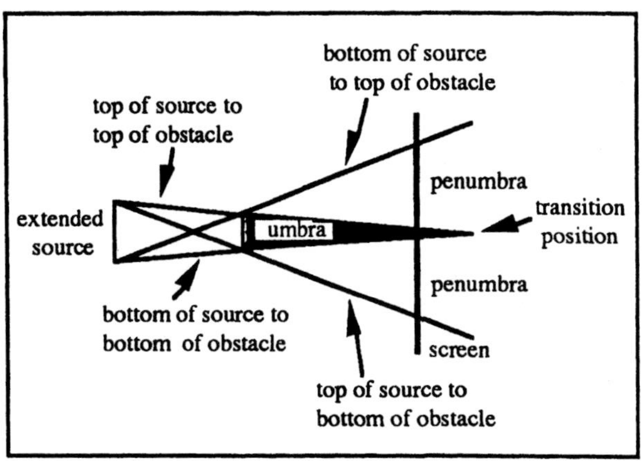

Figure 1-20. Rays from an extended source that are limited by an opaque obstacle.

When the source is larger than the opaque obstacle, the umbra shrinks in diameter and the penumbra increases in width as the screen is moved away from the obstacle (Figure 1-20). When the source is smaller than the obstacle, both the umbra and the outer diameter of the penumbra increase as the screen is moved away from the obstacle. (See *Example 1-m*.)

Example 1-m
A point source illuminates a 3 cm diameter opaque disk. If the disk is 10 cm from the source and 20 cm from an image screen, what is the diameter and type of the shadow formed on the screen?

Known	***Unknown***	***Equations/Concepts***
Disk diameter: 3 cm	Diameter and type of shadow	Point source yields a geometric shadow
Disk from source: 10 cm		Must diagram and setup similar triangles
Disk from screen: 20 cm		See Figure 1-21

The diagram shows the similar triangles, and the ratios are

$$\frac{3\text{ cm}}{10\text{ cm}} = \frac{y}{30\text{ cm}}$$

Solving for y

$$y = \frac{(3\text{ cm})(30\text{ cm})}{10\text{ cm}} = 9\text{ cm}$$

A geometric shadow with a diameter of 9 cm is formed on the screen.

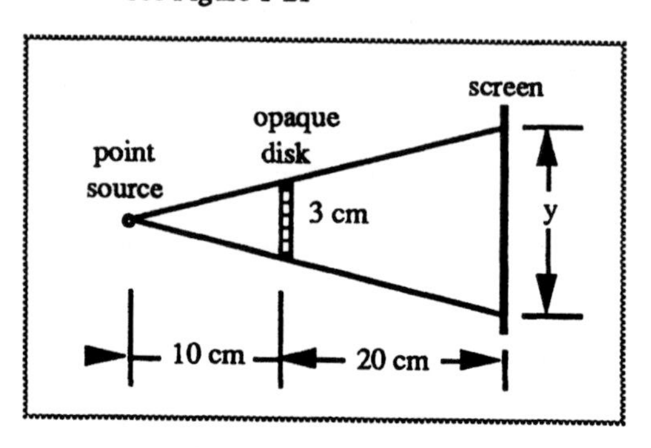

Figure 1-21. *Example 1-m.*

Example 1-n
A shadow is cast by a man standing 182 cm from a lamp on a post 458 cm above the ground. If the top of the shadow is 122 cm from the man, how tall is the man?

Known	***Unknown***	***Equations/Concepts***
Distance man to post: 182 cm	Man's height	Lamp acts as point source, yields a geometric shadow
Distance man to top shadow: 122 cm		Diagram and find similar triangles
Lamp above ground: 458 cm		

The problem is diagrammed in Figure 1-22. The similar triangles should be easily identified and t is found

$$\frac{458\text{ cm}}{304\text{ cm}} = \frac{t}{122\text{ cm}}$$

$$t = \frac{(458\text{ cm})(122\text{ cm})}{304\text{ cm}} = 183.8\text{ cm}$$

Is the answer feasible (i.e., can a man be this tall)? Converting to units of feet gives

$$(183.8\text{ cm})\left(\frac{1\text{ in}}{2.54\text{ cm}}\right)\left(\frac{1\text{ ft}}{12\text{ in}}\right) = 6.03\text{ ft}$$

6.03 ft is a reasonable height!

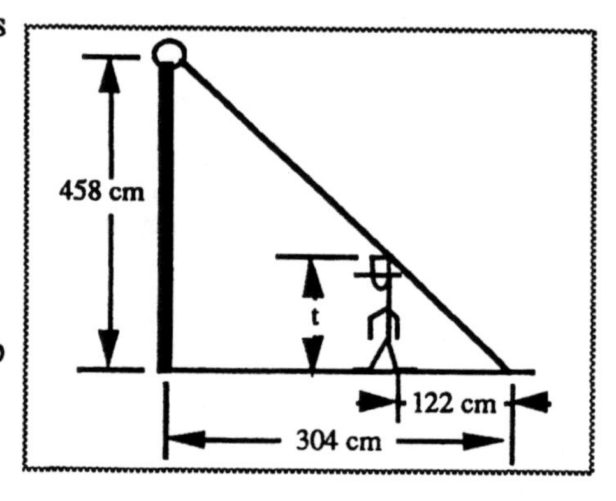

Figure 1-22. *Example 1-n.*

Example 1-o

An extended source, 9 cm in diameter, is centered above a circular table and forms a shadow on the floor consisting of a penumbra with an outer diameter of 46 cm and an umbra with a diameter of 34 cm. If the table is 36 cm above the floor, how far is the source above the table, and what is the diameter of the table?

Known	***Unknown***	***Equations/Concepts***
Source diameter: 9 cm	Distance source to table	Extended source
Penumbra outer diameter: 46 cm	Diameter of table	Diagram and set-up umbra and determine similar triangles
Umbra diameter: 34 cm	Size of source relative to table	See Figure 1-23
Table to floor: 36 cm		

An assumption is made that the light source is smaller than the table because the umbra has a large diameter compared to the width of the penumbra. Note that the dimensions and unknowns are labeled on the diagram with x as the distance from the source to the table and y as the diameter of the table.

The width of the penumbra, as illustrated in the figure, is calculated by:

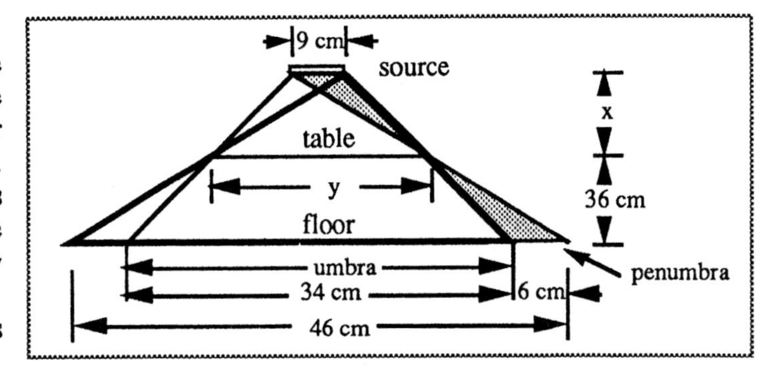

Figure 1-23. *Example 1-o.*

$$\text{penumbra width} = \frac{\text{penumbra diameter} - \text{umbra diameter}}{2} = \frac{46\text{ cm} - 34\text{ cm}}{2} = \frac{12\text{ cm}}{2} = 6\text{ cm}$$

Two sets of similar triangles are shown in the figure. The shaded set is used first.

$$\frac{9\text{ cm}}{6\text{ cm}} = \frac{x}{36\text{ cm}}$$

Solving for x, this becomes

$$x = \frac{(9\text{ cm})(36\text{ cm})}{6\text{ cm}} = 54\text{ cm} = \text{distance from source to table}$$

The trick here is to remember that the altitudes of similar triangles are similar (x and 36 cm). The next set of similar triangles is darkly outlined. The ratios are

$$\frac{54\text{ cm}}{y} = \frac{(54\text{ cm} + 36\text{ cm})}{(34\text{ cm} + 6\text{ cm})}$$

$$y = \frac{(40\text{ cm})(54\text{ cm})}{90\text{ cm}} = 24\text{ cm} = \text{diameter of table}$$

Note that the same similar triangles are used for all problems even when the source is larger than the obstacle or the aperture.

Example 1-p

A point source illuminates a 10 cm diameter frosted diffuser that is recessed 5 cm into a hanging lamp. The aperture opening in the lamp is the same diameter as the diffuser. The lamp is centered 35 cm above a table. What is the diameter of full illumination on the table? What is the largest diameter table that will just receive all the light from the lamp?

Known	***Unknown***	***Equations/Concepts***
Point source illuminates diffuser Diffuser diameter: 10 cm Diffuser illuminates aperture of: 10 cm Diffuser 5 cm from aperture Lamp 35 cm above table	Full illumination region Diameter table Really asking for limit of partial illumination.	Diffuser acts as extended source at position of diffuser Diagram and use similar triangles as with shadows

Figure 1-24 illustrates this problem. Note that the regions are labeled and the vertical and horizontal scales are different. The point source illuminates the diffuser, and the diffuser becomes an extended source that is used in the problem. Looking at the diagram it should be clear that the full illumination region is the same diameter as the diffuser and aperture: 10cm.

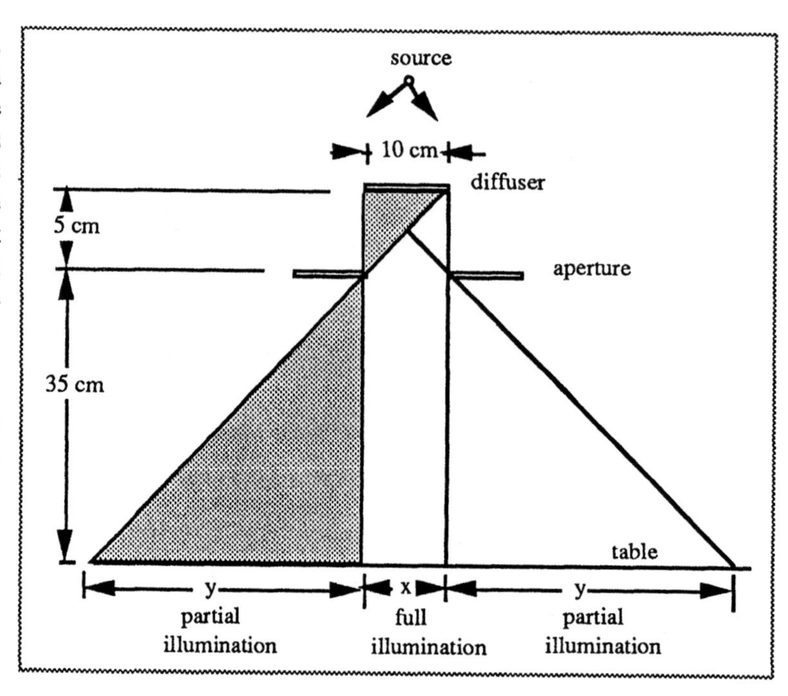

Figure 1-24. *Example 1-p.*

Use the similar triangles shaded in the figure to solve for the width of the partial illumination. Solving for y

$$\frac{5\text{ cm}}{10\text{ cm}} = \frac{35\text{ cm}}{y}$$

$$y = \frac{(35\text{ cm})(10\text{ cm})}{5\text{ cm}} = 70\text{ cm} =$$

width of partial illumination

Solve for the diameter of the table that will receive all the light from the lamp by adding the full illumination region to two times the width of the partial illumination (see Figure 1-24):

full diameter of illumination (full + 2 x partial) = 10 cm + 2(70 cm) = 150 cm = table diameter.

Supplemental Problems

Frequency, Wavelength, Velocity

1-1. What is the frequency associated with wavelengths of 400 nm, 500 nm, and 600 nm ?
ANS. For 400 nm: $f = 7.50 \times 10^{14}$; for 500 nm: $f = 6.00 \times 10^{14}$; for 600 nm: $f = 5.00 \times 10^{14}$

1-2. An electromagnetic wave travels at 1.73×10^8 m/sec in an ophthalmic medium. If the frequency of the wave is 5.70×10^{14} Hz, what is the wavelength of the light in a vacuum and in the medium.
ANS. In a vacuum: $\lambda = 526$ nm; In the medium: $\lambda = 306$ nm

Radius, Vergence

1-3. For each position in Figure 1-25, do the following:
a. Draw the wavefront on the diagram and determine the radius. Be sure to use the proper sign convention.
b. Determine the type of vergence (convergence, divergence, parallel).
c. Calculate the amount of vergence in diopters.

ANS. The wavefront is drawn for position (a) by placing the pointed end of the compass at the image location (c) and drawing an arc through position (a). The wavefronts for the other positions may be drawn using this procedure.

Position (a): radius = +30 cm or +0.30 m, and vergence = + 3.33 D convergence

Position (b): radius = + 10 cm = + 0.10 m, and vergence= +10.00 D convergence

Position (c): radius = 0 cm = 0.0 m, and vergence is infinite

Position (d): radius = – 20 cm = – 0.20 m, and vergence = – 5.00 D divergence

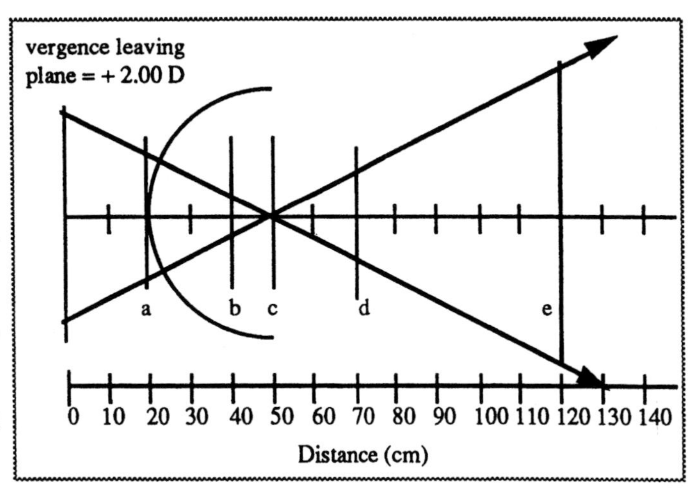

Figure 1-25. *Supplemental Problem 1-3.*

Position (e): radius = – 70 cm = – 0.70 m and vergence = – 1.43 D divergence

1-4. A point source is 15 cm in front (to the left) of a screen. What is the vergence incident on the screen?
ANS. – 6.67 D

1-5. An object is located 50 cm in front (to the left) of a lens. An image is formed 25 cm to the right of the lens. What must the lens power be to change the wavefronts accordingly?
ANS. Lens power = + 6.00 D

1-6. A lens is placed in a pencil that converges to a point 50 cm to the right of the lens. The vergence leaving the lens is + 7.50 D. What is the incident vergence and power of the lens?
ANS. Incident vergence = + 2.00 D; lens power = + 5.50 D

1-7. What is the vergence incident on a + 5.00 D lens that yields an emergent parallel pencil? Where must an object be placed to yield this vergence?
ANS. Vergence = – 5.00 D; object located – 20 cm or 20 cm in front of (to the left of) the lens

Rectilinear Propagation of Light, Pinhole Camera, Shadows, Illumination

1-8. A pinhole camera is used to photograph a model train. Where must the model be placed so that the image formed on the film 5 cm behind the pinhole is 0.1 times the size of the object?
ANS. Object is 50 cm in front of the pinhole

1-9. A 7 cm high object placed in front of a pinhole camera forms a 7 cm high image in the film plane. If the object is moved 10 cm closer to the pinhole, the image doubles in size. How far from the pinhole is the film plane?
ANS. Film plane is 20 cm behind the pinhole

1-10. On a screen 25 m from a point source, a shadow is formed by a 3 m diameter opaque disk. If the disk is 10 m from the screen, what is the diameter and type of the shadow formed?
ANS. 5 m in diameter; geometric or absolute shadow

1-11. If a 1 m diameter hole is drilled in the center of the disk in Problem 1-10, and the disk is moved 3 m closer to the point source, how large is the area of full illumination in the middle of the umbra?
ANS. 2.1 m in diameter

1-12. A boy holds a flashlight 305 cm from a tent. His brother holds a black frisbee with a diameter of 25 cm at 91 cm in front of the flashlight, and an umbra 71 cm in diameter is formed on the tent. What is the outer diameter of the shadow formed on the tent? What is the diameter of the flashlight?
ANS. Shadow 96.58 cm in diameter; flashlight 5.44 cm in diameter

1-13. A 10 cm diameter extended source illuminates a 3 cm diameter aperture. What is the size and type of illumination formed on a screen placed 8 cm, 21.3 cm, and 25 cm from the aperture if the source is 50 cm in front of the aperture?
ANS. Screen 8 cm from aperture: full illumination region with diameter of 1.88 cm; overall pattern (full plus partial illumination) is 5.08 cm
Screen 21.3 cm from aperture: partial illumination only with overall diameter of 8.57 cm; this is the position where the full illumination region approaches zero
Screen 25 cm from aperture: partial illumination only with overall diameter of 9.50 cm

1-14. When light from an extended source 15 cm in diameter is obstructed by an opaque disk, the umbra formed on a screen is 10 cm in diameter, and the outer limit of the penumbra is 50 cm in diameter. What is the diameter of the opaque disk, if the source is 25 cm from the disk?
ANS. 12.9 cm

1-15. Light from an extended source shines through a circular aperture with a diameter twice that of the source. A bright circle of light 50 cm in diameter is formed on the screen. If the aperture is placed in the exact middle between the screen and the source, what is the outer diameter of the partial illumination region formed on the screen?
ANS. 83.3 cm

Chapter 2

Laws of Reflection and Refraction

When light strikes a border or interface between two different media, changes in the velocity and direction of travel may occur. Depending on the media, light may be **reflected, refracted,** and/or **absorbed**. This chapter introduces reflection at a plane surface (i.e., the **Law of Reflection**). Other details about reflection at plane and curved surfaces are covered in Chapter 8. **Snell's Law** (the **Law of Refraction**) is presented here and applied to parallel plane surfaces. This information leads into Chapter 3, where refraction at nonparallel plane surfaces, such as prisms, is discussed.

Index of Refraction

In a vacuum, all electromagnetic waves travel at a velocity of 3×10^8 m/s. In other media, these waves travel at slower velocities; the exact velocity is a function of wavelength. For example, blue light with a wavelength of 350 nm travels in a medium at a slower velocity than red light with a wavelength of 750 nm. The velocity of light in any medium may be calculated with the following equation, first presented in Chapter 1:

$$v = f\lambda \tag{2-1}$$

Remember that the frequency (f) remains constant for all media; thus when the velocity (v) decreases, the wavelength (λ) must also decrease. (See *Examples 1-a, 1-b,* and *1-d*.)

The ratio of the velocity of light in a vacuum divided by the velocity of light in another medium is called the **index of refraction** and has the symbol **n**. It may be expressed by this equation:

$$n = \frac{\text{velocity in vacuum}}{\text{velocity in medium}} = \frac{c}{v_m} = \frac{3\text{x}10^8 \text{ m/s}}{v_m} \tag{2-2}$$

The velocity of light in a vacuum (3×10^8 m/s) is always greater than the velocity in other optical media (v_m). It should therefore be clear from Equation 2-2 that the index of refraction is always greater than or equal to one ($n \geq 1$). The velocity of light traveling in air is approximately equal to the velocity in a vacuum, therefore the index of refraction of air is usually considered to be unity. Indices of other optical materials are shown in Figure 2-1.

MATERIAL	INDEX
Vacuum	1.000
Air (clean, nonpolluted)	1.000
Water	1.333
Ophthalmic plastic (PMMA)	1.490
Ophthalmic crown glass	1.523
Dense flint glass	1.626
Diamond	2.417
Human cornea	1.376

Figure 2-1. Indices of common optical materials.

Example 2-a
In an unknown material, light with a wavelength of 589 nm travels at 2×10^8 m/s. What is the index of refraction of the material?

Known	***Unknown***	***Equations/Concepts***
Velocity of light in material: $v = 2 \times 10^8$ m/s	Index of refraction	Ratio of velocities
Velocity of light in vacuum : $c = 3 \times 10^8$ m/s		(2-2) $n = c / v$

Substitute the appropriate numbers into Equation 2-2. Remember that both velocities must have similar units.

$$n = \frac{c}{v_m} = \frac{3 \times 10^8 \text{ m/s}}{2 \times 10^8 \text{ m/s}} = 1.50$$

The index of refraction has no units because the m/s units cancel in the ratio.

Example 2-b
What is the wavelength of light in the material described in *Example 2-a* ?

Known	***Unknown***	***Equations/Concepts***
Velocity of light in vacuum: $c = 3 \times 10^8$ m/s	Wavelength of light in medium	(1-1) $v = f\lambda$
Wavelength of light in vacuum: $\lambda = 589$ nm		

The frequency of the light in a vacuum is first calculated using Equation 2-1 with the values of velocity and wavelength given.

$$v = f\lambda \qquad \text{or} \qquad f = \frac{v}{\lambda} = \frac{3 \times 10^8 \text{ m/s}}{589 \times 10^{-9} \text{ m}} = 5.09 \times 10^{14} \text{ Hz}$$

In the unknown material, the frequency remains constant and the velocity changes. Using the new velocity and the calculated frequency determine the new wavelength by solving Equation 2-1 for the wavelength:

$$\lambda = \frac{v}{f} = \frac{2 \times 10^8 \text{ m/s}}{5.09 \times 10^{14} \text{ Hz}} = 393 \times 10^{-9} \text{ m} = 393 \text{nm}$$

Dispersion

Since the velocity of light varies as a function of wavelength in media other than a vacuum, the index of refraction also varies as a function of wavelength in these media. This relationship is dependent upon the components that make up the optical material. A plot of index versus wavelength is shown in Figure 2-2 for a representative material. This plot is referred to as a **dispersion curve**. From Figure 2-2, the index for shorter wavelengths is greater than for longer wavelengths, i.e., the index of blue light (approximately 1.60 at 400 nm) is greater than the index for red light (approximately 1.47 at 700 nm).

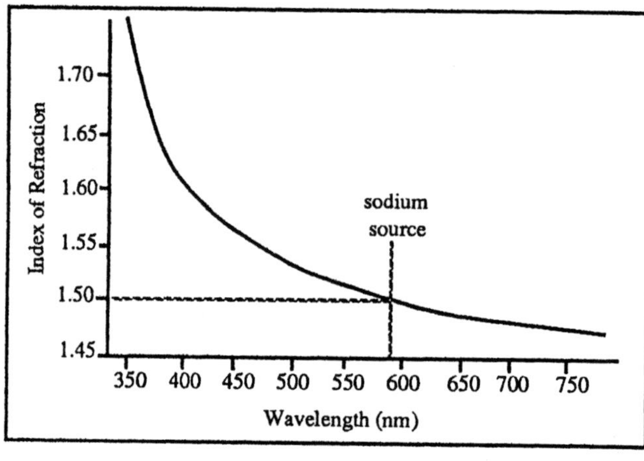

Figure 2-2. A dispersion curve showing the variation in index as a function of wavelength.

In defining the index of refraction of optical media, one wavelength (589 nm or the dominant wavelength of a sodium source) has been used as the standard. In Figure 2-2, the index of refraction at 589 nm is 1.50. In this and most other optics texts, the index of refraction is defined for the standard wavelength of 589 nm unless otherwise stated. The indices listed in Figure 2-1 are for this standard wavelength. A change in the standard wavelength to He D-line (587 nm) produced by a laser has been proposed recently.

Methods for measuring the index of refraction of unknown materials are discussed in later chapters.

Example 2-c

Using the dispersion curve in Figure 2-2, determine the velocities for 400 nm, 589 nm, and 700 nm. How does the wavelength relate to the velocity and index?

Known	***Unknown***	***Equations/Concepts***
From Figure 1-1 :	Velocity of 400nm, 589nm, and 700nm for material in figure.	(2-1) $v = f\lambda$
Index of 400 nm: $n = 1.60$	Relationship between wavelength and index	(2-2) $n = c/v$
Index of 589 nm: $n = 1.50$	Relationship between wavelength and velocity	
Index of 700 nm: $n = 1.47$		

Determine the index of refraction for each wavelength from Figure 2-1 (shown in the *Known* column above). The velocity for each wavelength of light may be calculated by solving Equation 2-1 for the velocity (v):

For 400 nm: $$v = \frac{c}{n} = \frac{3 \times 10^8 \text{ m/s}}{1.60} = 1.875 \times 10^8 \text{ m/s}$$

For 589 nm: $$v = \frac{c}{n} = \frac{3 \times 10^8 \text{ m/s}}{1.50} = 2.000 \times 10^8 \text{ m/s}$$

For 700 nm: $$v = \frac{c}{n} = \frac{3 \times 10^8 \text{ m/s}}{1.47} = 2.041 \times 10^8 \text{ m/s}$$

These calculations show that as the wavelength increases, the velocity increases and the index of refraction decreases.

Interface - the boundary (plane, curved, aspheric, etc.) between two media with different indices. In Figure 2-3, light travels in medium 1 (left) before striking the interface of medium 2 (right). Usually n is the label for the index of the medium which the incident ray travels in before striking the interface. This is the first medium listed when describing an interface. The other medium of the interface is labeled n' and is the second media listed. Figure 2-3 presents several plane or flat interfaces with the proper labels.

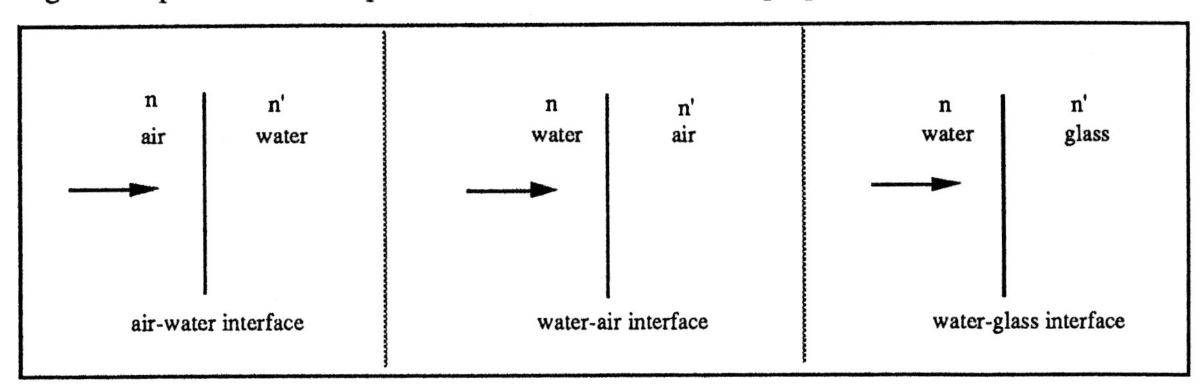

Figure 2-3. Examples of interfaces between two media.

Reflection and Refraction at a Plane Surface

When you throw a rubber ball at a wall, the direction in which the ball bounces back depends on several factors. First, if you throw the ball so that it hits normal (at a right angle) to the surface, the ball will come back toward you along the same path. On the other hand, if you bounce the ball off the ground first and then onto the wall, the ball will return as a pop-up fly. (See Figure 2-4.) Therefore, the incident angle in which the ball hits the wall determines the direction the ball takes when leaving the wall. This also applies to reflected and refracted light.

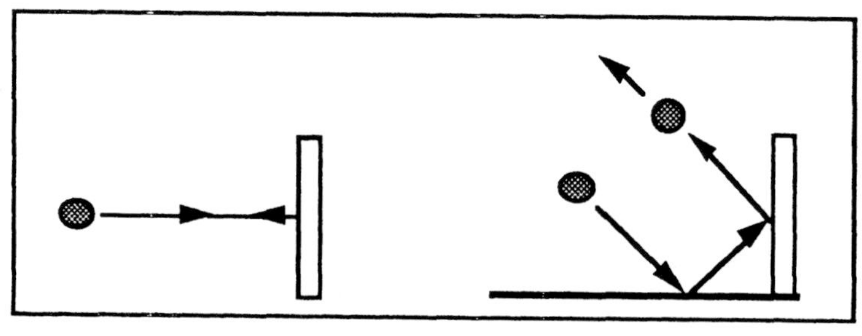

Figure 2-4. Ball normal to the surface and incident at some angle.

In the context of this and most other optics texts, light incident on an interface is represented as a ray. The incident angle the ray makes with the interface is measured from a normal (perpendicular line) to the surface. The emergent ray, be it reflected or refracted, also has an angular subtense measured from the normal. It will be assumed, unless otherwise stated, that an incident, reflected, or refracted angle is measured from the normal. *Read all questions carefully to be sure that angles are defined normal to the surface.*

Law of Reflection

Light that is reflected at an interface leaves the surface traveling in the opposite direction (Figure 2-5). For any incident angle, the angle of reflection may be calculated using the **Law of Reflection**: *The angle of incidence is equal to the angle of reflection.*

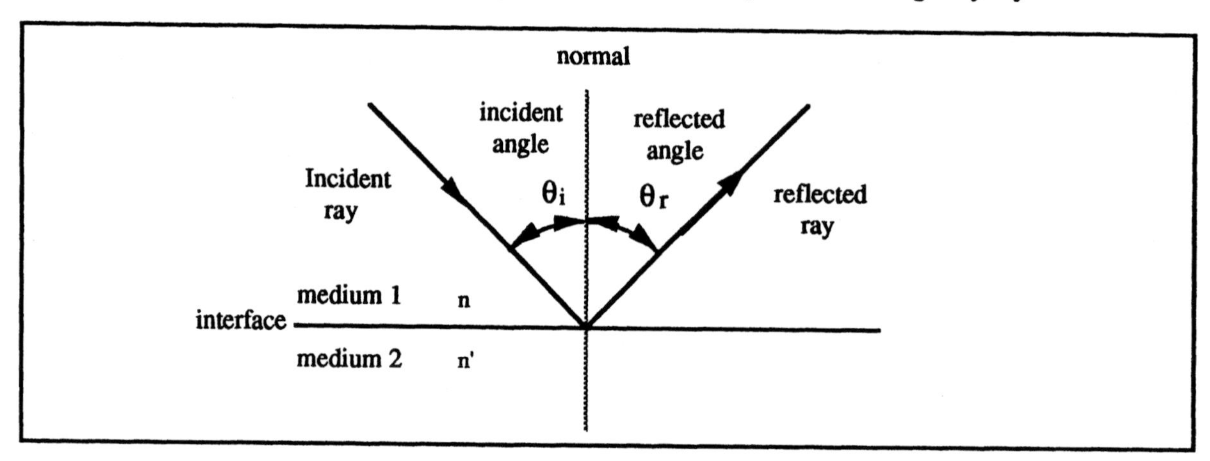

Figure 2-5. Incident and reflected rays and angles.

The Law of Reflection may be written as an equation:

$$\theta_i = \theta_r \tag{2-3}$$

where:

θ_i = the incident angle

θ_r = the reflected angle

Example 2-d
A ray is reflected at an interface between water and air at an angle of 25° from the normal. What is the angle of incidence?

Known	***Unknown***	***Equations/Concepts***
Index of refraction of water: $n = 1.33$	Incident angle: $\theta = ?$	(2-3) Law of Reflection
Index of refraction of air: $n = 1.00$		
Angle of reflection: $\theta_r = 25°$		

Simply knowing the relationship between the incident and reflected angles will yield the incident angle. No calculations are necessary.

$$\theta_i = \theta_r = 25°$$

Example 2-e
A ray of light is incident upon two interfaces that are at right angles to each other, as shown in Figure 2-6. Calculate the reflected angle at each interface, and trace the path of the ray.

Known	***Unknown***	***Equations/Concepts***
Figure 2-6	Reflected angles	(2-3) Law of Reflection
Incident angle: $\theta_i = 20°$		

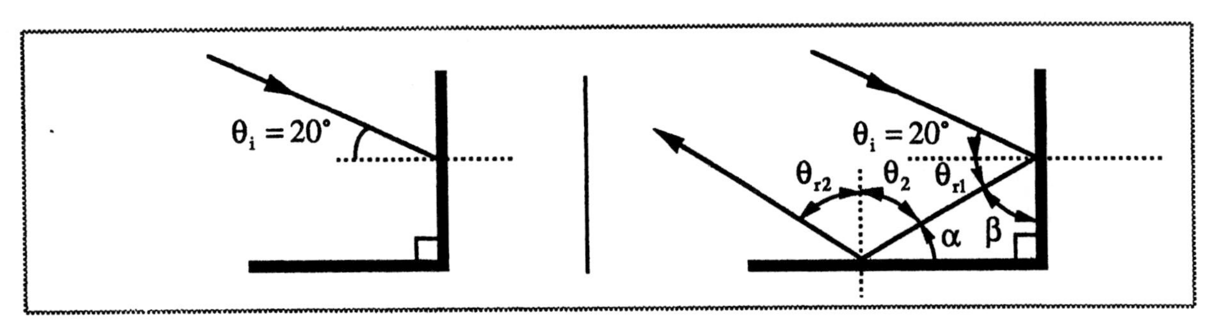

Figure 2-6. *Example 2-e.*

The left diagram in Figure 2-6 shows the incident ray striking the interface at an angle of 20° from the normal. According to the Law of Reflection, the reflected ray (θ_{r1}) will leave this interface at 20° from the normal. Looking at the diagram on the right, the normal to the surface may be represented by

$$\beta + \theta_{r1} = 90° \text{ or } \beta = 90° - 20° = 70°$$

The sum of the internal angles of a triangle are equal to 180° ($\beta + \alpha + 90° = 180°$). Substituting for β and solving for α:

$$\alpha = 180° - 90° - 70° = 20°$$

Because the normal to the second surface is 90°, the incident angle at the second surface is equal to

$$\theta_2 = 90° - \alpha = 90° - 20° = 70°$$

Using the Law of Reflection, the final reflected angle (θ_{r2}) is equal to 70°.

Regular or specular reflection - reflected light leaves the surface in a definite beam that follows the Law of Reflection. This type of reflection occurs at smooth polished surfaces, such as a plane mirror or a piece of glass. (See Figure 2-7.)

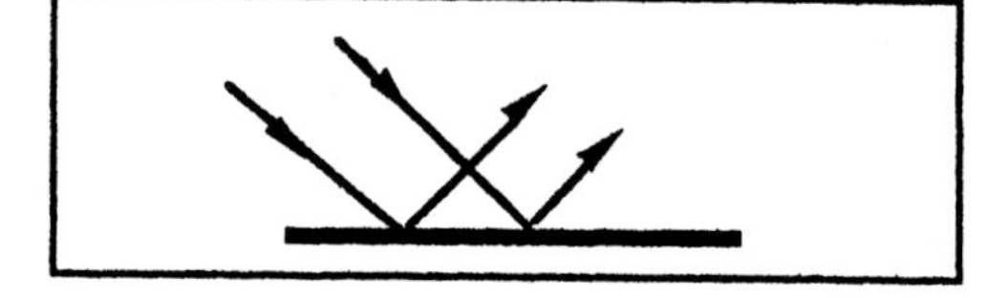

Figure 2-7. Specular reflection.

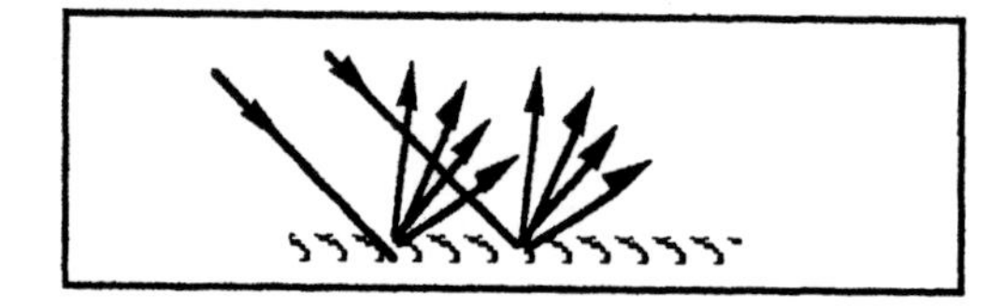

Figure 2-8. Diffuse reflection.

Irregular or diffuse reflection - incident light is reflected in all directions, and therefore the reflected rays can not be predicted with the Law of Reflection. This occurs at irregular surfaces, such as a projection screen or frosted glass. (See Figure 2-8.)

Fresnel's Law of Reflection

The actual amount of light that is reflected from a surface depends on several factors, including the following:

1. color - black absorbs, white reflects
2. surface - smooth, polished, rough, etc.
3. angle of incidence - for diffuse reflection but not for specular reflection
4. refractive index (type of medium)

For incident light that is normal or perpendicular to the surface, the percentage of light reflected from a transparent surface may be calculated using **Fresnel's Law of Reflection.** This law may be expressed as

$$\%\text{Reflected} = \left(\frac{n'-n}{n'+n}\right)^2 \text{x } 100 \qquad (2\text{-}4)$$

where:

n = the index of refraction to the left of the interface
n' = the index of refraction to the right of the interface

Example 2-f

How much light is reflected at air-water, water-air, and water-crown glass interfaces? How do the indices of the media that form the interfaces affect the amount of reflection? Assume the incident ray is normal to the interface.

Known
Air index of refraction: n = 1.00
Water index of refraction: n = 1.33
Crown glass index of refraction: n = 1.523

Unknown
% light reflected
Relationship between indices and reflection

Equations/Concepts
(2-4) Fresnel's Law

The interfaces are shown in Figure 2-3 with the proper labels. Using Fresnel's Law of Reflection, the amount of reflected light is calculated for each interface:

air(n) – water(n') $\quad \%R = \left(\frac{1.33-1.00}{1.33+1.00}\right)^2 \text{x } 100 = (0.142)^2 \text{x } 100 = 2.01\%$

water(n) – air(n') $\quad \%R = \left(\frac{1.00-1.33}{1.00+1.33}\right)^2 \text{x } 100 = (-0.142)^2 \text{x } 100 = 2.01\%$

water(n) – glass(n') $\quad \%R = \left(\frac{1.523-1.33}{1.523+1.33}\right)^2 \text{x } 100 = (0.068)^2 \text{x } 100 = 0.46\%$

From the calculations, the order of the media (air-water or water-air) does not affect the amount of reflected light. The closer the indices of the media that form the interface, the lower the amount of reflected light.

Example 2-g

According to Fresnel's Law of Reflection, how much light is transmitted through a piece of window glass (index of refraction = 1.50). Assume the thickness of the glass is negligible and that no light is absorbed.

Known	***Unknown***	***Equations/Concepts***
Glass index of refraction: n = 1.50	% light transmitted	(2-4) Fresnel's Law
Air index of refraction: n = 1.00		% Reflected + % Transmitted = 100%

To determine the total amount of light transmitted through the glass, one must first determine the amount of light lost by reflection at each surface (Figure 2-9). The first surface is an interface between air and glass, and the amount of light reflected is given by

$$\%R = \left(\frac{1.50 - 1.00}{1.50 + 1.00}\right)^2 \text{x } 100$$

$$= (0.2)^2 \text{x } 100 = 4.00\%$$

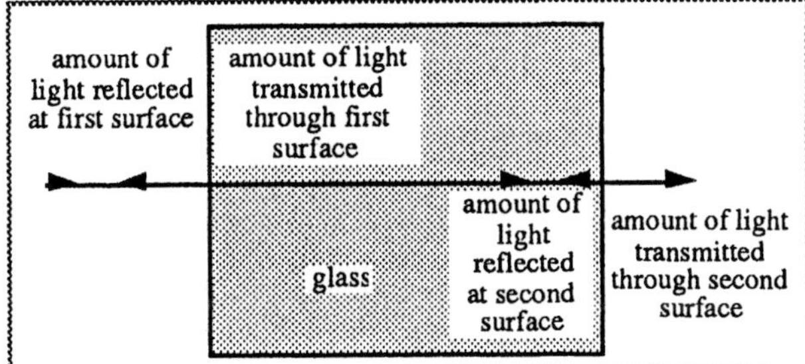

Figure 2-9. *Example 2-g*.

If we consider the incident light to be 100% and 4% of this light to be reflected, then the transmitted light at the first surface is:

$$100\% - 4\% = 96\%$$

As seen in the previous example, the amount of light reflected at the second surface (glass-air) is the same as the amount of light reflected at the first surface (air-glass) or 4% of the incident light. Since 4% of the 96% or 3.84% (0.96% x 0.04 = 0.038) of the incident light is reflected at the second surface, the amount of light transmitted is

$$96\% - 3.8\% = 92.16\%$$

Approximately 92% of the incident light is transmitted through the glass.

Snell's Law or Law of Refraction

Snell's Law (Law of Refraction) is used to calculate the direction of a ray refracted through an interface between two media. This law states that *the index of the medium in which the ray travels before the interface times the sine of the incident angle is equal to the index of the medium in which the ray travels after passing through the interface times the sine of the refracted angle.* The angle of refraction may be calculated if the indices of the interface and the incident angle are known. (See Figure 2-10.)

Snell's Law is written as

$$n \sin\theta = n' \sin\theta' \qquad (2\text{-}5)$$

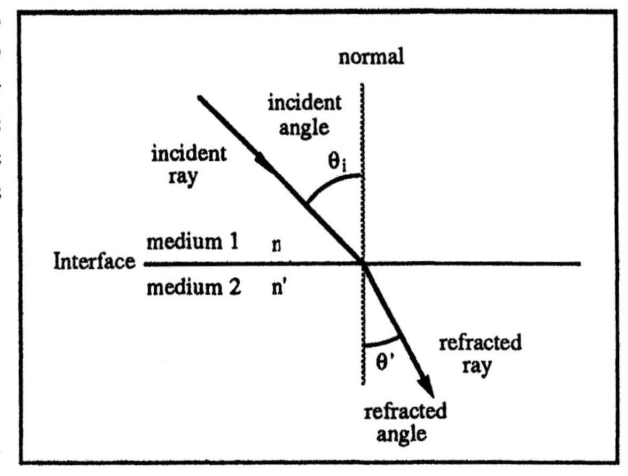

Figure 2-10. Refracted ray.

where:

n = the index of the medium before refraction
n' = the index of the medium after refraction
θ = the incident angle
θ' = the refracted angle

Example 2-h

A ray strikes an air-water interface at an incident angle of 30°. What is the angle of refraction? For the same incident angle, what would be the refracted angle at a water-air interface?

Known	***Unknown***	***Equations/Concepts***
Index of refraction of air: n = 1.00	Refracted angle θ' air-water interface	(2-3) Snell's Law
Index of refraction of water: n = 1.33	Refracted angle θ' water-air interface	n sin θ = n' sin θ'
Incident angle: θ = 30°		
Air-water interface		
Water-air interface		

Use Snell's Law to calculate the refracted angle. Be sure to determine first which medium is before refraction and which medium is after refraction. The variables before refraction are unprimed, and the variables after refraction are primed. The interface is written so that it indicates the media before and after refraction.
For the air-water interface

$$n \sin\theta = n' \sin\theta'$$
$$(1.00)\sin 30^\circ = (1.33)\sin\theta'$$
$$\sin\theta' = \frac{(1.00)(0.50)}{1.33} = 0.3759$$
$$\theta' = \sin^{-1}(0.3959) = 22.08^\circ$$

For the water-air interface

$$n \sin\theta = n' \sin\theta'$$
$$(1.33)\sin 30^\circ = (1.00)\sin\theta'$$
$$\sin\theta' = \frac{(1.33)(0.50)}{1.00} = 0.6650$$
$$\theta' = \sin^{-1}(0.6650) = 41.68^\circ$$

Example 2-i

A refracted ray leaves normal to a crown glass-air interface. What is the angle of incidence?

Known	***Unknown***	***Equations/Concepts***
Index before refraction - glass: n = 1.523	Angle of incidence: θ = ?	(2–5) Snell's Law
Index after refraction - air: n' = 1.00		Incident and refracted angles are measured from the normal
Refracted ray leaves normal to surface		
Angle of refraction: θ' = 0°		

Apply Snell's Law. Note that the sin 0° = 0°

$$n \sin\theta = n' \sin\theta'$$
$$(1.523)\sin\theta = (1.00)\sin 0^\circ$$
$$\sin\theta = \frac{(1.00)(0)}{1.523} = 0.00$$
$$\theta = \sin^{-1}(0.00) = 0.00^\circ$$

The incident angle is 0° and normal to the surface. A normal incident ray yields a normal refracted ray, and the ray does not change direction. Since the velocity changes, *there is refraction.*

Critical Angle and Total Internal Reflection

From *Examples 2-h* and *2-i*, some general rules about the refracted angle may be formulated. As shown in Figure 2-11, when light travels from a *low-high index* interface ($n < n'$), the *refracted ray bends towards the normal* and therefore the refracted angle is smaller than the incident angle. When light travels from a *high-low index* interface ($n > n'$), the *refracted ray bends away from the normal* and the refracted angle is greater than the incident angle.

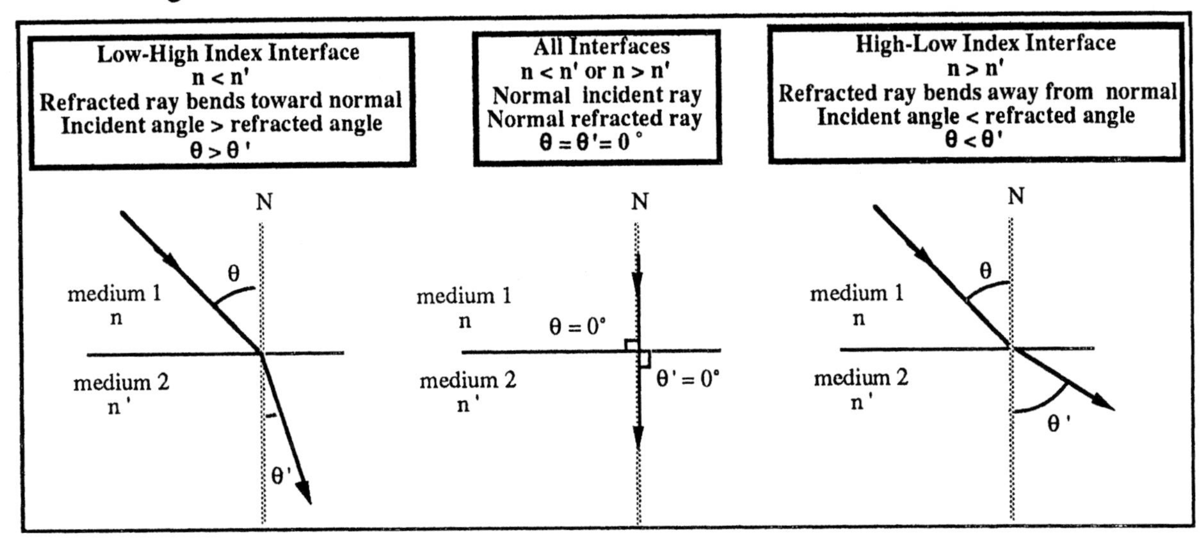

Figure 2-11. Refraction at a low-high and high-low index interface.

As shown in *Example 2-i*, for an incident ray normal to any interface (high-low or low-high index), the refracted ray leaves normal to the interface. That the ray does not change direction does not mean that the ray is not refracted. Refraction is related to the change in velocity, and the velocity of the refracted ray *does* change.

When a ray strikes a low-high index interface ($n < n'$), the ray refracts toward the normal. Figure 2-12 shows that as the incident angle increases, the refracted angle also increases but always remains smaller than the incident angle. The largest incident angle physically possible is 90°, i.e., the incident ray travels along the surface of the interface (see the black ray in Figure 2-12). This maximum incident angle may be used to find mathematical limits. However, a ray is rarely incident at 90°. Because for this interface the refracted angle is smaller than the incident angle, the refracted angle is less than 90°, and the ray is refracted through the interface.

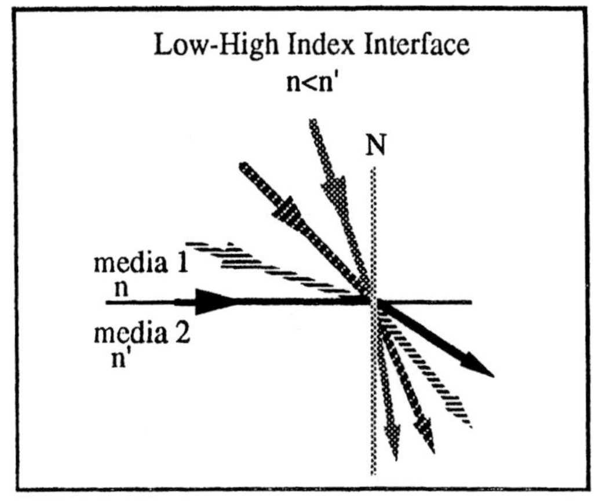

Figure 2-12. For low-high index interfaces, the refracted angle is smaller than the incident angle; all incident rays refract through the interface.

For a high-low index interface ($n > n'$), the refracted ray bends away from the normal, and therefore the refracted angle is greater than the incident angle. As the incident angle increases, the refracted angle also increases but always remains larger than the incident angle (Figure 2-13). Here there is a physical limit of 90° for the refracted angle (i.e., the refracted ray leaves along the surface

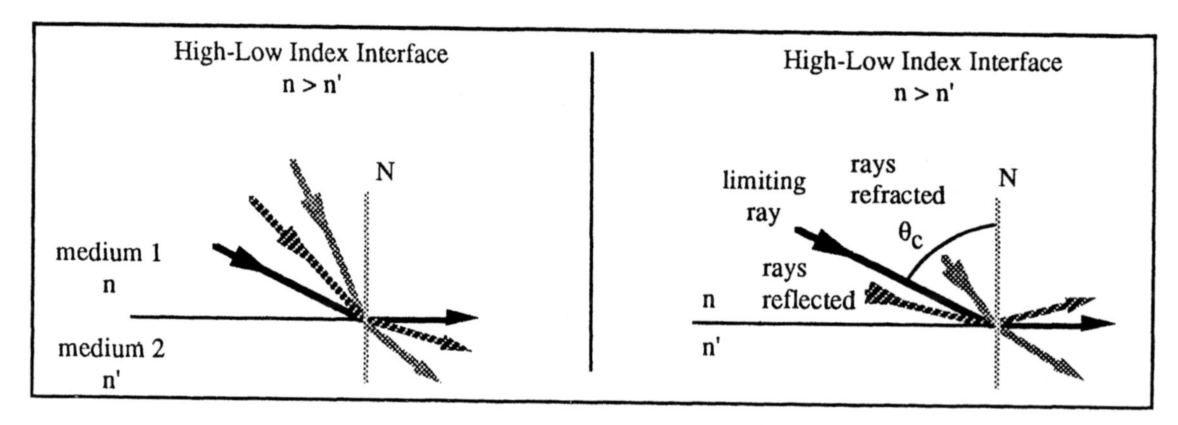

Figure 2-13. For high-low index interfaces, the refracted angle is greater than the incident angle. The limiting incident ray has an incident angle (called the critical angle) that yields a 90° refracted angle. Rays with incident angles greater that the critical angle are internally reflected.

of the interface). The incident angle that yields a refracted angle of 90° is called the **critical angle** and is labeled θ_c. For incident angles larger than the critical angle, the refracted angle is undefined. These incident angles must leave at angles greater than 90°, and therefore these rays are reflected back into the same medium as the incident ray. Snell's Law may be used to determine the angle of reflection. For incident angles greater than the critical angle, rays undergo **total internal reflection.** The phenomenon of internal reflection is often used in the design of reflecting prisms in low vision telescopes and in other ophthalmic devices. Internal reflection can also be used to calculate the index of refraction of an unknown material.

The critical angle may be calculated by using Snell's Law and solving for the incident angle (θ) that yields a refracted angle (θ') of 90°:

$$n \sin\theta = n' \sin\theta' \quad \text{or} \quad n \sin\theta_c = n' \sin 90°$$

$$\sin\theta_c = \frac{n'}{n} \qquad (2\text{-}6)$$

$$\theta_c = \sin^{-1}\left(\frac{n'}{n}\right)$$

Example 2-j

Find the critical angle at an interface between glass (n = 1.50) and air. Show by example that an incident angle smaller than the critical angle will refract through the interface. Solve Snell's Law for an incident angle greater than the critical angle.

Known	***Unknown***	***Equations/Concepts***
Index before refraction - glass: $n = 1.50$	Critical angle: $\theta_c = ?$	(2-6) critical angle
Index after refraction - air: $n' = 1.00$		$\sin\theta_c = n'/n$

The first step is to solve for the critical angle by substituting the indices into Equation 2-6. Remember that the unprimed variables are before refraction, the primed variables are after refraction, and a critical angle exists only when light travels from a high to a low index interface (n > n').

$$\sin\theta_c = \frac{n'}{n} \quad \text{or} \quad \sin\theta_c = \frac{1.00}{1.50} = 0.667$$

$$\theta_c = \sin^{-1}(0.667) = 41.80°$$

An incident angle of 41.80°will yield a refracted angle of 90°. Any incident angle less than 41.8° should be refracted through the interface. As an example of an incident angle less than the critical angle, solve Snell's Law for an incident angle of 20°:

$$\theta = 20°$$
$$n \sin\theta = n' \sin\theta'$$
$$1.50 \sin(20) = 1.00 \sin\theta'$$
$$\sin\theta' = \frac{(1.50)(0.3420)}{1.00} = 0.5130$$
$$\theta' = \sin^{-1}(0.5130) = 30.87°$$

Therefore a ray incident at an angle of 20° will have a refracted angle of 30.87°.
What happens for an incident angle greater than the critical angle? Use an incident angle of 60°:

$$\theta = 60°$$
$$n \sin\theta = n' \sin\theta'$$
$$1.50 \sin(60) = 1.00 \sin\theta'$$
$$\sin\theta' = \frac{(1.50)(0.8660)}{1.00} = 1.2990$$
$$\theta' = \sin^{-1}(1.2990) = \text{error on calculator}$$

What happened? The $\sin^{-1}$ function is only defined between 1 and −1, and therefore an error is displayed when the inverse sine is attempted for a number greater than ± 1. This occurs for incident angles greater than the critical angle because this is the limiting angle for refraction. Since the incident angle of 60° is greater than the critical angle of 41.8° there is internal reflection according to Snell's Law.

$$\theta = \theta_r = 60°$$

and the angle of incidence of 60°, is reflected at an angle of 60° into the initial medium n.

Example 2-k

You are told that a new optical material has a critical angle in air of 34.4°. What is the index of refraction of this material?

Known	***Unknown***	***Equations/Concepts***
Critical angle: $\theta_c = 34.4°$	Index of refraction of material	(2-6) $\sin\theta_c = n'/n$
Surrounding medium index - air: n = 1.00		
n > n' for critical angle		

Since a critical angle is defined, the index of the material must be higher than air. The medium before refraction is the material with an index labeled n, and the medium after refraction is air with an index labeled n'. Substituting the known values into Equation 2-6 and solving for n (the material before refraction) gives

$$\sin\theta_c = \frac{n'}{n} \quad \text{or} \quad n = \frac{n'}{\sin\theta_c}$$

$$n = \frac{1.00}{\sin(34.4)} = 1.770 = \text{unknown index}$$

Fermat's Principle

The time (t) for light to travel a distance (d) in a medium with an index (n) may be expressed by

$$\text{time (t)} = \frac{\text{distance (d) traveled in medium}}{\text{velocity (v) of light in medium}} = \frac{d}{v} = \frac{\text{meters}}{\text{meters / sec}} = \text{sec}$$

Time to travel distance (d) may also be expressed in terms of the index of refraction of the medium (n). Keep in mind the definition of the index of refraction ($n = c/v$):

$$t = \frac{nd}{c} = \frac{(c/v)d}{c} = \frac{cd}{cv} = \frac{d}{v} \qquad (2\text{-}7)$$

The term **nd** is referred to as the **optical distance** or **optical path length.** From Equation 2-7 it is seen that the time to travel through some medium is the optical path length divided by the speed of light in a vacuum.

To calculate the time to travel through several media, sum the time to travel in each medium:

$$t = \frac{1}{c}\sum_{i=1}^{m} n_i d_i = \frac{1}{c}\{n_1 d_1 + n_2 d_2 + n_3 d_3 + \ldots n_m d_m\} \qquad (2\text{-}8)$$

Some relationships between optical path and time are summarized.

Principle of Least Time - the path traveled by light from one point to another in a medium will be the path that requires the least amount of time. This is also called **Fermat's Principle**. There are exceptions to this principle when curved surfaces are involved, but this is beyond the scope of this workbook.

Optical path - the path in which light travels within the medium.

Optical path length (nd) - the actual distance light travels within a medium times the index of the medium. Equal optical path lengths require equal travel time.

Example 2-1

What is the velocity of Na light in glass with an index of 1.50? How far will light travel in air in 10^{-8} seconds? Compare this distance to the distance light will travel in glass for the same time interval. What is the optical path length in each case?

Known	*Unknown*	*Equations/Concepts*
Index of refraction of glass: $n = 1.50$	Velocity of ray in media	(2-2) $v_m = c/n$
Velocity of light in air: $c = 3 \times 10^8$ m/s	Distance traveled in time	Velocity x time = distance
Time: $t = 10^{-8}$ seconds	Optical path length	(2-7) optical path length = nd

The velocity of the light in glass may be calculated by simply substituting the values into Equation 2-2 and solving for the velocity:

$$v_m = \frac{c}{n} = \frac{3 \times 10^8 \text{ m/s}}{1.50} = 2 \times 10^8 \text{ m/s}$$

The distance light travels in 10^{-8} seconds is calculated by multiplying the velocity times the time:

$$d_{air} = (c)(t) = (3 \times 10^8 \text{ m/s})(10^{-8} \text{ s}) = 3 \text{ meters}$$

$$d_{glass} = (v_{glass})(t) = (2 \times 10^8 \text{ m/s})(10^{-8} \text{ s}) = 2 \text{ meters}$$

Thus the light in this glass slows to 2/3 the velocity and travels 2/3 the distance when compared to light in air. The optical path length for these distances is

$$\text{optical path length} = n_{air}d_{air} = (1.00)(3.00 \text{ m}) = 3.00 \text{ m}$$
$$\text{optical path length} = n_{glass}d_{glass} = (1.50)(2.00 \text{ m}) = 3.00 \text{ m}$$

Thus for the same time interval, light travels the same optical path length.

Example 2-m

As shown in Figure 2-14, light travels normal through several plane interfaces of different indices: 3cm in index 1.33, 5cm in index 1.55, 10cm in index 1.70. How much time is required to travel from the first through the last interface?

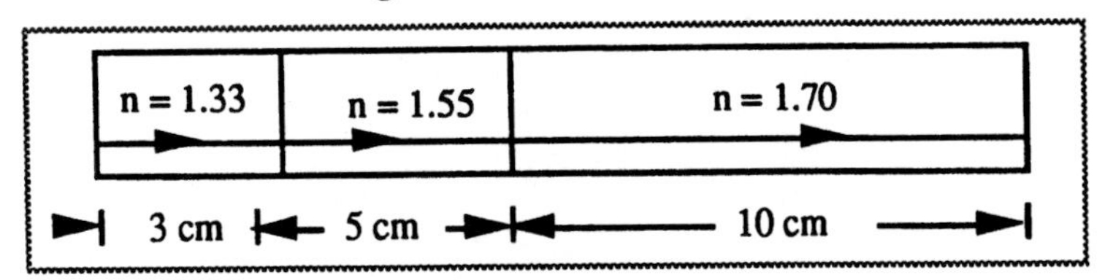

Figure 2-14. *Example 2-m.*

Known
d1 = 3cm; n1 = 1.33
d2 = 5cm; n2 = 1.55
d3 = 10cm; n3 = 1.70

Unknown
Time required to travel through the interfaces.

Equations/Concepts
(2-8) $t = (1/c) \sum n d$
$c = 3 \times 10^8$ m/s

Substitute the distances and indices given and diagrammed in Figure 2-14 into Equation 2-8 and solve for the time:

$$t = \frac{1}{c}\sum_{i=1}^{m} n_i d_i = \frac{1}{3\text{x}10^8}\sum_{i=1}^{3} n_i d_i$$

$$= \left(\frac{1}{3 \times 10^8}\right)\{n_1 d_1 + n_2 d_2 + n_3 d_3\}$$

$$= \left(\frac{1}{3 \times 10^8}\right)\{(1.33)(0.03 \text{ m}) + (1.55)(0.05 \text{ m}) + (1.70)(0.10 \text{ m})\}$$

$$= \left(\frac{1}{3 \times 10^8}\right)\{0.2874\} = 9.580 \times 10^{-10} \text{ seconds}$$

Refraction through Parallel-Sided Elements

Refraction through a parallel-sided element such as a block of glass, is shown in Figure 2-15. There are three media labeled in the diagram: n_1, the medium in which the incident ray travels or the incident medium; n_2, the internal medium; and n_3, the medium in which the ray travels after leaving the last interface or the emergent medium. Just as with a single refracting surface, the ray that stikes the first interface is called the **incident ray**, and the angle measured from the normal to the ray is called the **angle of incidence** or **incident angle.** With parallel-sided surfaces, the ray that leaves the last interface is called the **emergent ray**, and the angle this ray make with the normal to the surface is the **emergent angle.** These definitions would also hold true if there were a multiple number of parallel-sided interfaces (for example an aquarium that has interfaces of air-glass, glass-water, water-glass, and glass-air, as shown in *Example 2-n*).

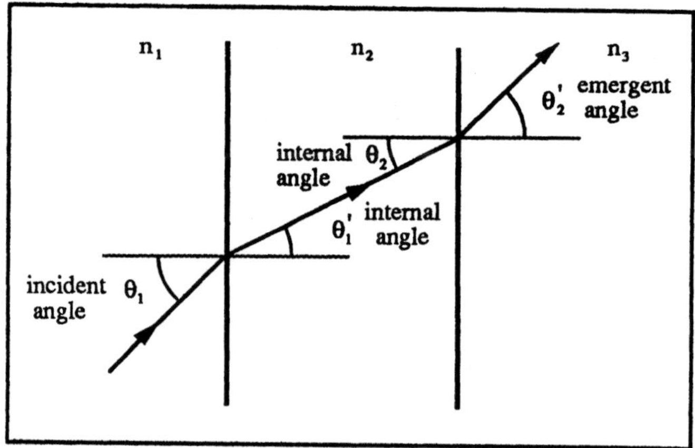

Figure 2-15. Refraction through a parallel sided surface.

The relationship between the internal refracted angles may be demonstrated with simple geometry. As illustrated in Figure 2-16, a line that intersects two parallel lines forms equal opposite internal angles. Using this premise, the internal angles within a parallel-sided medium are equal. In Figure 2-15, θ'_1 and θ_2 are equal.

When the incident and the emergent media (i.e., the media surrounding the parallel-sided surfaces) *are the same, the incident and emergent angles are equal.* This means that if the internal medium n_2 in Figure 2-15 were surrounded by air, for example ($n_1 = n_3 = 1.00$), θ_1 would equal θ'_2.

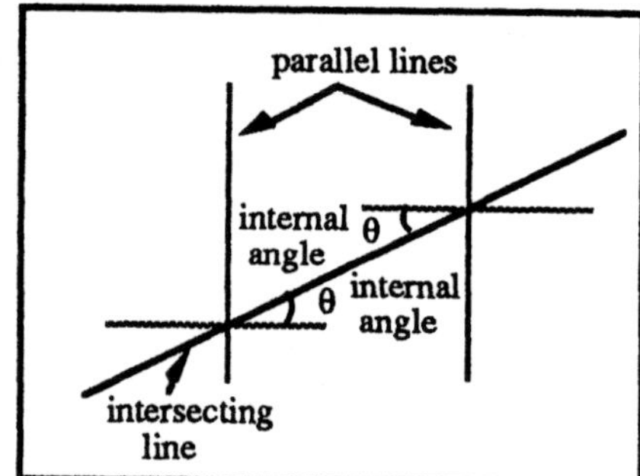

Figure 2-16. Opposite internal angles are equal.

Example 2-n

An incident ray traveling in air strikes a glass (n = 1.50) aquarium filled with water (n = 1.33). If the incident angle is 25°, trace the ray through the aquarium and solve for each of the refracted angles. Show that the emergent angle is equal to the incident angle. Draw to scale a ray traveling through the aquarium.

Known	*Unknown*	*Equations/Concepts*
Incident angle: $\theta = 25°$	Refracted angles	(2-5) Snell's Law $n \sin\theta = n' \sin\theta'$
Index of refraction of air: $n = 1.00$	Emergent angle	Snell's Law at each surface
Index of refraction of glass: $n = 1.50$		Incident and emergent angles are equal
Index of refraction of water: $n = 1.33$		Internal refracted angles are equal

The diagram in Figure 2-17 will be useful in showing the ray as it refracts through the aquarium. The indices and the incident ray and angle are labeled. The problem should be straightforward. Use Snell's Law at each interface to find the refracted angle. Use this angle at the next interface as the incident angle, and so on. To keep the calculation generic, the angles and indices will be relabeled according to the general Snell's Law

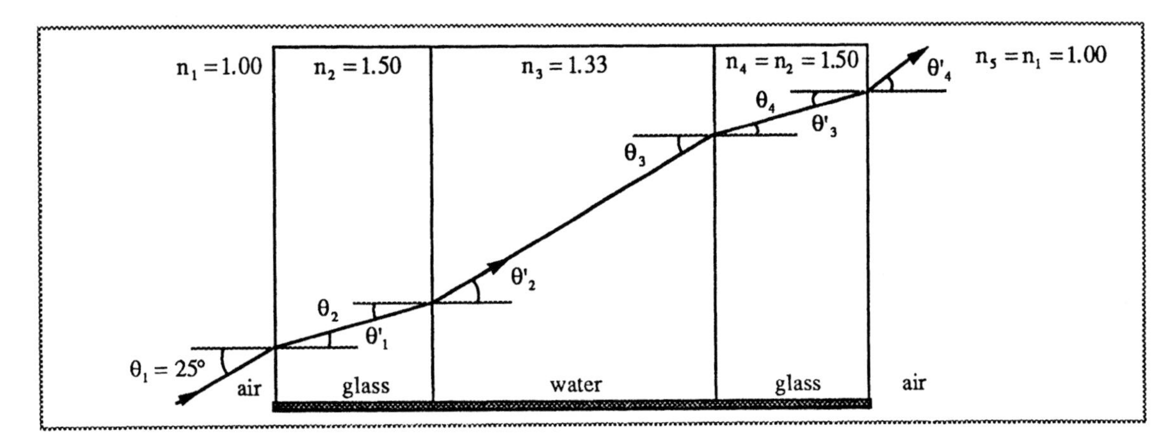

Figure 2-17. *Example 2-n.* Ray traced through an aquarium. The glass is thick to show the rays.

equation and then later relabeled to follow the diagram. The calculations follow:

At the first interface $n = n_1 = 1.00$, $n' = n_2 = 1.50$, and $\theta = \theta_1 = 25°$. To calculate the refracted angle θ'_1, use Snell's Law and solve for θ':

$$n \sin\theta = n' \sin\theta'$$
$$1.00 \sin(25°) = 1.50 \sin\theta'$$
$$\sin\theta' = \frac{(1.00)(0.4226)}{1.50} = 0.2817$$
$$\theta'_1 = \theta' = \sin^{-1}(0.2817) = 16.36°$$

Because the internal refracted angles are equal, $\theta'_1 = \theta_2 = 16.36°$ and angle θ_2 is the new incident angle at the next interface between n_2 and n_3. Using Snell's Law once again with $n = n_2 = 1.50$, $n' = n_3 = 1.33$, and $\theta = \theta_2 = 16.36°$, θ'_2 may be calculated as follows:

$$1.50 \sin(16.36°) = 1.33 \sin\theta'$$
$$\sin\theta' = \frac{(1.50)(0.2818)}{1.33} = 0.3178$$
$$\theta'_2 = \theta' = \sin^{-1}(0.3178) = 18.53°$$

Again because the internal refracted angles are equal, $\theta'_2 = \theta_3 = 18.53°$, and θ_3 becomes the new incident angle at the next interface between n_3 and n_4. Using Snell's Law once again with $n = n_3 = 1.33$, $n' = n_4 = 1.50$, and $\theta = \theta_3 = 18.53°$, θ'_3 may be calculated as follows:

$$1.33 \sin(18.53) = 1.50 \sin\theta'$$
$$\sin\theta' = \frac{(1.33)(0.3178)}{1.50} = 0.2818$$
$$\theta'_3 = \theta' = \sin^{-1}(0.2818) = 16.36°$$

Note that this is exactly the same angle formed between the glass-water interface. This indicates that the refracted angles will be equal regardless of the direction of light. This concept is discussed in the next section.

Continuing on through the last interface: $\theta'_3 = \theta_4 = 16.36°$, and θ_4 becomes the new incident angle at the next interface between n_4 and n_5.

Using Snell's Law once again with $n = n_4 = 1.50$, $n' = n_5 = 1.00$, and $\theta = \theta_4 = 16.36°$, θ'_4 may be calculated as follows:

$$1.50\sin(16.36°) = 1.00\sin\theta'$$

$$\sin\theta' = \frac{(1.50)(0.2818)}{1.00} = 0.4227$$

$$\theta'_4 = \theta' = \sin^{-1}(0.4227) = 25°$$

Therefore the emergent angle ($\theta'_4 = \theta' = 25°$) is equal to the incident angle. You could easily draw a ray traveling through the aquarium with a protractor and ruler. Note that the thickness of the glass of the aquarium does not affect the angles of refraction.

Example 2-o

Show that a ray will emerge from air-glass-water interfaces at the same angle as a ray that emerges from a single air-water interface.

Known	***Unknown***	***Equation/Concepts***
Index of refraction of air: n = 1.00	Emergent angle	(2-5) Snell's Law
Index of refraction of glass: n = 1.50		Internal refracted angles are equal
Index of refraction of water: n = 1.33		

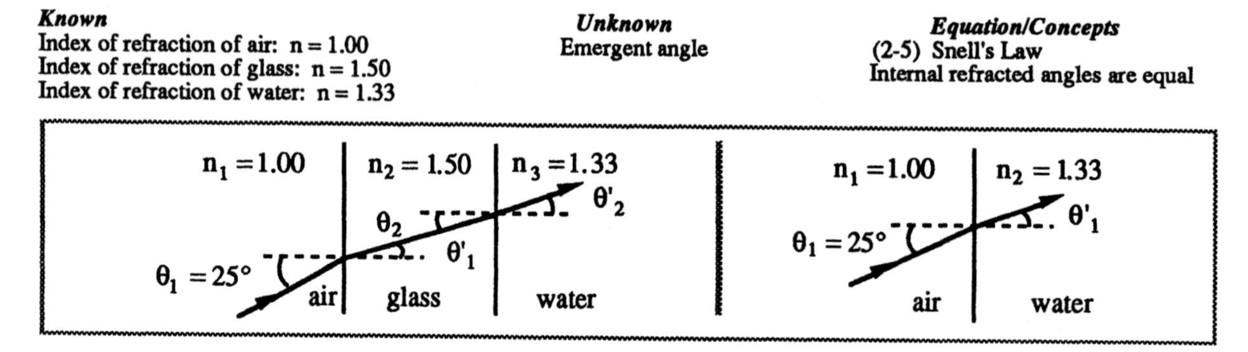

Figure 2-18. *Example 2-o.* The emerging angle (18.53°) will be the same in both cases above.

Use any incident angle and calculate the refracted angle at each interface. If we use the incident angle in *Example 2-n,* most of the calculations have been completed. For an incident angle of 25°, the refracted angle at the air-glass interface is 16.36°, and at the glass-water interface, it is 18.53°. This is the emergent angle for the first part of this problem. Now solve for the emergent angle for an air-water interface using Snell's Law:

$$n\sin\theta = n'\sin\theta'$$

$$1.00\sin(25°) = 1.33\sin\theta'$$

$$\sin\theta' = \frac{(1.00)(0.4226)}{1.33} = 0.3177$$

$$\theta' = \sin^{-1}(0.3177) = 18.53°$$

From this example it should be clear that the emergent angle is dependent on the medium in which the incident ray travels and the medium in which the emergent ray travels. The media between these interfaces do not change the angle of emergence. The ray will, however, be displaced a different amount. This is illustrated in later problems.

Lateral Displacement - the perpendicular distance between an incident and emerging ray after traveling through parallel sided interfaces. See Figure 2-19a, where d is the lateral displacement.

Example 2-p
Show the relationship between the thickness of a parallel-sided plate, the incident and refracted angles, the indices, and the lateral displacement. If a glass plate (n = 1.56) is 15 cm thick, what is the lateral displacement for a ray incident at 45°?

Known	***Unknown***	***Equations/Concepts***
Glass plate thickness: t = 15cm	Lateral displacement	(2-5) Snell's Law
Incident angle = 45°		Refraction through parallel-sided plate

Start with a diagram of a ray being refracted through a parallel sided plate (Figure 2-15, for example), extend the incident ray through the plate, and draw a line perpendicular from the incident to the emergent ray. In Figure 2-19b, several angles and triangles are labeled. The angle that encompasses both triangles is labeled θ_1, the refracted angle in the lower striped triangle is labeled θ'_1, and the angle in the shaded triangle is the difference between these two angles or $\{\theta_1-\theta'_1\}$ as indicated. The tangent of this angle is defined as

$$\tan\{\theta_1 - \theta'_1\} = \frac{d}{ab} \qquad (2\text{-}9)$$

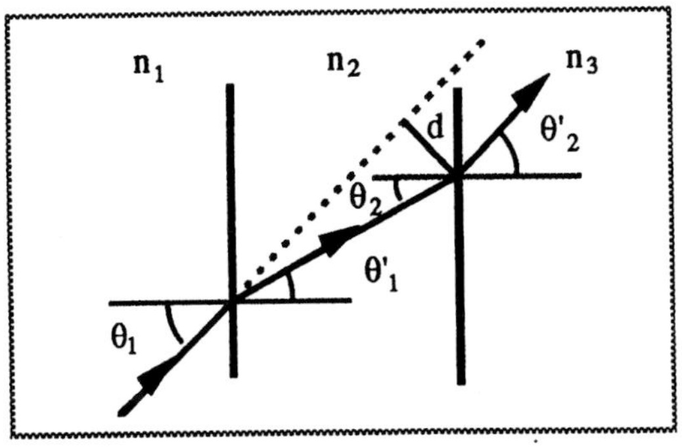

Figure 2-19a. *Example 2-p.* Parallel-sided plate with the lateral displacement labeled d.

where ab is the hypotenuse of the triangle, and d is the lateral displacement. From the striped triangle, calculate the value ab using the cosine function:

$$\cos\theta'_1 = \frac{ab}{t} \quad \text{or} \quad ab = t\cos\theta'_1 \qquad (2\text{-}10)$$

where: t = the thickness of the plate

Substituting for ab in Equation 2-9 and solving for d, this becomes

$$\tan\{\theta_1 - \theta'_1\} = \frac{d}{t\cos\theta'_1}$$
$$d = (t\cos\theta)\tan\{\theta_1 - \theta'_1\} \qquad (2\text{-}11)$$

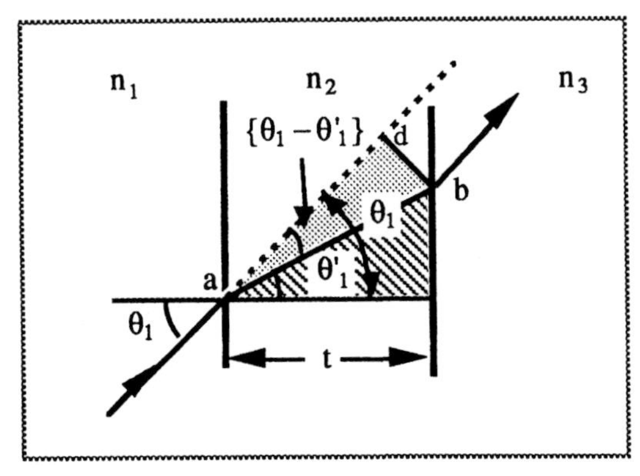

Figure 2-19b. *Example 2-p.* Triangles required for the derivation of lateral displacement.

For the incident angle of 45°, the refracted angle must be determined using Snell's Law:

$$n\sin\theta = n'\sin\theta' \qquad 1.00\sin 45° = 1.56\sin\theta' \qquad \theta' = 27°$$

Using Equation 2-11, solve for the displacement:

$$d = (t\cos\theta)\tan\{\theta_1 - \theta'_1\}$$
$$d = 15\cos(45°)\tan\{45° - 27°\} = 3.45 \text{ cm}$$

Reversibility of optical path - light travels along the same path regardless of the direction in which it is traveling.

Homocentric bundles - all rays in a pencil converge or diverge from a single point. This assumes that there are no aberrations in a system. For small pencils (small angles), light approximates homocentric bundles.

Apparent Position or Thickness

When an object in one medium is viewed from another medium, the apparent position of the object differs from its actual position. You may have experienced this phenomenon when viewing an under water object from above the surface of the water. This **apparent position** phenomenon may be explained by exploring the refraction of rays at the interface. As shown in Figure 2-20, rays from an object are refracted at a high-low index interface. The refracted angle is larger than the incident angle. Since we do not see refraction or the bending of the rays at the surface, the emergent rays appear to be coming (projecting back) from a position along the midline. Assuming that the bundle of rays is small and homocentric, all the rays appear to be coming from a position closer than the actual object. In trying to grab the object, one would undershoot the actual position.

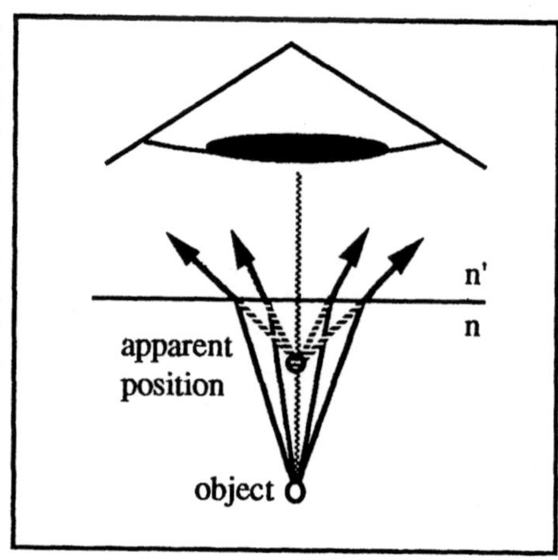

Figure 2-20. Apparent position of object in medium with higher index than medium viewed from.

Equation 2-12, may be used to calculate the relationship between the actual and apparent position of an object if the indices and actual object position are known:

$$\frac{n}{\ell} = \frac{n'}{\ell'} \tag{2-12}$$

where:

n = Index of space where real object is located
n' = Index of space from which object is viewed
ℓ = Actual distance of object from interface
ℓ' = Apparent distance of object (image) from interface

The apparent position may be closer or farther than the actual object position depending on the indices. This is shown in the following examples, along with a more formal explanation of Equation 2-12.

Example 2-q

You view from the air an object that is under water. Diagram and show the relationship between the actual and apparent position of the object.

This example is intended to demonstrate the apparent position relationship and to show a simplified derivation of Equation 2-12. First, the relationship between tangent, sine, and radians will be explored (Figure 2-21). Using this relationship, the derivation may be easily shown with the aid of Figure 2-22a, where one set of rays from the object is drawn. In this figure, the incident angle is θ, and the refracted angle is θ'; the apparent and actual positions of the object are ℓ and ℓ', respectively. Note that $n' < n$. Snell's Law is used to determine the refracted angle:

$$n \sin \theta = n' \sin \theta'$$

For small angles, the sine is approximately equal to the tangent. An easy way to demonstrate this relationship is to put angles into your calculator and determine the sine and tangent. For example, sin 1° and tan 1° are 0.174524 and 0.174550, respectively. Another way to approach this relationship is to look at the definition of the tangent:

$$\tan\beta = \frac{\sin\beta}{\cos\beta}$$

For 0°, the sine is zero, the cosine is one, and thus the tangent is zero. For small angles, the sine is approximately equal to the tangent because the cosine is approximately equal to one (i.e., the cos 1° is 0.999885), and dividing the sine by one yields the sine. The graph in Figure 2- 21, compares the sine, tangent, and radian measure. Note here again that for small angles, the sine, tangent, and radian measure are all equal. Therefore in this problem, if θ and θ' are small angles, the tangent may be substituted for the sine in Snell's Law:

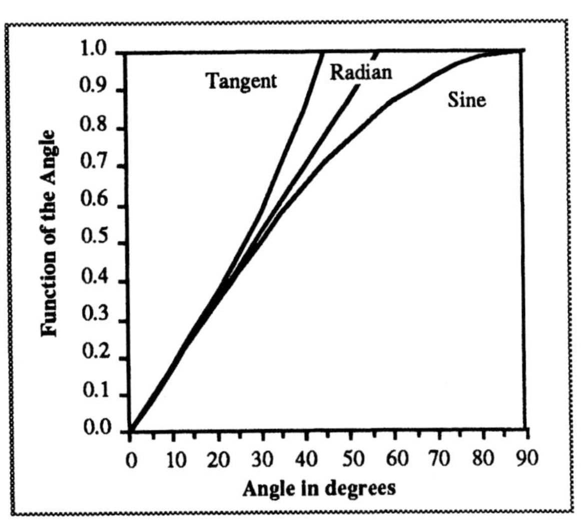

Figure 2-21. *Example 2-q.* Relationship of sine, tangent, and radians.

$$n \tan\theta = n' \tan\theta' \qquad (2\text{-}13)$$

Figure 2-22b shows triangles that may be used in the development of the apparent position relationship. The tangents may be rewritten in terms of distance by using the definition of tangent:

$$\tan\beta = \frac{\text{opposite side}}{\text{adjacent side}}$$

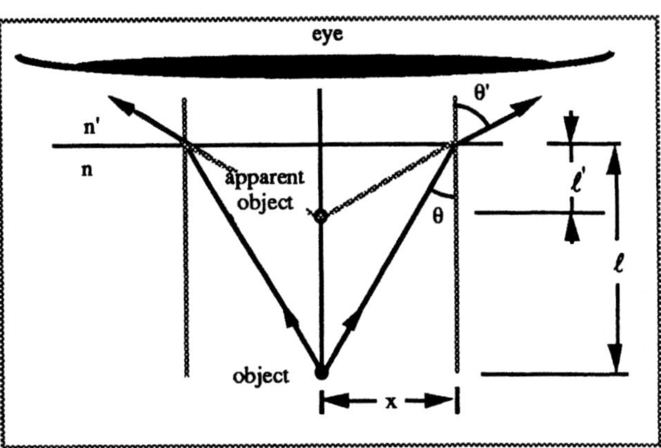

Figure 2-22a. *Example 2-q.* Apparent position of an object.

For the dashed triangle, the angle labeled θ' is equal to the refracted angle (opposite internal angles), and the tangent is equal to

$$\tan\theta' = \frac{x}{\ell'}$$

For the shaded triangle the tangent is equal to

$$\tan\theta = \frac{x}{\ell}$$

Substituting for the tangent Equation 2-13 becomes

$$n\left\{\frac{x}{\ell}\right\} = n'\left\{\frac{x}{\ell'}\right\}$$

Dividing both sides of the equation by x yields

$$\frac{n}{\ell} = \frac{n'}{\ell'} \qquad (2\text{-}12)$$

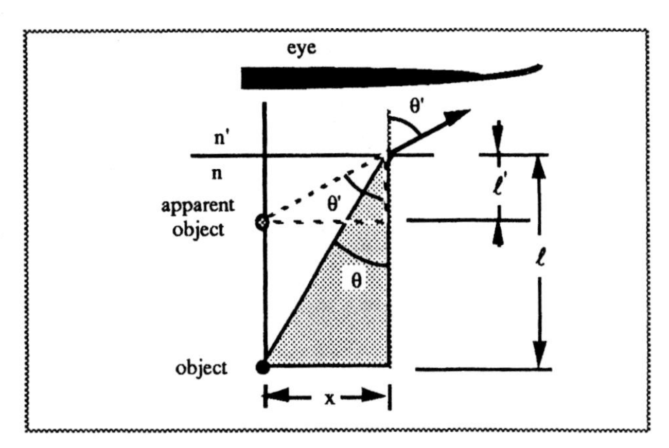

Figure 2-22b. *Example 2-q.* Derivation of the apparent position.

Example 2-r
A pebble located at the bottom of a fish tank appears to be 22.55 cm from the surface. What is the depth of the water in the tank?

Known
Index of refraction of water: n = 1.33
Index of refraction of air: n = 1.00
Apparent position of pebble: $\ell' = 22.55$ cm
Pebble located in water

Unknown
Depth of water
Actual position of pebble

Equations/Concepts
(2-12) $n/\ell = n'/\ell'$
Apparent position
Figure 2-3

Use the general rules for the variables: n = 1.33 (object in water); n' = 1.00 (viewing from air); ℓ = ? (actual distance of object from interface is the depth of the water); ℓ' = 22.55 cm (apparent position of object from interface). Substituting into the Equation 2-12 gives

$$\frac{n}{\ell} = \frac{n'}{\ell'} \quad \text{or} \quad \frac{1.33}{\ell} = \frac{1.00}{22.55}$$

Solving for ℓ:

$$\ell = (1.33)(22.55) = 29.99 \text{ cm} \approx 30.00 \text{ cm} = \text{depth of water}$$

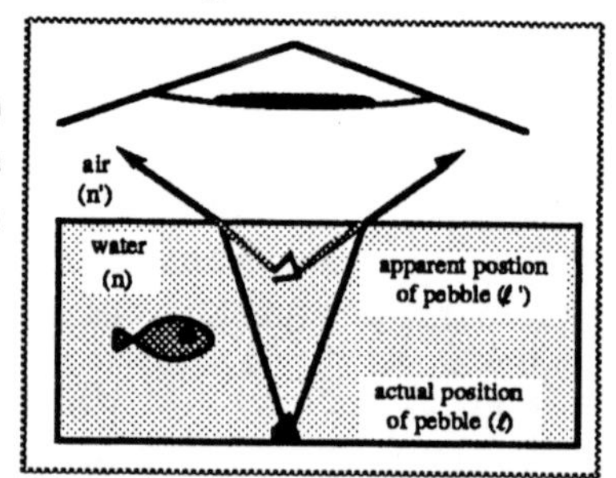

Figure 2-23. *Example 2-r.*

The apparent position is 7.45 cm (30.00 cm – 22.55 cm = 7.45 cm) closer than the actual pebble.

Example 2-s
A fish in the tank in *Example 2-r* views a fly that appears to be 7 cm above the water surface. How far above the surface is the fly?

Known
Index of refraction of water: n=1.33
Index of refraction of air: n = 1.00
Apparent position of fly in air: $\ell' = 7$ cm

Unknown
Actual position of fly: ℓ =?

Equations/Concepts
(2-12) $n/\ell = n'/\ell'$
Apparent position
Figure 2-4

Again use the definitions for the equation: n = 1.00 (index object located in); n' = 1.33 (index viewing from); ℓ = ? (actual distance object from interface); ℓ' = 7 cm (apparent distance of object from interface). Substituting into Equation 2-12 and solving for ℓ yields

$$\frac{n}{\ell} = \frac{n'}{\ell'} \quad \text{or} \quad \frac{1.00}{\ell} = \frac{1.33}{7.00}$$

$$\ell = \frac{7.00}{1.33} = 5.26 \text{ cm} = \text{actual position of fly}$$

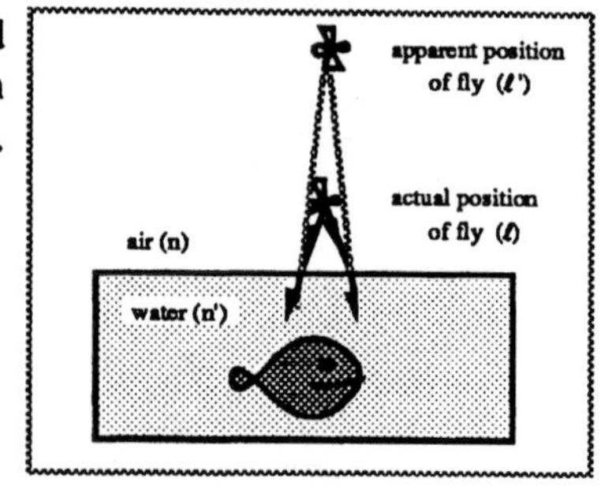

Figure 2-24. *Example 2-s.*

Note that the actual position of the fly is closer than the apparent position in this case.

From these examples, general rules about apparent position may be developed:

When viewing from a low index (n') into a higher index (n), the apparent position (ℓ') is closer than the actual position (ℓ).
When viewing from a high index (n') into a lower index (n), the apparent position (ℓ') is farther than the actual position (ℓ).

Apparent longitudinal displacement - the apparent position of an object viewed normally through a parallel-sided plate surrounded by the same medium. The displacement may be calculated using this equation:

$$d = \frac{t(n'-n)}{n'} \qquad (2\text{-}14)$$

In Equation 2-14, t is the thickness of the plate, n' is the index of the plate, n is the index of the surrounding medium, and d is the apparent displacement of the object. The displacement d is relative to the actual position of the object, *not the interface*. (See Figure 2-25.)

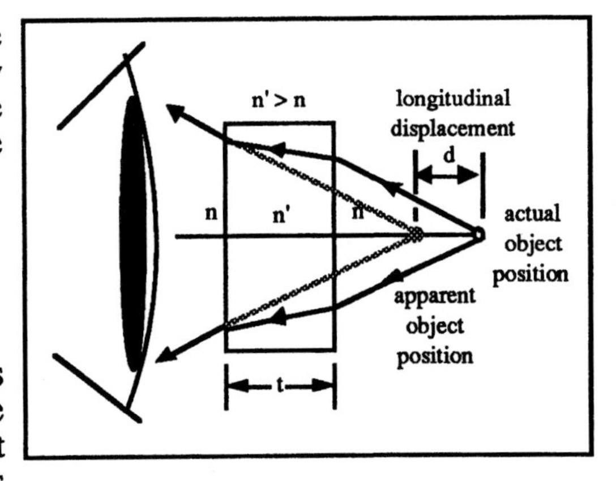

Figure 2-25. Displacement of object viewed through a parallel plate.

Apparent thickness - the apparent depth or thickness of a parallel-sided plate viewed normal to its surface.

Example 2-t
An object is viewed through a 5cm thick glass plate (n = 1.45) normal to the surface. The object is 20 cm from the front of the plate. Where is the apparent position of the object relative to the actual object position and to the back surface of the glass plate? What is the apparent thickness of the glass plate?

Known	***Unknown***	***Equations/Concepts***
Thickness of plate: t = 5 cm	Apparent position of object relative to actual object and back surface plate	(2-14) Displacement formula
Index of the glass: n = 1.45	Apparent thickness of plate	(2-12) Apparent position
Index of air: n = 1.00		
Object 20 cm from front of plate		

From Figure 2-25, the glass plate with index n' is surrounded by air (n = 1.00), and the displacement along the axis is calculated using Equation 2-14:

$$d = \frac{t(n'-n)}{n'} = \frac{(5)(1.45-1.00)}{1.45} = 1.55 \text{ cm}$$

The apparent position of the object is 1.55 cm from the actual object. In this example, the surrounding medium has a lower index than the parallel-sided plate, and therefore the apparent position is closer to the plate than the actual object (see Figure 2-25). The apparent position of the object relative to the back of the plate may be calculated by:

distance to object from front of plate – thickness of plate – displacement

20 cm – 5 cm – 1.55 cm = 13.45 cm

Label these distances in Figure 2-25, and show that this is the proper distance from the back of the plate to the apparent position.

The apparent thickness of the plate may be calculated using Equation 2-12. Here the object is the back surface of the plate so the actual object distance (ℓ) is equal to the thickness of the plate (t). The back of the glass plate has an index (n) of 1.45, and the viewing position is in air (n' = 1.00). Substituting into Equation 2-12 yields

$$\frac{n}{\ell} = \frac{n'}{\ell'} \quad \text{or} \quad \frac{1.45}{5.00} = \frac{1.00}{\ell'} \qquad \ell' = \frac{5.00}{1.45} = 3.45 \text{ cm}$$

The glass plate appears to be thinner than it actually is.

Apparent Position: Multiple Parallel-Sided Surfaces

Reduced distance - the thickness of a parallel sided plate or the path a ray travels divided by the index of the medium in which it travels:

$$\text{reduced distance} = \frac{t}{n} = \frac{\ell}{n} \tag{2-15}$$

The apparent position of an object viewed through multiple parallel-sided interfaces may be calculated by adding the reduced distance of each interface. The final position is determined relative to the last interface. This is written as

$$\frac{\ell'}{n'} = \sum_{i=1}^{m} \frac{\ell_i}{n_i} = \frac{\ell_1}{n_1} + \frac{\ell_2}{n_2} + \frac{\ell_3}{n_3} \cdots + \frac{\ell_m}{n_m} \tag{2-16}$$

where n' is the index from which the object is viewed, and ℓ' is the apparent position relative to the last interface.

Example 2-u

In Figure 2-26, an object is located under a tank filled with three media. The indices and thickness of the media are labeled. Where is the apparent position of the object viewed from the air above the tank?

Known
Thickness 1 = 3 cm; index n1 = 1.523
Thickness 2 = 2 cm; index n2 = 1.33
Thickness 3 = 5 cm; index n3 = 1.62
Object at bottom of tank
Viewed from air n' = 1.00

Unknown
Apparent position of object

Equations/Concepts
(2-16) Apparent position is sum of reduced distances
(2-15) Reduced distance

Solve for the apparent position using Equation 2-16:

$$\frac{\ell'}{n'} = \sum_{i=1}^{3} \frac{\ell_i}{n_i} = \frac{\ell_1}{n_1} + \frac{\ell_2}{n_2} + \frac{\ell_3}{n_3}$$

$$\frac{\ell'}{1.00} = \frac{3.00}{1.523} + \frac{2.00}{1.33} + \frac{5.00}{1.62} = 6.56 \text{ cm}$$

In this case, the apparent position of the object is 6.56 cm from the last interface of oil-air. The actual position of the object is 10 cm from the last interface. The apparent position is significantly closer than the actual object.

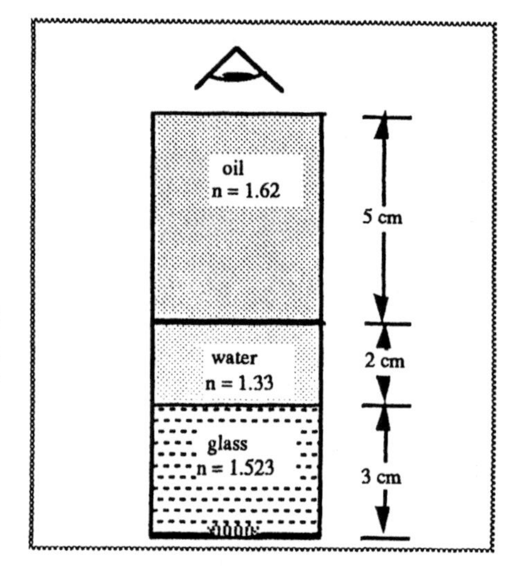

Figure 2-26. *Example 2-u.*

Supplemental Problems

Index of Refraction, Velocity of Light

2-1. What is the velocity of light from a sodium source in media with indices of 1.33, 1.60, and 1.80? How does the frequency and velocity vary as a function of the index?
ANS. 2.26 x 108 m/s, 1.88 x 108 m/s, 1.67 x 108 m/s. The velocity decreases as the index of refraction increases. The frequency remains constant.

2-2. An electromagnetic wave travels at 1.73×10^8 m/s. Calculate the index of refraction of the medium in which the wave is traveling.
ANS. $n = 1.73$

2-3. Light travels in a medium at 1.5×10^8 m/s. What is the index of the medium?
ANS. $n = 2.00$

2-4. If the wavelength of electromagnetic radiation in water is 562 nm, what is the wavelength in air?
ANS. $\lambda = 747$ nm

2-5. The velocity of Na light in a medium is 2.1×10^8 m/s. What is the index of refraction of the material?
ANS. $n = 1.43$

2-6. What is the approximate velocity of light in alcohol ($n = 1.36$)?
ANS. $v = 2.21 \times 10^8$ m/s

2-7. Light with a frequency of 4×10^{14} Hz travels in a medium with an index of 1.55. What is the wavelength of light in this medium?
ANS. $\lambda = 484$ nm

Reflected Light

2-8. Light is reflected normally from a water-glass interface. If 1.64% of the light is reflected, what is the index of the glass?
ANS. $n = 1.72$

2-9. What is the percentage of incident light reflected from a surface between air and plastic with a refractive index of 1.49?
ANS. 3.87% reflected

Snell's Law

2-10. A ray traveling in air is incident upon an interface of unknown medium at an angle of 36° from the normal. The refracted angle is half the reflected angle. What is the index of the unknown medium?
ANS. $n = 1.90$

2-11. When light strikes a glass-water ($n = 1.33$) interface at an angle of 49°, the ray emerges parallel to the interface. What is the index of the glass?
ANS. $n = 1.76$

2-12. Light is incident onto an interface of air to water ($n = 1.33$) at an angle of 15°. What is the angle of refraction?
ANS. 11.2°

2-13. An incident ray traveling in air strikes a flat plastic surface ($n = 1.49$) at an angle of 27° from the normal. What is the angle of refraction?
ANS. 17.74°

2-14. A ray is refracted at an angle of 35° from an oil-glass interface ($n = 1.73$ and 1.58, respectively). What is the incident angle?
ANS. 31.59°

Critical Angle

2-15. What is the critical angle at a glass-water interface ($n=1.67$ and 1.33, respectively)?
ANS. 52.8°

2-16. The critical angle at an interface between two media is 37°. Light travels in the second medium at a velocity of 3×10^8 m/s. What is the index of refraction of the first medium?
ANS. $n = 1.66$

2-17. Light traveling in a medium at 2.25×10^8 m/s strikes a second medium at an angle of 48.5° and emerges parallel to the interface. What is the index of refraction of the second medium?
ANS. $n = 1.00$

Apparent Position

2-18. Find the index of refraction of a piece of glass 3 cm thick if the apparent thickness under water is 2.22 cm.
ANS. $n = 1.80$

2-19. The olive in a martini cocktail ($n = 1.35$) appears to be 2 cm below the surface. What is the actual depth of the olive?
ANS. 2.70 cm

2-20. You view an object from the air through two pieces of glass ($n_1=1.45$ and $n_2=1.80$) with the same thickness. If the object appears to be 5.6 cm away, what is the thickness of each piece of glass?
ANS. 4.50 cm

2-21. You view a ring at the bottom of a beaker. The beaker contains two liquids: a bottom layer of water ($n = 1.33$) 3 cm deep and a top layer of oil ($n = 1.61$) 5 cm deep. Where does the ring appear to be relative to the top of the oil-air interface?
ANS. 5.4 cm

Chapter 3

Prisms

A prism, unlike a glass plate, does not have parallel sides, and therefore the internal refracted angles are typically *not* equal (except for a special case). Prisms may be used for a variety of purposes including changing the apparent position of an object, reinverting an image, and altering the direction that light travels. In an optometric practice, prisms are useful in measuring and correcting relative deviations (phorias and tropias) of the two eyes. Although prisms may also be designed to reflect light at certain surfaces, the emphasis of this chapter is on topics related to the refraction of light through prisms. For more information on reflecting and other types of prisms, consult your optics textbook.

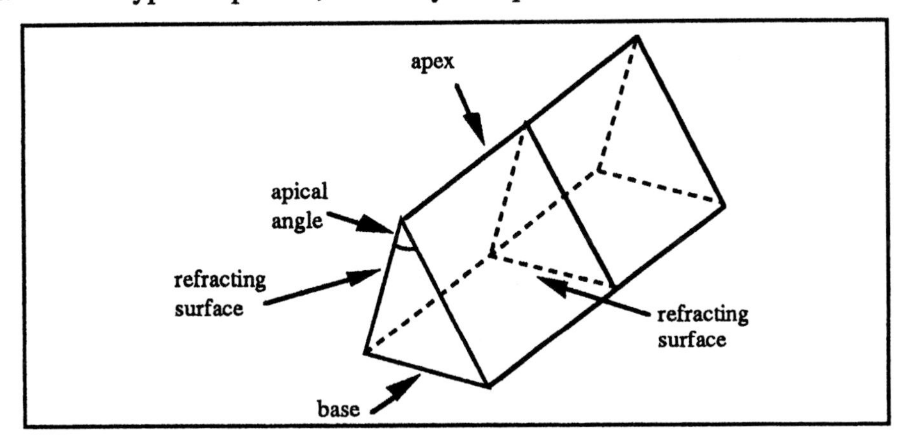

Figure 3-1. A prism with surfaces and angles.

The names of the surfaces and angles of a prism that are illustrated in Figure 3-1 are defined here:

Refracting surface - the incident ray strikes the first refracting surface, and the emergent ray leaves the other refracting surface.

Reflecting surface - (not shown in Figure 3-1) in certain prisms, the internal rays hit surfaces at angles greater than the critical angle and therefore reflect. In some reflecting prisms, there may be several reflecting surfaces.

Apical angle (a, γ) - the angle between the two refracting surfaces in a standard refracting prism. This angle may also be called the **Refracting angle**.

Apex - the tip of the prism where the two refracting surfaces meet. The apical angle is at the apex of the prism.

Base - the bottom of the prism or the side opposite the apex or apical angle. The orientation of an ophthalmic prism is described relative to the base.

Refraction of a Ray through a Prism

The procedure for refracting light through a prism involves the same principles described for refraction through parallel-sided plates. A common path a refracted ray takes through a prism is shown in Figure 3-2. The incident ray usually starts below the base of the prism, strikes the first refracting surface, where it is refracted into the prism. The ray then travels in the prism to the second refracting surface and emerges after refracting at the second

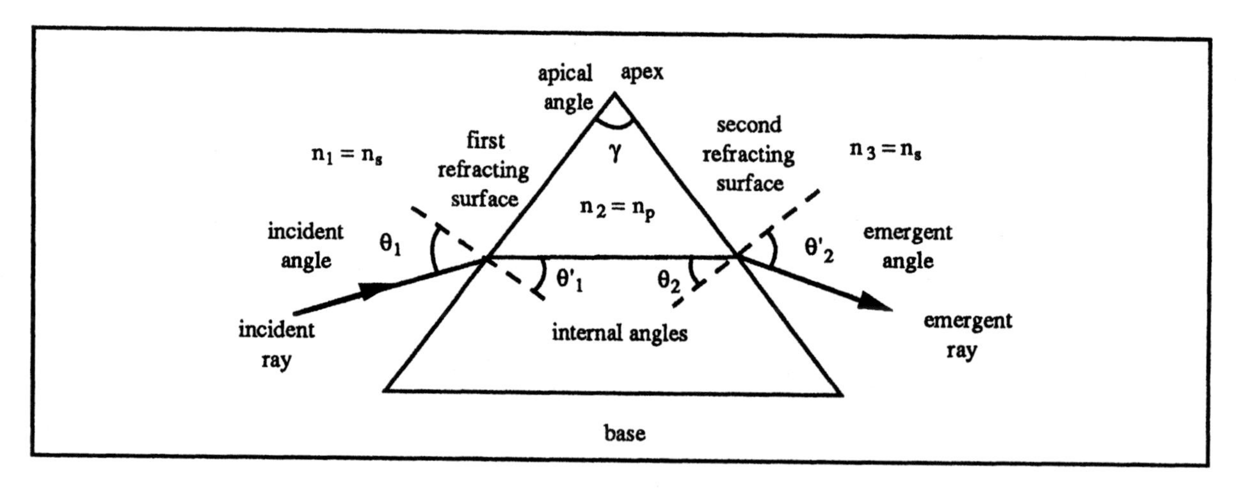

Figure 3-2. A ray refracted through a prism.

refracting surface. The incident and emergent angles formed by the ray are labeled θ_1 and θ'_2, respectively; the internal angles are labeled θ'_1 and θ_2; and the apical angle is γ. As previously defined (and shown in the figure), all incident and refracted angles are measured relative to a line normal to the refracting surface. At each refracting surface, Snell's Law is used to find the refracted angle. To calculate the first refracted angle θ'_1, use Snell's Law:

$$n_1 \sin\theta_1 = n_p \sin\theta'_1 \qquad (3\text{-}1a)$$

where: n_1 = the index of refraction of the medium for the incident ray
$n_p = n_2$ = the index of refraction of the prism
n_3 = the index of refraction of the medium for the emergent ray

The relationship between the two internal angles may be determined by

$$\gamma = \theta'_1 + \theta_2 \qquad (3\text{-}2)$$

Use Snell's Law at the second refracting surface to find the emergent angle θ'_2.

$$n_p \sin\theta_2 = n_3 \sin\theta'_2 \qquad (3\text{-}3a)$$

If the media surrounding the prism are equal ($n_1 = n_3 = n_s$), Equations 3-1a and 3-3a become

$$n_s \sin\theta_1 = n_p \sin\theta'_1 \qquad (3\text{-}1b) \qquad\qquad n_p \sin\theta_2 = n_s \sin\theta'_2 \qquad (3\text{-}3b)$$

Example 3-a
Using simple geometry, show that the sum of the internal refracted angles of a ray are equal to the apical angle (Equation 3-2).

From Figure 3-3, a triangle that contains the apical angle is outlined and labeled ΔABC. As shown, from the normal to the surface is a 90° angle. The internal angles that the ray makes with these normals are part of each right angle. The remainder of the right angles are contained within the outlined triangle angles and are represented by $(90^\circ - \theta'_1)$ and $(90^\circ - \theta_2)$. Using the simple geometric fact that the sum of the internal angles of a triangle is equal to 180°, the relationship between the internal angles and the apical angle is easily derived:

$$(90^\circ - \theta'_1) + (90^\circ - \theta_2) + \gamma = 180^\circ$$

Combining terms and solving for γ yield

$$180^\circ - \theta'_1 - \theta_2 + \gamma = 180^\circ$$

$$\gamma = \theta'_1 + \theta_2 \qquad (3\text{-}2)$$

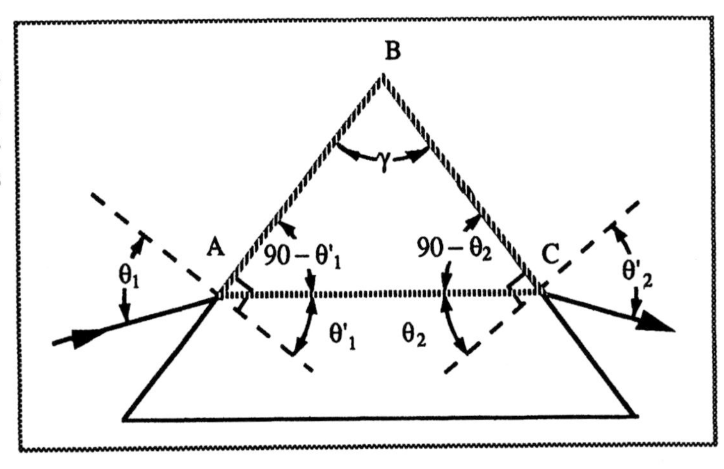

Figure 3-3. *Example 3-a.* Relationship between the internal the angles and apical angle.

Example 3-b
A prism has an apical angle of 35° and an index of refraction of 1.49. If the incident angle is 15°, trace the ray through the prism and calculate the emergent angle.

Known
Apical angle of prism: $\gamma = 35^\circ$
Index of refraction of prism: $n_p = 1.49$
Incident angle: $\theta_1 = 15^\circ$
Index of the surrounding media: $n_s = 1.00$

Unknown
Refracted angle: θ'_1
Internal angle: θ_2
Emergent angle: θ'_2

Equations/Concepts
(3-1) and (3-3) Prism equations
(3-4) $\theta_2 = \gamma - \theta'_1$

Start by solving for the first refracted angle using Snell's Law and the index of the prism (Equation 3-1). Assume the prism is surrounded by air when no surrounding medium is given.

$$n_s \sin\theta_1 = n_p \sin\theta'_1$$
$$1.00\sin(15^\circ) = 1.49\sin\theta'_1$$
$$\sin\theta'_1 = \frac{0.2588}{1.49} = 0.1737$$
$$\theta'_1 = \sin^{-1}(0.1737) = 10.0^\circ$$

Solve for the internal angle θ_2 using Equation 3-2:

$$\theta_2 = \gamma - \theta'_1 \qquad \theta_2 = 35^\circ - 10^\circ = 25^\circ$$

Use this angle to calculate the emergent angle θ'_2 using Equation 3-3:

$$n_p \sin\theta_2 = n_s \sin\theta'_2$$
$$1.49\sin(25^\circ) = 1.00\sin\theta'_2$$
$$\sin\theta'_2 = (1.49)(0.4226) = 0.6297$$
$$\theta'_2 = \sin^{-1}(0.6297) = 39^\circ = \text{Emergent angle}$$

Example 3-c

A ray emerges from a 40° plastic (n = 1.49) prism at an angle of 25°. If the prism is under water, what is the incident angle at the first refracting surface of the prism?

Known

Emergent angle: $\theta'_2 = 25°$

Index of water: $n_s = 1.33$

Index of prism: $n_p = 1.49$

Apical angle: $\gamma = 40°$

Unknown

Incident angle: $\theta_1 = ?$

Equations/Concepts

(3-1), (3-2), (3-3)

Refraction through a prism

The incident angle is determined by tracing the ray backward through the prism (i.e., using Equations 3-3, 3-2, and 3-1). First the internal refracted angle (θ_2) at the second refracting surface is calculated. Be sure to note that the surrounding medium is water.

$$n_p \sin\theta_2 = n_s \sin\theta'_2$$

$$1.49\sin(\theta_2) = 1.33\sin(25°)$$

$$\sin\theta_2 = \frac{0.5621}{1.49} = 0.3772$$

$$\theta_2 = \sin^{-1}(0.3772) = 22.16°$$

Calculate the internal refracted angle at the first refracting surface (θ'_1) using Equation 3-2:

$$\theta'_1 = \gamma - \theta_2 \qquad \theta'_1 = 40° - 22.16° = 17.84°$$

Solve for the incident angle (θ_1) using Equation 3-1:

$$n_s \sin\theta_1 = n_p \sin\theta'_1$$

$$1.33\sin\theta_1 = 1.49\sin(17.84°)$$

$$\sin\theta_1 = \frac{0.4564}{1.33} = 0.3432$$

$$\theta_1 = \sin^{-1}(0.3432) = 20.07°$$

Example 3-d

An incident ray is refracted at the first refracting surface of a prism at an angle of 17° and is internally incident on the second refracting surface at an angle of 17°. What is the refracting angle of the prism?

Known

Internal refracted angle, first surface: $\theta'_1 = 17°$

Internal angle, second surface: $\theta_2 = 17°$

Unknown

Refracting angle: $\gamma = ?$

Equations/Concepts

(3-2) Internal angles of prism

Refracting angle = apical angle

The refracting or apical angle of the prism is equal to the sum of the internal angles made by the ray (Equation 3-2):

$$\gamma = \theta'_1 + \theta_2 \qquad \gamma = 17° + 17° = 34° = \text{refracting angle}$$

Deviation of a Prism

The angular bending that the incident ray undergoes after refraction through the prism is called the **deviation δ**. The deviation is used to describe the power of a prism. The larger the deviation, the more the incident light is bent and the higher the prismatic power. Figure 3-4 shows the deviation an incident ray undergoes at each refracting surface (δ_1, δ_2). The sum of the deviations at each surface is the total deviation:

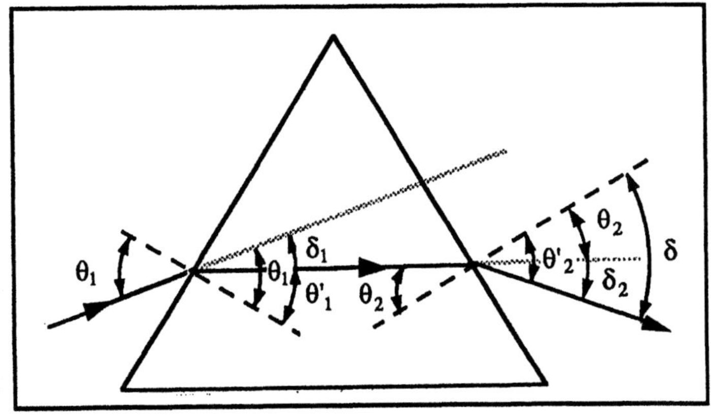

Figure 3-4. Deviation of a prism.

$$\delta = \delta_1 + \delta_2 \qquad (3\text{-}4)$$

The total deviation of a prism δ may be calculated in terms of the incident, emergent, and apical angles:

$$\delta = \theta_1 + \theta'_2 - \gamma \qquad (3\text{–}5)$$

Example 3-e

Using the fact that the total deviation of a prism is the sum of the deviation at each surface, show the relationship between deviation and the incident, emergent, and apical angles.

The deviation at the first refracting surface may be expressed in terms of the incident and refracted angles θ_1 and θ'_1. Note that in Figure 3-4, θ_1 is labeled inside the prism (because opposite internal angles are equal) and that the sum of the deviation δ_1 and the refracted angle θ'_1 is equal to the incident angle θ_1:

$$\theta_1 = \delta_1 + \theta'_1 \quad \text{or} \quad \delta_1 = \theta_1 - \theta'_1 \qquad (3\text{-}6)$$

At the second refracting surface, the internal angle θ_2 is labeled outside the prism. The sum of θ_2 and the deviation at the second surface δ_2 is equal to the emergent angle θ'_2:

$$\theta'_2 = \theta_2 + \delta_2 \quad \text{or} \quad \delta_2 = \theta'_2 - \theta_2 \qquad (3\text{-}7)$$

Substituting for δ_1 and δ_2 into Equation 3-4, you can solve for the total deviation:

$$\delta = \delta_1 + \delta_2 = (\theta_1 - \theta'_1) + (\theta'_2 - \theta_2)$$

Collecting terms this becomes:

$$\delta = \theta_1 + \theta'_2 - \theta'_1 - \theta_2 \quad \text{or} \quad \delta = \theta_1 + \theta'_2 - (\theta'_1 + \theta_2)$$

Because $\theta'_1 + \theta_2 = \gamma$ or the apical angle of the prism, the equation becomes

$$\delta = \theta_1 + \theta'_2 - \gamma \qquad (3\text{-}5)$$

Example 3-f
What is the total deviation and the deviation at each refracting surface for the prism in *Example 3-b*?

Known	***Unknown***	***Equations/Concepts***
See *Example 3-b* for information See Figure 3-4	Deviation at each surface Total deviation	(3-4), (3-5), (3-6), (3-7) *Example 3-e*

Using the information from *Example 3-b* ($\theta_1 = 15°$, $\theta'_1 = 10.0°$, $\theta_2 = 25°$, $\theta'_2 = 39°$) and substituting in Equations 3-6 and 3-7, solve for the deviations:

$$\delta_1 = \theta_1 - \theta'_1 \qquad \delta_1 = 15° - 10° = 5°$$

$$\delta_2 = \theta'_2 - \theta_2 \qquad \delta_2 = 39° - 25° = 14°$$

Total deviation may be solved either with Equation 3-4 or Equation 3-5. Using Equation 3-4

$$\delta = \delta_1 + \delta_2 \qquad \delta = 5° + 14° = 19°$$

Using Equation 3-5

$$\delta = \theta_1 + \theta'_2 - \gamma \qquad \delta = 15° + 39° - 35° = 19°$$

Example 3-g
A glass prism has an apical angle of 38°. If a ray is incident on the first refracting surface at an angle of 10°, the total deviation of the prism is 21°. What is the emergent angle at the second refracting surface?

Known	***Unknown***	***Equations/Concepts***
Incident angle: $\theta_1 = 10°$ Apical angle: $\gamma = 38°$ Total deviation: $\delta = 21°$	Emergent angle: $\theta'_2 = ?$	(3-5) Deviation of a prism

Here is one of those problems that students think lack sufficient information. If you try to refract through the prism to find the emergent angle, indeed there is not enough information. To solve the problem, you must use the deviation information given, and solve for the emergent angle using Equation 3-5:

$$\delta = \theta_1 + \theta'_2 - \gamma \quad \text{or} \quad \theta'_2 = \delta + \gamma - \theta_1 \qquad \theta'_2 = 21° + 38° - 10° = 49°$$

In Figure 3-5, a plot of deviation as a function of apical angle is shown for a constant incident angle. It can be seen that the deviation of a prism increases as the apical angle increases. The amount of deviation and the resulting change in the direction of the emergent ray may be controlled by altering the apical angle of the prism. This concept is used in the design of a special lens called a Fresnel lens where more deviation is required for off-axis lens positions than for on-axis lens positions (off-axis prisms have larger apical angles than on-axis prisms). Consult your textbook for more details on Fresnel lenses.

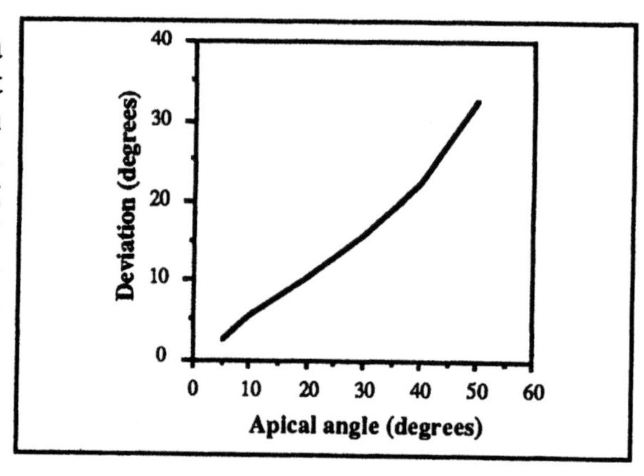

Figure 3-5. Relation of deviation and apical angle.

Normal Incidence and Emergence

The equations for refraction through a prism and the resulting deviation are simplified if the incident or emergent ray leaves along the normal to the surface or at 0°. (See Figure 3-6.)

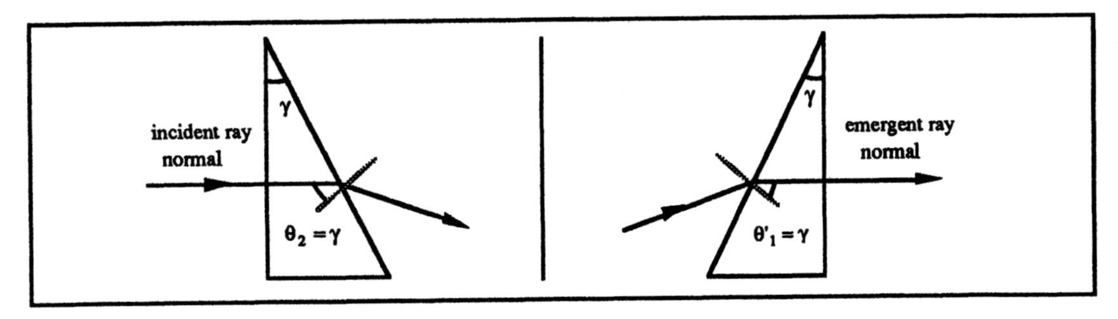

Figure 3-6. Normal incidence on and normal emergence from a prism.

For *normal incidence*, the incident ray strikes the prism perpendicular to the first refracting surface ($\theta_1 = 0°$). Substituting $\theta_1 = 0°$ into the formulas for refraction through a prism and deviation (Equations 3-1, 3-2, 3-3, 3-5), the following relationships are found:

$$\theta_1 = 0° \qquad \theta'_1 = 0° \qquad \gamma = \theta_2 \qquad \delta = \theta'_2 - \gamma$$

For *normal emergence* from a prism, the emergent ray leaves the prism perpendicular to the second refracting surface ($\theta'_2 = 0°$), resulting in these relationships:

$$\theta'_2 = 0° \qquad \theta_2 = 0° \qquad \gamma = \theta'_1 \qquad \delta = \theta_1 - \gamma$$

These relationships are extremely helpful in solving certain prism problems. Memorizing these relationships is not recommended, however. Be familiar with the general prism equations, and these relationships are easily derived, as shown in the following example.

Example 3-h
Using the refracting and the deviation equations (3-1, 3-2, 3-3, and 3-5), derive the relationships for normal incidence.

Substituting $\theta'_1 = 0°$ into the equations, one can easily derive the relationships described:

(3-1) $\qquad n_s \sin\theta_1 = n_p \sin\theta'_1 \qquad \sin\theta'_1 = \dfrac{n_s \sin 0°}{n_p} \qquad \sin\theta'_1 = 0°$

$$\theta'_1 = \sin^{-1}(0°) \qquad \theta'_1 = 0°$$

(3-2)

$$\gamma = \theta'_1 + \theta_2 = 0° + \theta_2$$
$$\gamma = \theta_2$$

(3-5)

$$\delta = \theta_1 + \theta'_2 - \gamma = 0° + \theta'_2 - \gamma$$
$$\delta = \theta'_2 - \gamma \qquad \text{or} \qquad \theta'_2 = \delta + \gamma$$

Example 3-i
Light is incident on the front refracting surface of a prism such that the refracted angle is 25° and the emergent ray leaves normal to the back refracting surface. Under these conditions, the deviation of the prism is 22.7°. What is the index of refraction of the prism?

Known	***Unknown***	***Equations/Concepts***
Refracted angle at first surface: $\theta'_1 = 25°$	Index of refraction of prism:	(3-2), (3-3), (3-4), (3-5)
Emergent ray normal: $\theta'_2 = 0°$	$n_p = ?$	Refraction through a prism
Deviation of prism: $\delta = 22.7°$		Normal emergent ray relationships

The relationships between the incident and refracted angles and the fact that the ray leaves normal to the second refracting surface is required to solve this problem. Since the ray leaves normal to the second surface, we know

$$\theta_2 = \theta'_2 = 0°$$

$$\gamma = \theta'_1 + \theta_2 \text{ and substituting } 0° \text{ for } \theta_2;\ \gamma = \theta'_1 = 25°$$

Solving the Equation 3-5 for the incident angle by substituting the known values of deviation, emergent angle, and apical angle gives

$$\delta = \theta_1 + \theta'_2 - \gamma$$

$$\theta_1 = \delta + \gamma = 22.7° + 25° = 47.7°$$

Use Equation 3-1 to solve for the index of the prism by substituting the values for θ_1 and θ'_1:

$$n_s \sin\theta_1 = n_p \sin\theta'_1 \qquad 1.00\sin(47.7°) = n_p \sin(25°)$$

$$n_p = \frac{1.00\sin(47.7°)}{\sin(25°)} = \frac{0.7396}{0.4226} = 1.75 = \text{index of prism}$$

Eample 3-j
What is the apical angle of a plastic prism (n = 1.49) if a normal incident ray yields an emergent angle of 32°?

Known	***Unknown***	***Equations/Concepts***
Index of refraction of prism: $n_p = 1.49$	Apical angle: $\gamma = ?$	(3-2), (3-3), (3-4), (3-5)
Incident angle at first surface: $\theta_1 = 0°$	Deviation of prism: $\delta = ?$	Refraction through a prism
Emergent angle: $\theta'_2 = 32°$		Normal incident ray relationships

For normal incidence, the apical angle is equal to the internal angle at the second surface ($\gamma = \theta_2$). Because the emergent angle is known ($\theta'_2 = 32°$), use Equations 3-3 and 3-2 to calculate the internal angle and then the apical angle:

$$n_p \sin\theta_2 = n_s \sin\theta'_2 \qquad 1.49\sin\theta_2 = 1.00\sin(32°) \qquad \sin\theta_2 = \frac{0.5299}{1.49} = 0.3557$$

$$\gamma = \theta_2 = \sin^{-1}(0.3557) = 20.83° = \text{apical angle}$$

Limitations on Refraction through a Prism

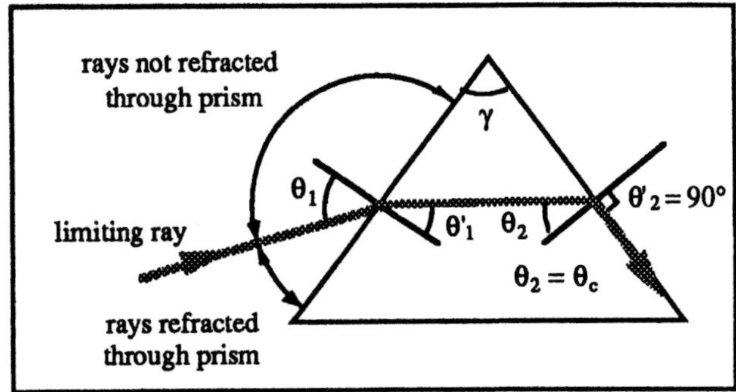

Figure 3-7. Limitations on refraction through a prism.

Light will not be refracted through a prism if the internal angle at the second refracting surface is greater than the critical angle. (See Figure 3-7.) Remember that the critical angle is the incident angle that yields a refracted angle of 90° and this angle only exists at a high-low index interface. In determining the limiting incident ray, we must ask this question: What incident angle at the first refracting surface of the prism will yield the critical angle at the second refracting surface? To answer this question, we must solve the formula for refracting through a prism backward, starting with an emergent angle of 90°. This process is shown in the following example.

Example 3-k

A plastic 60° prism with an index of refraction of 1.49 has an emergent angle of 90°. What is the critical angle of the prism? What is the incident angle at the first surface?

Known	***Unknown***	***Equations/Concepts***
Apical angle of prism: $\gamma = 60°$	Critical angle of material	(2-6) Critical angle: $\sin\theta = n'/n$
Index of prism: $n_p = 1.49$	Incident angle: $\theta_1 = ?$	(3-1), (3-2), (3-3)
Emergent angle: $\theta'_2 = 90°$		Refraction through prism
Index of surround: $n_s = 1.00$		backward using formula

First calculate the critical angle of the material (the critical angle has nothing to do with the prism itself, only the index of refraction). Use n as the index of the prism and n' as the index of air (high-low (n-n') interface):

$$\sin\theta_c = \frac{n'}{n} = \frac{1.00}{1.49} = 0.6711$$

$$\theta_c = \sin^{-1}(0.6711) = 42.16°$$

Because, in this problem, the emergent angle (θ'_2) is 90°, the incident angle at the second refracting surface (θ_2) is equal to the critical angle (θ_2). Using these values, calculate the first refracted angle with Equation 3-2:

$$\gamma = \theta'_1 + \theta_2$$

$$\theta'_1 = \gamma - \theta_2 = 60° - 42.16° = 17.84°$$

Use Equation 3-1 to solve for the incident angle ($n_s = 1.00$; $n_p = 1.49$; $\theta'_1 = 17.84°$):

$$n_s \sin\theta_1 = n_p \sin\theta'_1 \qquad 1.00 \sin\theta_1 = 1.49 \sin(17.84°)$$

$$\sin\theta_1 = \frac{(1.49)(0.3064)}{1.00} = 0.4565$$

$$\theta_1 = \sin^{-1}(0.4565) = 27.16°$$

An incident angle of 27.16° will have an emergent angle of 90°. This is the limiting incident angle.

What will happen to incident angles greater or less than this limiting incident angle? Let's try to refract through the prism with two incident angles: one larger (35°) and one smaller (15°) than this incident angle.
For $\theta_1 = 35°$

$$n_s \sin\theta_1 = n_p \sin\theta'_1 \qquad 1.00\sin(35°) = 1.49\sin\theta'_1$$

$$\sin\theta'_1 = \frac{(1.00)(0.5736)}{1.49} = 0.3850$$

$$\theta_1 = \sin^{-1}(0.3850) = 22.64°$$

$$\theta_2 = \gamma - \theta'_1 = 60° - 22.64° = 37.36°$$

$$n_p \sin\theta_2 = n_s \sin\theta'_2 \qquad 1.49\sin(37.36°) = 1.00\sin\theta'_2$$

$$\sin\theta'_2 = \frac{(1.49)(0.6068)}{1.00} = 0.9041$$

$$\theta'_2 = \sin^{-1}(0.9041) = 64.71°$$

For an incident angle greater than the limiting incident angle which yields the critical angle to the second refracting surface, light is refracted through the prism.

For $\theta_1 = 15°$

$$n_s \sin\theta_1 = n_p \sin\theta'_1 \qquad 1.00\sin(15°) = 1.49\sin\theta'_1$$

$$\sin\theta'_1 = \frac{(1.00)(0.2588)}{1.49} = 0.1737$$

$$\theta_1 = \sin^{-1}(0.1737) = 10.00°$$

$$\theta_2 = \gamma - \theta'_1 = 60° - 10.00° = 50.00°$$

$$n_p \sin\theta_2 = n_s \sin\theta'_2 \qquad 1.49\sin(50.00°) = 1.00\sin\theta'_2$$

$$\sin\theta'_2 = \frac{(1.49)(0.7660)}{1.00} = 1.1414$$

$$\theta'_2 = \sin^{-1}(1.1414) = \text{error}$$

Because the sine of the angle θ'_2 is undefined, the incident angle at the second surface is greater than the critical angle, and there is internal reflection. This indicates that rays at incident angles (at the first refracting surface) that are smaller than the limiting incident angle will not be refracted directly through the prism.

From this example it should be clear that there are limitations for refraction through a prism (Figure 3-7). To determine these limitations, first solve for the critical angle of the prism material. Keep in mind that the surrounding medium may not be air. Then solve the prism equations backward for the limiting incident angle θ_1. Incident angles greater than this angle will be refracted through the prism. Incident angles less than this limit will not be refracted through the prism and the ray will undergo internal reflection at the second surface.

The above relationship defines the minimum incident angle that can be refracted through the prism. What limits the maximum angle? The largest incident angle possible is 90°. As shown in Figure 3-8, this incident angle can be refracted through the prism with the same restrictions as in the previous example (i.e., the ray at the second refracting surface must be less than the critical angle). In this case, the apical angle of the prism determines what incident angles will be refracted through the prism. The relationship below can be seen in Figure 3-8:

$$\theta_1 = \theta'_2 = 90^\circ$$
$$\theta'_1 = \theta_2 = \theta_c$$

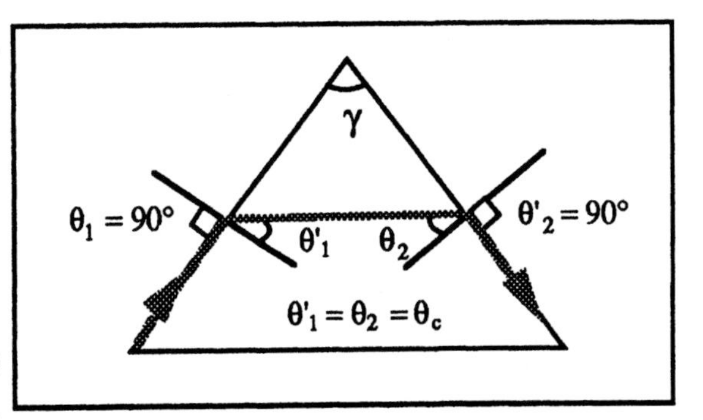

Figure 3-8. Apical angle limitations.

Note that θ'_1 is not really defined as the critical angle, but it has the same magnitude. Remember the reversibility of rays (i.e., a ray traveling backward through the prism will follow the same path as one incident at the first refracting surface at 90°).

To determine the apical angle, use Equation 3-2 and substitute the critical angle for the two internal angles:

$$\gamma = \theta'_1 + \theta_2 = \theta_c + \theta_c$$

$$\gamma = 2\theta_c \qquad (3\text{-}8)$$

If the apical angle is greater than two times the critical angle, no incident rays will be refracted directly through the prism.

$$\gamma > 2\theta_c \qquad (3\text{-}9)$$

No incident ray refracted through prism

Example 3-1
A prism is made of glass with an index of 1.75. What is the largest apical angle that the prism can have and still refract light through directly?

Known	***Unknown***	***Equations/Concepts***
Index of prism: $n_p = 1.75$	Apical angle of prism: $\gamma = ?$	(2-6) Critical angle
		(3-9) Limit for refraction through prism

The first step in this problem is to solve for the critical angle of the plastic material. If the apical angle is greater than two times the critical angle (Equation 3-10) then we know that no incident rays can be refracted through the prism directly. Solve for the critical angle:

$$\sin\theta_c = \frac{n'}{n} = \frac{1.00}{1.75} = 0.5714$$
$$\theta_c = \sin^{-1}(0.5714) = 34.85^\circ = \text{critical angle}$$

Multiply the critical angle times two to determine the apical angle that limits refraction:

$$\gamma > 2\theta_c \qquad \gamma > 2(34.85^\circ) \qquad \gamma > 69.70^\circ$$

Therefore the limiting apical angle is approximately 70°.

Example 3-m

What is the smallest incident angle that may be refracted through a prism that has an index of 1.75 and an apical angle of 20°?

Known	***Unknown***	***Equations/Concepts***
Index of prism: $n = 1.75$	Smallest incident angle refracted through prism	(3-1), (3-2), (3-3) Prism formulas
Apical angle of prism: $\gamma = 20°$		Limitation for refraction through prism
Critical angle: $\theta_c = 34.85°$		

Since the apical angle is less than two times the critical angle, as in *Example 3-k* , solve the prism formula backward starting with an emergent ray of 90° and an incident ray at the second refracting surface equal to the critical angle ($\theta_2 = \theta_c = 34.85°$). Solve for the refracted angle (θ'_1) at the first refracting surface using Equation 3-2.

$$\theta'_1 = \gamma - \theta_2 = 20° - 34.85° = -14.85°$$

Don't panic because the refracted angle is negative. Just use your rules and formulas in the normal manner. The sine of a negative angle is valid and can be determined using your calculator. Remember the relationship:

$$\sin(-\beta) = -\sin(\beta)$$

Now solve for the incident angle using Equation 3-1:

$$n_s \sin\theta_1 = n_p \sin\theta'_1 \qquad 1.00 \sin\theta_1 = 1.75 \sin(-14.85°)$$

$$\sin\theta_1 = \frac{(1.75)(-0.2563)}{1.00} = -0.4485$$

$$\theta_1 = \sin^{-1}(-0.4485) = -26.65°$$

Thus, the smallest incident angle is −26.65°. All incident angles less negative or more positive will be refracted through the prism.

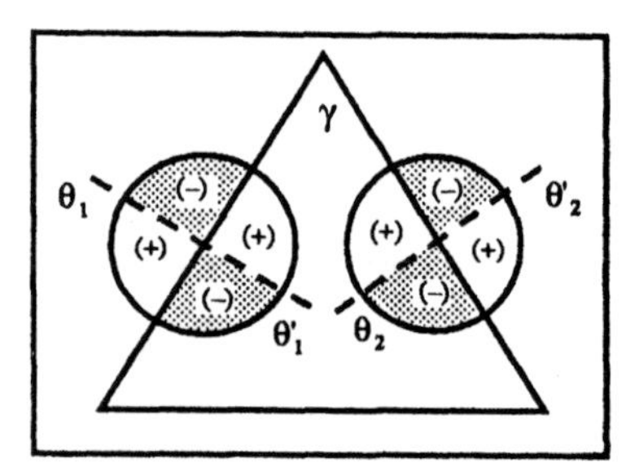

Figure 3-9. Sign convention for refraction through a prism.

The negative angles found in *Example 3-m* may not be clear in the context of refraction through a prism. Figure 3-9 indicates the sign convention for incident and refracted angles for a ray traced through a prism. For the incident and emergent angles, angles above the normal are negative, and those below are positive. For the internal angles, below the normal is negative, and above is positive. In most applications, all the angles will be positive (i.e., a ray is incident below the normal, refracted above the normal, strikes the second surface above the normal, and emerges below).

Example 3-n

Trace the incident ray found in *Example 3-m* through the prism.

Start with the incident ray above the normal with an angle of −26.65°. The refracted angle at the first surface is negative (−14.85°), so it is below the normal. The ray incident on the second surface is positive (+34.85°) and thus above the normal and the emerging ray is + 90°, which is traced below the normal along the surface of the prism. (See Figure 3-10.)

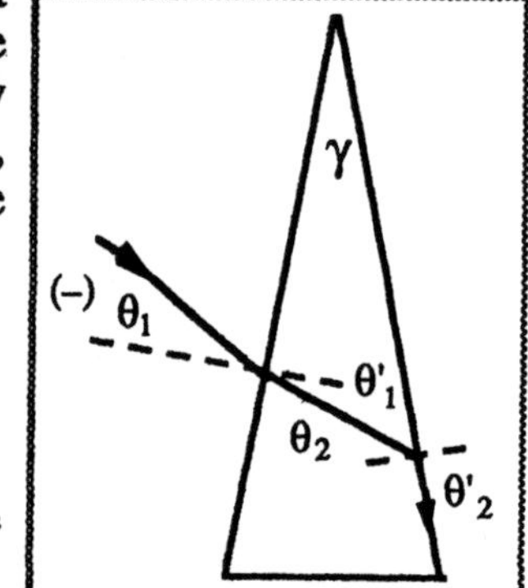

Figure 3-10. *Example 3-n.*

Minimum Deviation of a Prism

Deviation of a prism has been shown to be related to the incident, emergent, and apical angles. Figure 3-11 is a plot of deviation as a function of incident angle for a 30° glass prism. Note that the deviation plotted in this manner has a minimum of approximately 16°. This minimum deviation condition holds when the two external angles are equal and when the two internal angles are equal:

$$\theta_1 = \theta'_2 \qquad (3\text{-}10)$$

$$\theta'_1 = \theta_2 \qquad (3\text{-}11)$$

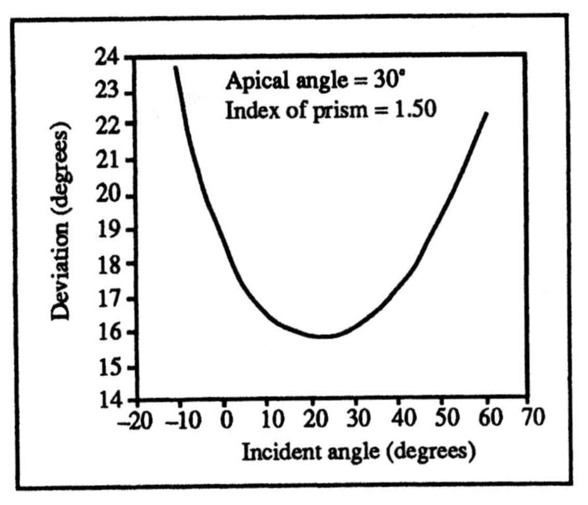

Figure 3-11. Deviation as a function of incident angle.

Substituting these relationships into the equations for the apical angle (Equation 3-2) and deviation (Equation 3-5) yields

$$\gamma = \theta'_1 + \theta_2 = 2\theta'_1 = 2\theta_2$$

$$\theta'_1 = \theta_2 = \frac{\gamma}{2} \qquad (3\text{-}13)$$

$$\delta_m = \theta_1 + \theta'_2 - \gamma = 2\theta_1 - \gamma = 2\theta'_2 - \gamma$$

$$\theta_1 = \theta'_2 = \frac{\delta_m + \gamma}{2} \qquad (3\text{-}14)$$

where: δ_m = the minimum deviation

Example 3-o

The minimum deviation of a prism can be measured in the laboratory using an instrument called a *spectrometer*. From this measurement, the index of the prism may be calculated. Derive an equation for the index of the prism (in air) as a function of the minimum deviation.

Use either Equation 3-1 or 3-3, substitute the minimum deviation relationships for angles shown in Equations 3-12 and 3-13, and solve for the prism index. Using Equation 3-1

$$n_s \sin\theta_1 = n_p \sin\theta'_1$$

$$\theta'_1 = \theta_2 = \frac{\gamma}{2} \qquad \theta_1 = \theta'_2 = \frac{\delta_m + \gamma}{2} \qquad n_s = 1.00$$

$$1.00 \sin\left(\frac{\delta_m + \gamma}{2}\right) = n_p \sin\left(\frac{\gamma}{2}\right)$$

$$n_p = \frac{\sin\left(\frac{\delta_m + \gamma}{2}\right)}{\sin\left(\frac{\gamma}{2}\right)} \qquad (3\text{-}15)$$

Although found in many textbooks, this equation is not easy to solve in this form. I recommend solving for the incident and refracted angles directly using Equations 3-12 and 3-13 and then substituting into Equation 3-1 or 3-2 to solve for the prism index. See your textbook for more details on using the spectrometer to measure index the of refraction.

Example 3-p
A 48° prism has a minimum deviation of 27°. What is the index of refraction of the prism?

Known	***Unknown***	***Equations/Concepts***
Apical angle of the prism: $\gamma = 48°$	Index of prism: $n_p = ?$	(3-12), (3-13) Minimum deviation
Minimum deviation: $\delta_m = 27°$		(3-1) or (3-3)

Use the procedure shown in *Example 3-o* to calculate the index of refraction. First determine either θ'_1 or θ_2 using Equation 3-12:

$$\theta'_1 = \theta_2 = \frac{\gamma}{2} = \frac{48°}{2} = 24°$$

Next calculate either θ_1 or θ'_2 by substituting the apical angle and the minimum deviation into Equation 3-13:

$$\theta_1 = \theta'_2 = \frac{\delta_m + \gamma}{2} = \frac{27° + 48°}{2} = \frac{75°}{2} = 37.5°$$

Substitute the angles into Equation 3-1:

$$n_s \sin\theta_1 = n_p \sin\theta'_1$$

$$1.00 \sin(37.5°) = n_p \sin(24°)$$

$$n_p = \frac{0.6018}{0.4067} = 1.48 = \text{index of prism}$$

Example 3-q
What is the minimum deviation of a 35° glass prism (n = 1.523)?

Known	***Unknown***	***Equations/Concepts***
Apical angle of the prism: $\gamma = 35°$	Minimum deviation: $\delta_m = ?$	(3-12), (3-13) Minimum deviation
Index of prism: $n_p = 1.523$		(3-1) or (3-3)

Solve for either internal angle using Equation 3-12:

$$\theta'_1 = \theta_2 = \frac{\gamma}{2} = \frac{35°}{2} = 17.5°$$

Calculate the emergent angle using Equation 3-3:

$$n_p \sin\theta_2 = n_s \sin\theta'_2$$

$$1.523 \sin(17.5°) = 1.00 \sin\theta'_2$$

$$\sin\theta'_2 = 0.4580 \qquad \theta'_2 = \sin^{-1}(0.4580) = 27.26°$$

Rearrange Equation 3-14 and solve for the minimum deviation:

$$\delta_m = 2\theta'_2 - \gamma = 2(27.26°) - 35° = 19.51° \qquad \text{minimum deviation}$$

Ophthalmic Prisms

Ophthalmic prisms are usually considered to be thin prisms (apical angles less than 15°) in which either the incident or emergent ray leaves normal to the surface. As shown in Figure 2-21 in Chapter 2, the sine of a small angle is approximately equal to the angle in radians. Assuming that the incident and refracted angles are small (angles in radians), the sines in Equation 3-1 and 3-3 may be dropped. This results in a simplification of the deviation Equation 3-4:

$$\delta = \left(\frac{n_p}{n_s} - 1\right)\gamma \quad (3\text{-}15)$$

If this surrounding medium is air ($n_s = 1.00$), this may be further simplified:

$$\delta = (n_p - 1)\gamma \quad (3\text{-}16)$$

When the apical angle in Equation 3-15 or 3-16 is in degrees, the deviation is in degrees; and when the apical angle is expressed in radians, the deviation is expressed in radians. These equations indicate that the deviation of a thin prism is independent of the incident or emergent angles.

Example 3-r

Show how Equations 3-15 and 3-16 are derived by dropping the sines in the prism equations (3-1 and 3-3) and substituting values into Equation 3-4.

Let's start this problem by reviewing one relationship that can be used to convert between radians and degrees (i.e., π radians/180°). Use this conversion factor to test the idea that for small angles the sine is approximately equal to the angle in radians; try sin (5°):

$$\sin(5^\circ) = 0.8716$$

Convert 5° to radians:

$$(5^\circ)(\pi \text{ rad}/180^\circ) = (5^\circ)(0.01745 \text{ rad/deg}) = 0.8727 \quad \text{(approximately equal)}$$

Now simplify Equations 3-1 and 3-3 by dropping the sine:

$$n_s \sin\theta_1 = n_p \sin\theta'_1 \qquad n_s\theta_1 = n_p\theta'_1 \qquad \theta_1 = \frac{n_p\theta'_1}{n_s} \quad (3\text{-}17)$$

$$n_p \sin\theta_2 = n_s \sin\theta'_2 \qquad n_p\theta_2 = n_s\theta'_2 \qquad \theta'_2 = \frac{n_p\theta_2}{n_s} \quad (3\text{-}18)$$

The deviation of the prism can be simplified by substituting these values for θ_1 and θ'_2 into Equation 3-5:

$$\delta = \frac{n_p\theta'_1}{n_s} + \frac{n_p\theta'_2}{n_s} - \gamma$$

From Equation 3-2, substitute γ for $(\theta'_1 + \theta_2)$:

$$\delta = \left(\frac{n_p}{n_s}\right)\gamma - \gamma \qquad \delta = \left(\frac{n_p}{n_s} - 1\right)\gamma \quad (3\text{-}15)$$

Example 3-s
What is the deviation under water and in air of a plastic ophthalmic prism (n = 1.49) that has an apical angle of 8°? Express your answers in degrees and radians.

Known
Apical angle of the prism: $\gamma = 8^\circ$
Index of prism: $n_P = 1.49$
Index of water surround: $n_S = 1.33$
Index of air surround: $n_S = 1.00$

Unknown
Deviation of prism: $\delta = ?$

Equations/Concepts
(3-15), (3-16) Ophthalmic prism equations
Conversion of degrees to radians
Example 3-r

Calculate the deviation (in degrees) under water using Equation 3-15:

$$\delta^\circ = \left(\frac{n_P}{n_s} - 1\right)\gamma^\circ = \left(\frac{1.49}{1.33} - 1\right)(8^\circ) = 0.96^\circ$$

Convert the apical angle to radians, and solve for the deviation in radians:

$$(8^\circ)\left(\frac{\pi \text{ radians}}{180^\circ}\right) = 0.1396 \text{ radians}$$

$$\delta^{rad} = \left(\frac{n_P}{n_s} - 1\right)\gamma^{rad} = \left(\frac{1.49}{1.33} - 1\right)(0.1396 \text{ rad}) = 0.01679 \text{ rad}$$

Check to see if the deviation is the same by converting the radian measure into degrees:

$$(0.01679 \text{ rad})\left(\frac{180^\circ}{\pi \text{ radians}}\right) = 0.96^\circ$$

Repeat the procedure and determine the deviation for the prism in air using Equation 3-16:

$$\delta^\circ = (n_p - 1)\gamma^\circ = (1.49 - 1)(8^\circ) = 3.92^\circ$$

$$\delta^{rad} = (n_p - 1)\gamma^{rad} = (1.49 - 1)(0.1396 \text{ rad}) = 0.0684 \text{ rad}$$

Prism diopter (pd; Δ) - The unit of deviation used for ophthalmic applications. One prism diopter is a deviation that yields a lateral displacement of 1 cm at 1 m (Figure 3-12). Prism diopters are proportional to displacement (e.g. a 2 cm displacement at 1 meter is 2^Δ, etc.)

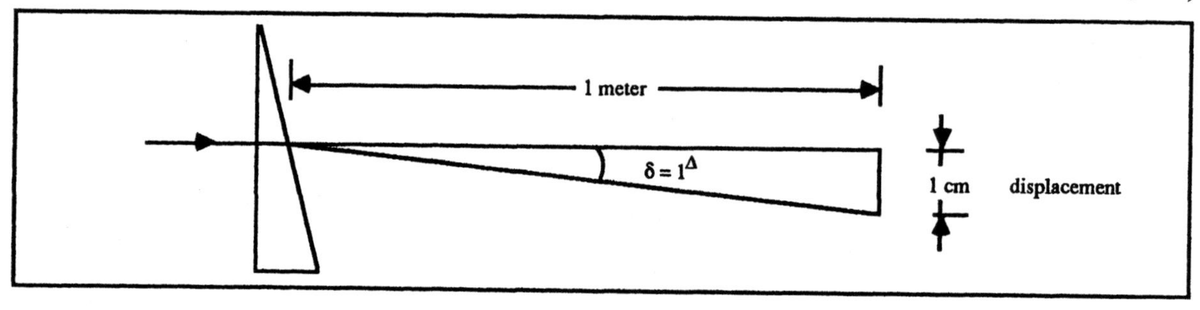

Figure 3-12. Definition of a prism diopter.

Equation 3-16 may be modified to calculate directly the deviation in prism diopters. As shown in Figure 3-12, the deviation angle of the prism may be defined in terms of the tangent. For prism diopters, this represents a one to one hundred ratio (cm versus m).

$$\delta^{\Delta} = \tan\delta^{\circ} = \delta^{rad} = \frac{1}{100} \qquad \delta^{\Delta} = 100\tan\delta^{\circ} = 100\delta^{rad}$$

Substituting this relationship into Equation 3-16, the equation becomes

$$\delta^{\Delta} = 100\,(n_p - 1)\tan\gamma^{\circ} = 100\,(n_p - 1)\gamma^{rad} \qquad (3\text{-}19)$$

Centrad (∇) - A unit that was used to express deviation. One centrad is equal to 0.01 radian deviation:

$$\delta^{\nabla} = (0.01)\tan\delta^{\circ} = 0.01\delta^{rad} \qquad (3\text{-}20)$$

Example 3-t
Calculate the deviation in prism diopters and centrads for the prism in *Example 3-s* in air.

Known	***Unknown***	***Equations/Concepts***
Apical angle of the prism: $\gamma = 8^{\circ} = 0.14$ rad	Deviation of prism in	(3-15), (3-16), (3-19), (3-20)
Index of prism: $n_p = 1.49$	prism diopters: $\delta^{\Delta} = ?$	Prism diopters, centrads
Index of air surround: $n_s = 1.00$	centrads: $\delta^{\nabla} = ?$	*Example 3-r*
Deviation of the prism: $\delta = 3.92^{\circ} = 0.0684$ rad		

Substitute the given values into the equations:

$$\delta^{\Delta} = 100(n_p - 1)\tan\gamma^{\circ} = 100(1.49 - 1)\tan(8^{\circ}) = 6.89^{\Delta}$$

$$\delta^{\Delta} = 100(n_p - 1)\gamma^{rad} = 100(1.49 - 1)(0.14 \text{ rad}) = 6.86^{\Delta}$$

$$\delta^{\nabla} = (0.01)\tan\delta^{\circ} = (0.01)\delta^{rad} = (0.01)\tan(3.92^{\circ}) = (0.01)(0.0684 \text{ rad}) \approx 6.84^{\nabla} \text{ x } 10^{-4}$$

The values for the deviation in prism diopters are approximately equal using either degrees or radians for the apical angle. The small discrepancy is due to the radian - tangent approximation.

Chromatic Dispersion

Images formed by a prism may also have chromatic dispersion. Because the index of refraction for each wavelength varies, so does the deviation. Therefore incident white light spreads out into its component colors, as shown in Figure 3-13. This effect may add fringes of color around objects viewed through a prism. Other distortions, such as curvature of straight edges, are also seen when viewing through prisms.

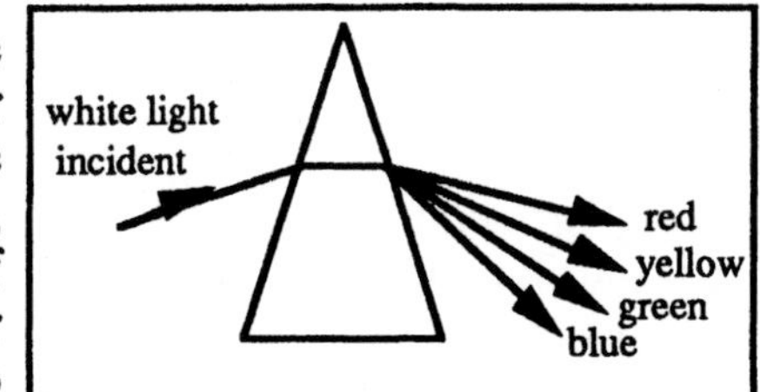

Figure 3-13. Prismatic dispersion.

Images Formed by a Prism

Light that emanates from a point source and is refracted through a prism will be seen as a single image point if the incident angles are small. Two types of images may be formed by a prism. If convergent light is incident on the prism, the convergent rays will be deviated toward the base, and a real image will be formed that is also deviated toward the base (Figure 3-14).

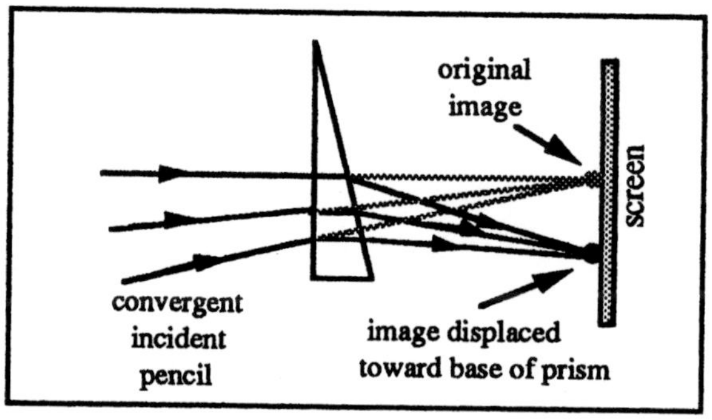

Figure 3-14. Image displacement through a prism.

In an ophthalmic application, one views through the prism (Figure 3-15). Here again the incident rays are deviated toward the base of the prism. However, the rays appear to be emanating from the apex of the prism. Therefore the object viewed through the prism appears to be deviated toward the apex. The apparent position of the object may be found by extending the deviated rays back through the prism.

Prisms are often used in correction of phoria and tropia. In tropia, one eye is deviated in (eso) or out (exo) so that the two eyes are not viewing the same position, and diplopia (double images) may result. A prism, used to correct a tropia, deviates light so that the apparent position of an object is in front of the eye. The base of the prism is appropriately placed (base in, base out, base up, or base down). In certain cases, a cosmetic correction is desirable. Here the eye acts as the object, and the base of the prism is positioned so that the image of the eye is deviated to "look" proper. This base position, however, is the opposite of a functional correction. You will, no doubt, receive much more information on the clinical uses of prisms in your courses on binocular vision.

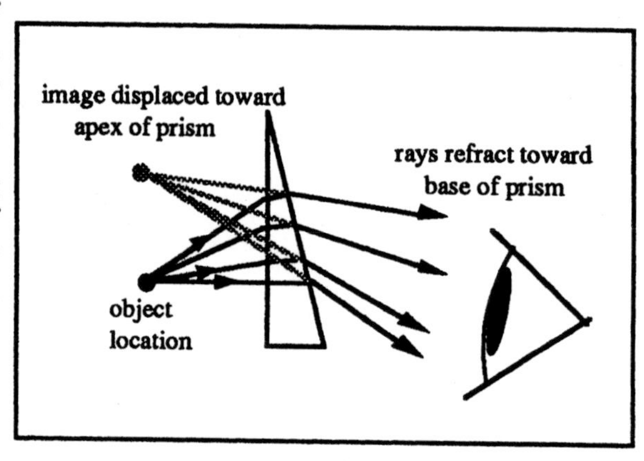

Figure 3-15. Deviation of the image toward the apex of an ophthalmic prism.

Example 3-u

A patient has a constant left esotropia (left eye deviated toward nose). What is the orientation of the base of the prism placed in front of the left eye to correct this problem? What would be the orientation for a cosmetic correction? Explain your answers.

Known	***Unknown***	***Equations/Concepts***
Left constant esotropia	Base of prism for functional correction	Image formed by prism
	Base of prism for cosmetic correction	Deviated toward apex

Draw a diagram of the eyes with the left eye deviated toward the nose (Figure 3-16). A functional correction will move the "straight ahead" position in front of the deviated eye. Because the image is moved toward the apex and the image is to be moved toward the nose, the apex is at the nose. Thus a base-out prism is required.

For the cosmetic correction, the image of the left eye must be out, and the apex of the prism must be out. This means a base-in prism is required.

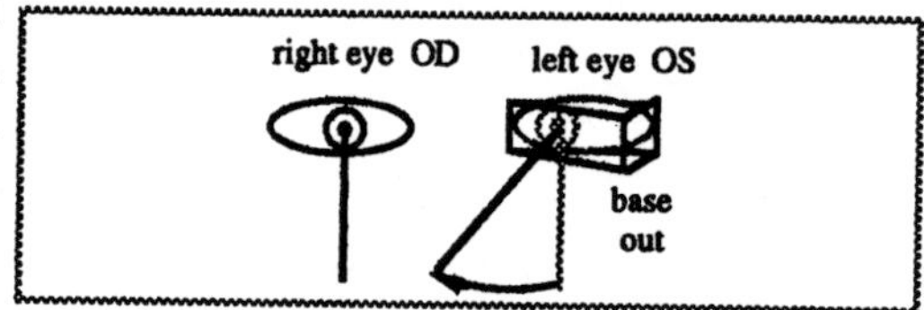

Figure 3-16. Functional correction of left constant esotrope with base out prism.

Reflecting Prisms

A prism in which one or more surfaces reflect light is called a *reflecting prism*. These types of prisms are designed so that the internal angles are incident at an angle greater than the critical angle. This is illustrated by *Example 3-v*.

Example 3- v

A ray of light incident on the front refracting surface of a 45°-90°-45° plastic prism (n=1.49), emerges out of the base. For an incident angle of 48.2°, find the emergent angle and the total deviation of the ray. The critical angle for the material is 42.2°.

Known	***Unknown***	***Equations/Concepts***
Index of prism: $n_p = 1.49$	Emergent angle	(2-6) Critical angle
Apical angle of prism: $\gamma = 90°$	Deviation	(3-1), (3-2), (3-3) Refraction through a prism
Incident angle: $\theta_1 = 48.20°$		
Index of surround: $n_s = 1.00$		
Critical angle or material: $\theta_c = 42.2°$		

Use Equations 3-1 and 3-2 to determine the incident angle at the second refracting surface:

$$n_s \sin\theta_1 - n_p \sin\theta'_1 \qquad (1.00)\sin(48.20°) = (1.49)\sin\theta'_1 \qquad \theta'_1 = \sin^{-1}(0.50) = 30°$$

$$\theta_2 = \gamma - \theta'_1 = 90° - 30° = 60° \qquad \text{Greater than the critical angle}$$

The Law of Reflection is used to determine the internal reflected angle at the second interface:

$$\theta_i = \theta_r = 60°$$

The incident angle at the base of the prism is found using simple geometry. The triangle formed by the reflected ray and the second surface has internal angles of 30°, 45°, and α. Calculate the value of α by using the fact that the sum of the internal angles of a triangle is equal to 180°. (See *Example 2-e* for another example of this calculation.)

$$\alpha + 30° + 45° = 180° \qquad \alpha = 180° - 30° - 45° = 105°$$

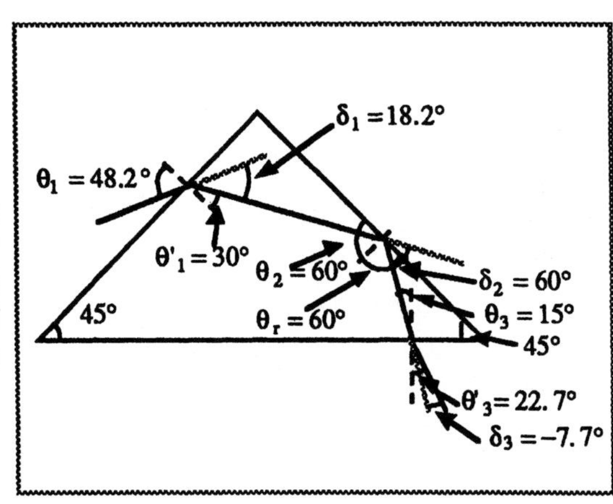

Figure 3-17. *Example 3-u.* Ray refracting and reflecting through a prism.

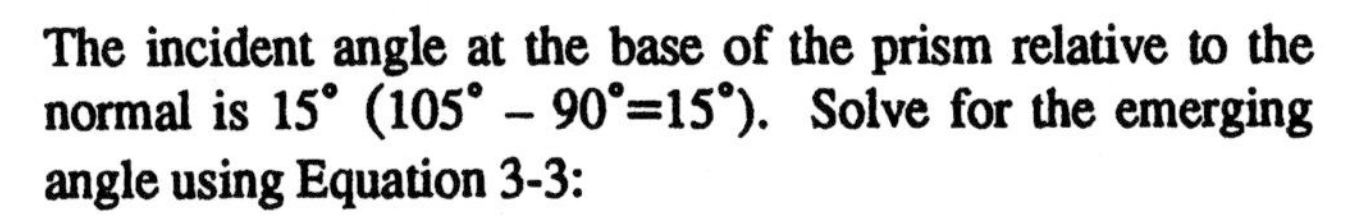

The incident angle at the base of the prism relative to the normal is 15° (105° – 90°=15°). Solve for the emerging angle using Equation 3-3:

$$n_p \sin\theta_3 = n_s \sin\theta'_3 \qquad (1.49)\sin(15°) = (1.00)\sin\theta'_3 \qquad \theta'_3 = \sin^{-1}(0.3856) = 22.7°$$

Calculate and sum the deviation at each interface to find the total deviation (see Figure 3-17):

$$\delta_1 = \theta_1 - \theta'_1 = 48.2° - 30° = 18.2°$$

$$\delta_2 = 180° - \theta_2 - \theta_r = 180° - 60° - 60° = 60°$$

$$\delta_3 = \theta_3 - \theta'_3 = 15° - 22.7° = -7.7°$$

$$\delta = \delta_1 + \delta_2 + \delta_3 = 18.2° + 60° - 7.7° = 70.5°$$

Supplemental Problems

Refraction and Deviation of a Prism

3-1. A ray is incident on the front surface of a 28° glass prism (n = 1.5) at an angle of 55°. What is the angle of emergence and the deviation of the prism?
ANS. Emergent angle = –7.66°; deviation = 19.3°

3-2. A prism has a deviation of 34° when a ray is incident at 30° and emerges at 44°. What is the apical angle of the prism?
ANS: Apical angle ~ 40°

Normal Incidence and Emergence

3-3. When the incident angle is 22°, light is refracted through a prism (n = 1.69) such that the emergent ray is normal to the second refracting surface. What is the apical angle of the prism?
ANS. Apical angle = 12.8°

3-4. Light is incident on the front refracting surface of a thick prism such that the refracted angle is 25° and the emergent ray leaves normal to the back refracting surface. Under these conditions, the deviation of the prism is 22.7°. What is the index of refraction of the prism?
ANS. n = 1.75

3-5. When an incident ray strikes the first refracting surface of a prism at an angle of 50.77°, all the deviation occurs at the first refracting surface. When an internal ray strikes the second refracting surface of the same prism at an angle of 28° all the deviation occurs at the second surface. Calculate the apical angle and index of refraction of the prism.
ANS. Apical angle = 28°; n = 1.65

3-6. A ray strikes a prism normal to the first refracting surface. The ray is incident on the second refracting surface at an angle of 26°. The prism is made of a material with a critical angle of 53.71° when surrounded by water (n = 1.33). Find the index of refraction of this prism, its apical angle, its emergent angle in air, and the deviation in degrees and prism diopters.
ANS. n = 1.65; apical angle = 26°; emergent angle = 46.33°; deviation = 20.33° or 31.7^{Δ}

Limitations on Refraction through a Prism

3-7. If the apical angle of a prism (n = 1.6) is 32°, what is the smallest incident ray that can be refracted through the prism?
ANS. Incident ray = –10.73° (critical angle = 38.68°)

3-8. What is the smallest incident ray that can be refracted through a 20° glass prism (n = 1.70)?
ANS. Incident ray = –28.00° (critical angle = 36.03°)

Minimum Deviation

3-9. A 20° prism has a refractive index of 1.6. What is the minimum deviation of this prism?
ANS. Minimum deviation = 12.26°

3-10. A 60° prism has a minimum deviation of 41°. What is the index of refraction of the prism?
ANS. n = 1.543

3-11. A plastic prism with an apical angle of 45° has a refractive index of 1.44. Find the angle of incidence that yields the minimum deviation.
ANS. Angle of incidence = 33.4°

3-12. When a ray is incident onto the first refracting surface of a glass prism (n = 1.50) at an angle of 39.34°, the deviation is a minimum. What is the apical angle of the prism?
ANS. Apical angle = 50°

3-13. A 30° glass prism has a minimum deviation of 15° when submerged under water (n = 1.33). What is the index of refraction of the prism?
ANS. n = 1.97

3-14. When a 20° prism is placed in an optical medium of gas (not air), the minimum deviation is measured at 9.51°. What is the critical angle at the prism-gas interface?
ANS. Critical angle = 42.99°

3-15. A glass prism in air yields a 23° emergent angle when the minimum deviation condition exists. If this prism were made of plastic (n = 1.49), an incident ray of 16.21° would yield the same emergent angle. Find the index of the glass prism, its apical angle, and the minimum deviation.
ANS. n = 1.74; apical angle = 26°; minimum deviation = 20°

Ophthalmic Prisms

3-16. What orientation (base out our base in) prism would a constant right exotrope require for a cosmetic correction? Draw a diagram of the eye and the base of the prism.
ANS. Base out

3-17. A patient has a temporal right eye turn of 5° from the straight ahead. What prismatic power is required to compensate for this condition? Use an index of 1.50, if required.
ANS. Power of prism = 8.75 $^{\Delta}$ BI

3-18. A glass ophthalmic prism (n = 1.52) displaces the image of an object 0.037 m at 2.2 m. What is the deviation of the prism in prism diopters?
ANS. Deviation = 1.68 $^{\Delta}$

3-19. An observer located 50 cm behind a 2 $^{\Delta}$ ophthalmic prism views an object located 30 cm in front of the prism. By how much does the object appears to be displaced?
ANS. 6 mm displacement

3-20. Find the deviation and the apical angle for a 5 $^{\Delta}$ ophthalmic prism (n = 1.49). At a distance of 1.5 m, how far is a spot of light displaced by an 8 $^{\Delta}$?
ANS. Deviation = 2.86°; apical angle = 5.83°; 12 cm displacement

3-21. An object 500 cm from an ophthalmic prism appears to be displaced 18 cm. What is the deviation of the prism?
ANS. Deviation = 3.6$^{\Delta}$ or 2.1°

Chapter 4

Curved Refracting Surfaces

In most applications in optics, the vergence of the incident wavefront is altered by refraction at a curved surface. Curved refracting surfaces may have many different shapes (spherical, cylindrical, aspheric, etc.). Although any point on the curve will refract light according to Snell's Law, the shape of the surface will determine the position, size, type, and quality of the image produced. The spherical surface, discussed in this chapter, is the most common ophthalmic refracting surface used for spectacle, contact, and intraocular lenses. Although these elements have two curved surfaces, the discussion in this chapter is simplified by describing the imaging properties of one **single refracting curved surface.** In the next chapter, curved surfaces are combined to form lenses.

A single refracting surface is represented by a boundary between two media with different indices. (See Figure 4-1.) If this boundary is spherical, it may be defined by its **radius, curvature,** or **power.**

Spherical surface - a surface generated by a single arc with equal distance from any point on the surface to the reference point (called the *center of curvature*).

Radius (r) - the distance from the generating arc to any point on the surface. Spherical surfaces can be defined by their radii.

Curvature (R) - the reciprocal of the radius.

Center of Curvature (C) - the reference point equal distance from any point on a spherical surface.

Sag (s) - the perpendicular distance from the center of any chord to the surface.

Chord - a straight line connecting any two points on an arc or circumference of a circle.

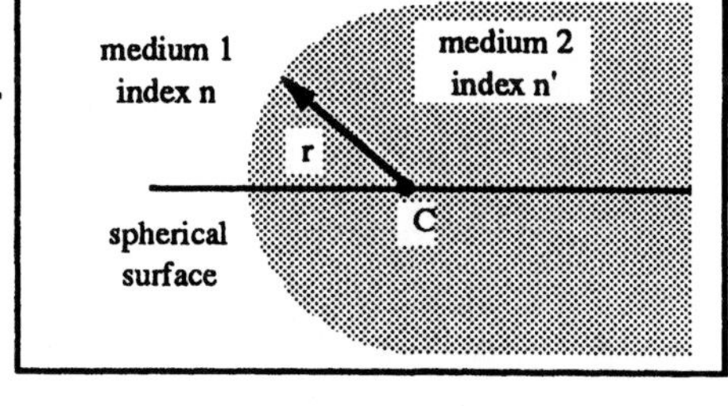

Figure 4-1. Spherical refracting surface.

As shown in Figure 4-2, a two dimensional representation of a spherical surface can be drawn with a compass. The center of curvature is where the pointed end of the compass is placed in order to generate the arc. The radius, center of curvature, chord, and sag are labeled in the figure.

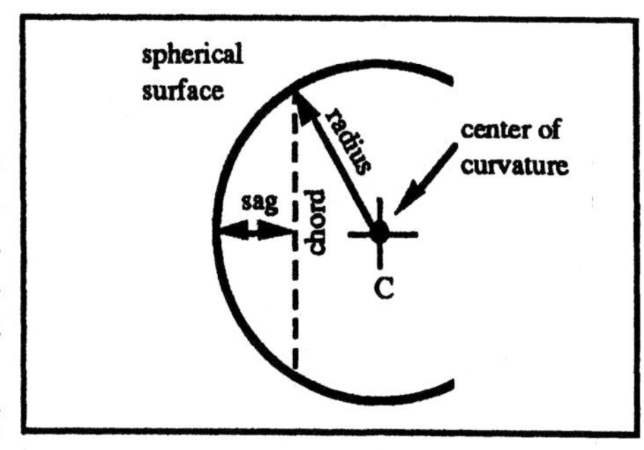

Figure 4-2. Spherical surface.

The relationship between the radius and the sag may be expressed as

$$r = \frac{y^2}{2s} + \frac{s}{2} \qquad (4\text{-}1)$$

where: r = the radius of the surface
s = the sag of the surface to the chord
y = one half of the chord length (see Figure 4-3)

When working with contact lenses (small diameters, short radii), Equation 4-1 must be used to determine the radius-sag relationship. However, for conventional lenses, the sag is usually much smaller than the radius, and therefore the second term in Equation 4-1 does not change the value of the radius significantly. Thus if $s \ll r$, then Equation 4-1 becomes

$$r = \frac{y^2}{2s} \qquad (4\text{-}2)$$

When using Equations 4-1 and 4-2, be sure that all the variables have the same units, and remember that y represents *half* of the chord length.

Example 4-a

Show the relationship between the sag and the radius using the basic equation for a circle.

The basic equation for a circle relates any point (x,y) on the circumference to the radius (r):

$$r^2 = x^2 + y^2 \qquad (4\text{-}3)$$

As shown in Figure 4-3, the expression (r – s) can be substituted for x in Equation 4-3, and this becomes

$$r^2 = (r - s)^2 + y^2$$

Expanding the equation gives

$$r^2 = r^2 - 2rs + s^2 + y^2$$

Subtracting r^2 from both sides of the equation and solving for the radius yield

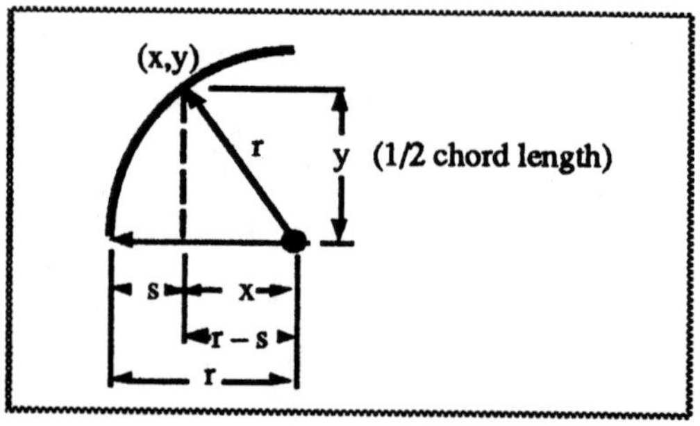

Figure 4-3. *Example 4-a.* Radius - sag relationship.

$$0 = y^2 - 2rs + s^2 \quad \text{or} \quad 2rs = y^2 + s^2$$

$$r = \frac{y^2 + s^2}{2s} = \frac{y^2}{2s} + \frac{s^2}{2s}$$

$$r = \frac{y^2}{2s} + \frac{s}{2} \qquad (4\text{-}1)$$

Example 4-b

A lens has one curved surface with a radius of 50 cm and one flat surface, as shown in Figure 4-4a. The blank diameter of the lens is 60 mm. What is the minimum center thickness of the lens?

Known
Surface radius: r = 50 cm = 500 mm
Diameter of lens: d = 60 mm
Chord length: 60 mm
Half chord length: y = 30 mm

Unknown
Center thickness: ct = ?

Equations/Concepts
From Figure 4-4a, the relationship between the sag, radius, and center of curvature
(4-2) Sag formula

In Figure 4-4a, the sag (s) is equal to the center thickness (ct) of the lens. This is true when the lens has a knife edge or zero edge thickness. The diameter (d) of the blank and half the chord length (y) are also labeled. Because the sag for the entire lens is required, the chord is half the diameter of the lens. Using Equation 4-2 and rearranging terms, you can solve for the sag:

$$s = \frac{y^2}{2r} = \frac{(30)^2}{(2)(500)} = \frac{900}{1000} = 0.90 \text{ mm}$$

Therefore the center thickness is 0.90 mm.

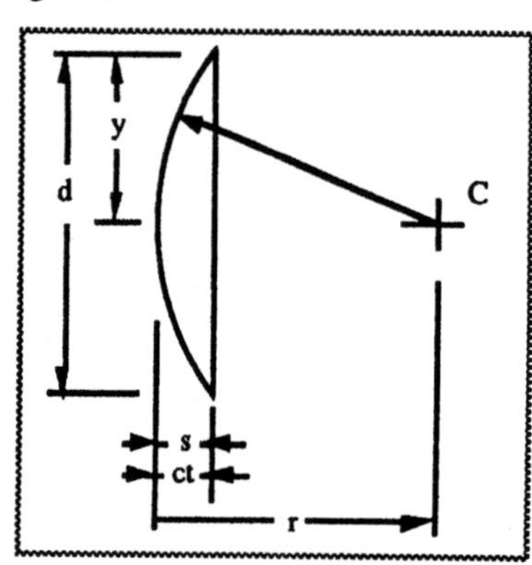

Figure 4-4a. *Example 4-b.*

Let's go one step farther with this type of problem. What if the lens had an edge thickness? This would mean that the edge thickness must be added to the *minimum center thickness* to determine the total center thickness. This is shown in Figure 4-4b. Notice from the diagram that the relationship between the sag (s), center thickness (ct), and edge thickness (et) can be written as a simple equation:

$$ct = et + s$$

Let's assume an edge thickness of 2 mm for this example. The sag calculated above would represent the minimum center thickness that must be added to the edge thickness to obtain the actual center thickness:

$$ct = 2 \text{ mm} + 0.9 \text{ mm} = 2.9 \text{ mm}$$

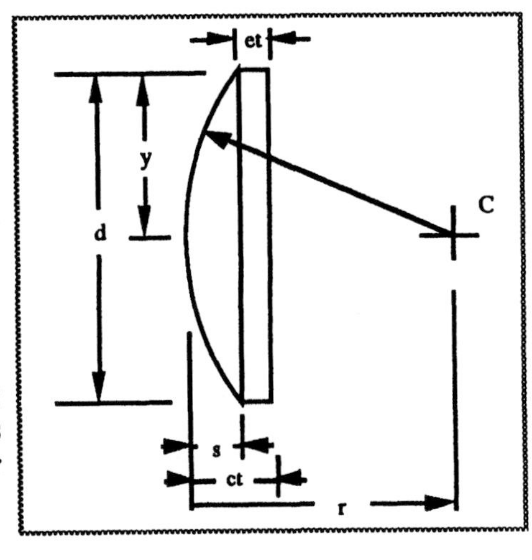

Figure 4-4b. *Example 4-b.* Lens with finite edge thickness.

This procedure may be used for surfaces that are curved inward (concave surfaces) and for lenses with two curves. However, each problem must be diagrammed and labeled so that the relationship between sag, center thickness and edge thickness can be determined. *Don't memorize the formula, learn the technique.*

Radius-Power Relationship

When an incident wavefront is refracted at a curved surface, there is a change in the radius of the wavefront that results in a change in the vergence. This change in vergence is directly related to the **power** of the surface. There is a simple relationship between the radius, the indices that make up the interface, and the power of a surface:

$$F = \frac{n' - n}{r} \quad (4\text{-}4)$$

where:
- F = surface power in diopters
- n = index of refraction of space before refraction (left of interface)
- n' = index of refraction of space after refraction (right of interface)
- r = surface radius in meters (the sign convention is important)

When more convergence is added to the wavefront, the surface is considered to have positive power; when more divergence is added to the wavefront, the surface is considered to have negative power. Always put a plus or minus sign in front of the power to denote the type of surface. Care must be taken with the sign of the radius in the equation because this will change the calculated power. Remember that the radius is measured from the surface to the center of curvature. If this measurement is in the direction that light travels (left to right), it has a positive value; if it is opposite the direction that light travels (right to left), it has a negative value.

The radius is not the only factor that determines the sign of the power; curved surfaces may also be converging or diverging depending on the relative indices. Be sure that you identify n and n' properly. *Fewer errors are made when you draw the surface and label n, n', and r before calculating the power. Use an arrow to indicate the direction of the radius measure.*

Example 4-c

The end of a long glass rod (n = 1.50) is polished with a spherical curve that has a radius of 5 cm. What is the power of the spherical surface?

Known	***Unknown***	***Equations/Concepts***
Index of rod: n = 1.50	Power of surface	(4-4) F = (n'−n)/r
Radius of curved surface: r = 5 cm = 0.05 m	Shape curved surface	
Index of air: n = 1.00	Positive or negative	

This question is somewhat ambiguous in that the sign of the radius is unknown. If the surface is extended out of the rod, it may be considered to be convex or positive; if the surface is extended into the rod, it may be concave or negative. Both power calculations will be made here. In Figure 4-5a, the two surfaces are shown. The radii are labeled with arrows indicating the direction. The indices are labeled in the diagram with the primed and unprimed notation (i.e., n' is to the right of the interface, and n is to the left). For the power to be in diopters, the radius must be changed to meters. For the concave (inward) and convex (outward) surfaces, the calculations are as follows:

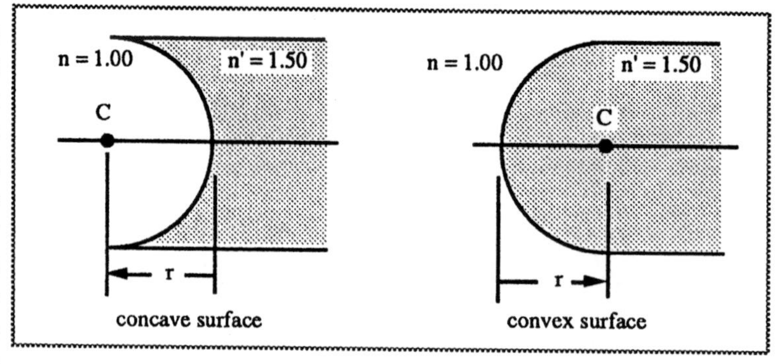

Figure 4-5a. *Example 4-c.* Concave and convex surfaces with labels.

$$F = \frac{n'-n}{r} = \frac{1.50-1.00}{-0.05} = \frac{0.50}{-0.05} = -10.00 \text{ D} \qquad \text{concave}$$

$$F = \frac{n'-n}{r} = \frac{1.50-1.00}{+0.05} = \frac{0.50}{+0.05} = +10.00 \text{ D} \qquad \text{convex}$$

The concave surface has a negative power, and the convex surface has a positive power.

These surfaces could have been drawn another way (with the same power), as shown in Figure 4-5b. Note that n and n' have different values (n is the medium to the left of the interface, and n' is the medium to the right) and that the radius has the opposite sign. The surface power calculations are as follows:

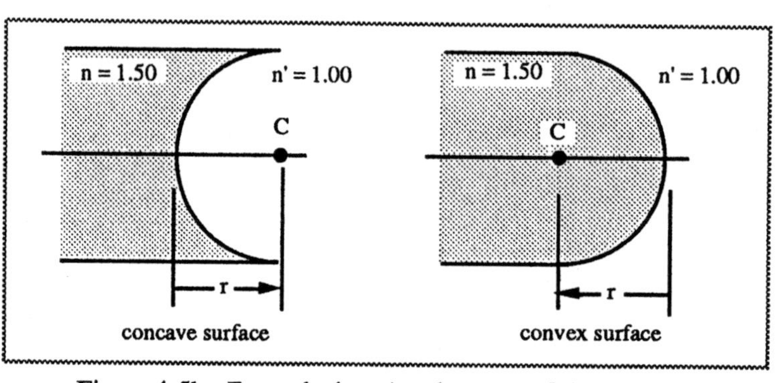

Figure 4-5b. *Example 4-c.* Another way of drawing the surfaces. The power is the same as in Figure 4-5a.

$$F = \frac{n'-n}{r} = \frac{1.00-1.50}{+0.05} = \frac{-0.50}{+0.05} = -10.00 \text{ D} \qquad \text{concave}$$

$$F = \frac{n'-n}{r} = \frac{1.00-1.50}{-0.05} = \frac{-0.50}{-0.05} = +10.00 \text{ D} \qquad \text{convex}$$

Surfaces have only one power independent of orientation.

Laboratory Measurements of Radius and Sag

The radius of an unknown spherical surface may be measured in the laboratory. The "brute force" method involves matching the surface with a template or surface profile cut in metal. This involves trial and error. It is not a practical method because a large number of templates and a great amount of time are required.

The radius of the surface may be determined indirectly by measuring the sag. For these measurements, the sag is considered to be the difference between a flat surface and the curved surface. A spherometer (Figure 4-6) may be used to measure directly the sag of a spherical surface. The spherometer has three stationary outer legs and a middle inner leg that moves vertically. A scale on the instrument indicates the position of the middle leg. The spherometer is first zeroed by placing it on a flat surface and adjusting the middle leg until all four legs touch the surface (Figure 4-6a). The scale reading of the middle leg is noted. The spherometer is then placed on the curved surface, and the middle leg is moved again so that all four legs contact the surface (Figure 4-6b). The scale position of the middle leg is again noted. The difference between the two readings is the sag. Once the sag is determined, the radius and the power of the surface may be calculated using Equations 4-2 and 4-4.

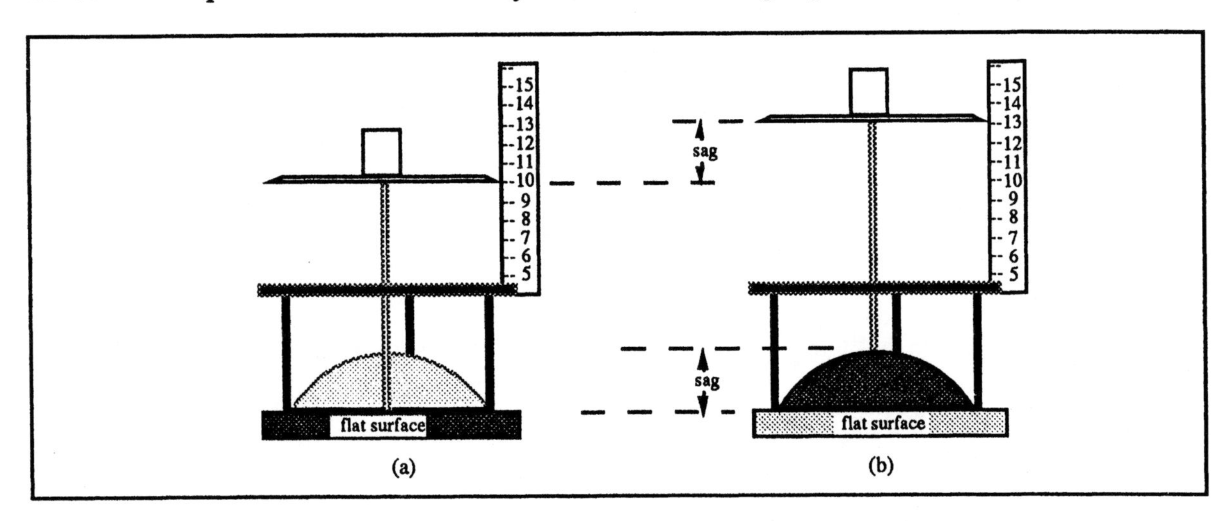

Figure 4-6. (a) The spherometer in the zero position on a flat surface. Note that the scale position of the middle leg is approximately 10 units. (b) The spherometer on a convex surface with the middle leg adjusted (13 units) so that all four legs touch the surface. The difference in the scale readings (13 – 10 = 3 units) for the two positions is the sag.

The lens clock is another device that may be used to determine the radius of a spherical surface. The lens clock has three legs: two stationary outer legs and one movable middle leg. The middle leg moves in or out of the base of the instrument. Just as with the spherometer, the difference between the middle leg's position on a flat surface and its position on a curved surface is the sag (Figure 4-7). The lens clock, however, is geared internally so that the sag measurement is converted directly to a power. This direct power measurement is based on these assumptions:

1. The surface material is crown glass with an index of 1.53.
2. The surrounding media is air with an index of 1.00.
3. The chord length (i.e., the distance between the two outer legs). For most lens clocks the chord length is a fixed distance of approximately 2.0 cm.
4. The shape of the surface (i.e., convex or concave) is known.

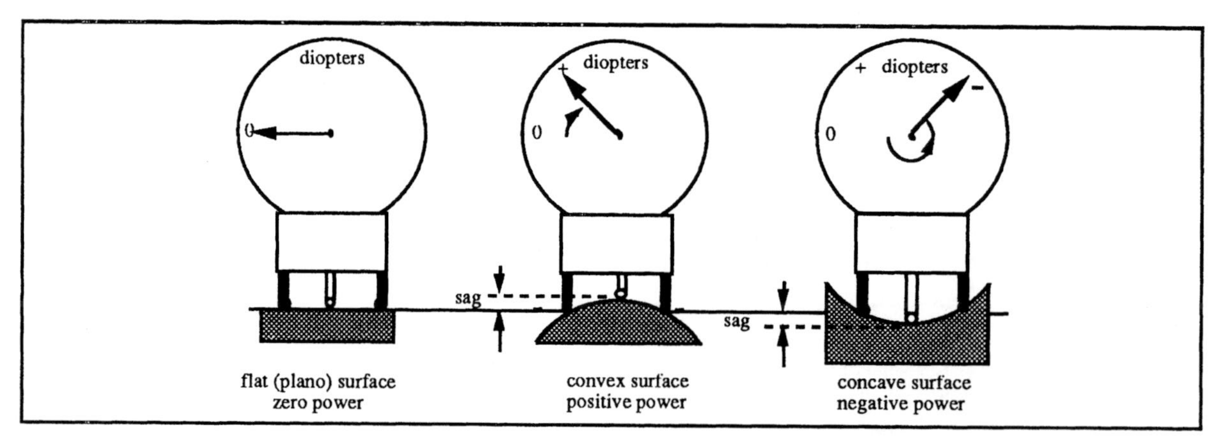

Figure 4-7. The lens clock on a flat surface, a convex surface, and a concave surface. The sag is directly converted to power (in diopters) by an internal gearing system. The shape of the surface must be known to read the proper scale (i.e., a convex surface is positive; a concave surface is negative).

Example 4-d

Show the mathematics involved in the lens clock by developing a formula that can be used to solve directly for the power given the sag.

Known	***Unknown***	***Equations/Concepts***
Index of surface: n' = 1.53	Radius of surface: r = ?	(4-2) Relationship of sag to radius
Index of surround: n = 1.00	Power of surface: F = ?	(4-4) Relationship of radius to power
Half chord length: y = 1.00 cm = 0.01 m		
Sag is known: s		

Substitute Equation 4-2 into Equation 4-4 for the radius and substitute the known values:

$$F = \frac{(n'-n)}{r} = \frac{(n'-n)}{\left(\frac{y^2}{2s}\right)} = \frac{2s(n'-n)}{y^2} = \frac{2s(1.53-1.00)}{0.01\text{ m}} = \frac{2s(0.53)}{0.01\text{ m}} = 106s \qquad \text{(power = 106 times sag)}$$

Example 4-e

A lens clock measures a power of – 8.50 D on a plastic surface (n = 1.49). What is the radius of the surface? What is the actual power of the surface?

Known	***Unknown***	***Equations/Concepts***
Lens clock power: F= –8.50 D	Radius of surface	(4-2) and (4-4)
Index of material: n = 1.49	Actual power of surface	*Example 4-d* Lens clock theory

If a lens clock is used to measure a surface consisting of a material other than crown glass, the power readings will be in error. This is easily corrected if the material has a known index. Given the power reading, the radius of the measured surface may be determined using Equation 4-4. Using the calculated radius, substitute back into Equation 4-4 with the appropriate index of refraction, and calculate the power. From the lens clock reading, using the assumed values for the index, solve Equation 4-4 for the radius:

$$r = \frac{n'-n}{F} = \frac{1.53-1.00}{-8.50} = \frac{0.53}{-8.50} = -0.0624\text{ m} = -6.24\text{ cm}$$

Using the calculated radius and the index of the surface, the actual power is calculated using Equation 4-4:

$$F = \frac{n'-n}{r} = \frac{1.49-1.00}{-0.0624} = \frac{0.49}{-0.0624} = -7.85\text{ D}$$

As shown in the previous example, a significant power error may result when the lens clock is used to measure the power of materials other than crown glass. You must also be aware of the shape of the refracting surface when making these measurements. The sign of the sag is not important as long as the sign convention is maintained when using the calculated radius. Diagram the surface, and label the indices and radius.

Image-Object Relationships: Position

Once the power of the surface is determined, it should remain constant for all imaging conditions. For this reason, it is suggested that you solve for the power of the surface before considering the image-object relationships. Vergence is an important concept used in deriving image-object relationships. Vergence, as shown in the previous chapters, is related to the radius of the wavefront. If the radius is increasing as the wave propagates, the wavefront is divergent and is denoted as having a negative vergence. If the radius of the wavefront is decreasing as the wave propagates, the wavefront is converging and has a positive vergence. In Chapter 1, the vergence is defined as the reciprocal of the radius of the wavefront. This is true in most cases; however, if the wave is traveling in a medium other than air, the decrease in velocity (and the change in vergence) must be taken into account. This is called the **reduced vergence,** and is defined as follows:

$$\text{Reduced Vergence} = \frac{\text{index of media}}{\text{radius of wavefront}}$$

For objects, the radius of the wavefront is the same as the object distance (ℓ) and is defined as the distance from the refracting surface to the object. For images, the radius of the wavefront is the same as the image distance (ℓ') and is defined as the distance from the refracting surface to the image. When measuring left to right, the distance is positive; measuring right to left, the distance is negative. Arrows should be used to indicate the direction of measurement.

Using this notation, the reduced vergence for the (incident) object and (emergent) image is

Incident vergence or reduced object vergence: $$L = \frac{n}{\ell} \quad (4\text{-}5)$$

Emergent vergence or reduced image vergence: $$L' = \frac{n'}{\ell'} \quad (4\text{-}6)$$

As shown in Figure 4-8, the space in front of the surface (to the left of) may be called **object space*** and the space behind the surface (to the right of) may be called **image space.*** To work problems properly, sign conventions must be followed. This convention requires that "real objects" or objects that are emitting divergent wavefronts (which is the case for all physical objects we see) are to be placed to the left of the interface in object space. Many students overlook this when working these problems. The question that must be asked when setting up problems is: *What is the object? Is it to the left of the interface?* If not, redraw the system.

*Note that the object space and image space definitions used in this workbook may not conform to those in your text. In the strictest sense, any place an object is positioned may be called object space, and any place an image is located may be called image space. I feel that this leads to unnecessary confusion, and unless you are confronted with question that asks for a strict definition, the definitions presented here are simpler to remember and adequately explain the image-object relationships.

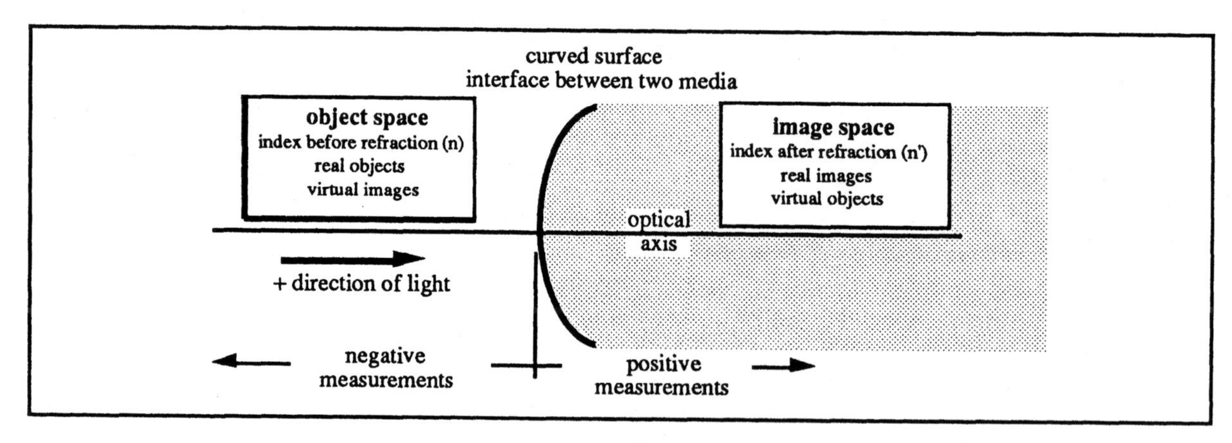

Figure 4-8. Space, image, and object definitions.

Real object - an object that emits divergent or expanding wavefronts. The radius of the wavefront and the vergence have negative values. Real objects must be to the left of the interface.

Real image - an image that is formed with convergent or constricting wavefronts. The radius of the wavefront and the vergence have positive values. Real images are located to the right of the interface. Only real images can be formed on a screen.

Virtual object - when a convergent wavefront is obstructed by an obstacle (such as a refracting surface), the position to which the wavefront is headed is considered to be the virtual object position. The radius of the wavefront is the distance from the obstacle to the virtual object position. The convergent wavefront is formed with some other element that does not need to be known. Virtual objects are to the right of the interface.

Virtual image - when a divergent wavefront leaves a refracting surface, the center of curvature of the wavefront is considered to be the virtual image position. The radius of the wavefront is the distance from the surface to the virtual image position. Virtual images are located to the left of the interface.

Optical axis - the line joining the center of curvature to the vertex or pole of a curved refracting surface. In most optical systems, all surfaces have a common optical axis that is centered symmetrically.

Gaussian Optics - rays strike the lens about a region close to the optical axis so that all rays form a perfect image without aberrations. This may also be referred to as *first order optics* because the angles the ray makes with the axis are small and the sine of the angle equals the angle in radians. (See *Paraxial region* on page 81.) Higher order optics (3rd, 5th, 7th, etc.) have aberrations.

A refracting surface changes the vergence incident on it. The amount of the change is equal to the power of the surface. Since vergence and power have the same units (diopters), you need only sum the power of the surface and the incident vergence to obtain the emerging vergence. This may be expressed as the **Gaussian Imaging Equation**:

$$L' = L + F \tag{4-7}$$

where: L' = the emerging reduced vergence leaving the surface after refraction
L = the incident reduced vergence stricking the surface
F = the power of the surface

From this equation, the relationship between the object position, image position, and power of the surfaces may be calculated. The type of object or image (real or virtual) and the resulting type of vergence (parallel, convergent, divergent) must be known before the equation can be used properly. You must be careful to follow the established sign convention. The table in Figure 4-9 may be helpful in setting up problems.

OBJECT/IMAGE	SYMBOL	WAVEFRONT	LOCATION	SIGN DISTANCE/VERGENCE
real object	RO	Divergent incident	Left of interface	Negative
real image	RI	Convergent emerging	Right of interface	Positive
virtual object	VO	Convergent incident	Right of interface	Positive
virtual image	VI	Divergent emerging	Left of interface	Negative

Figure 4-9. The relationship between object location, image location, and vergence.

Example 4-f

The reduced eye is a perfect example of a single refracting surface. The spherical refracting surface is the cornea, with air on one side and aqueous fluid (n = 1.33) on the other side. If the radius of the cornea is 7.5 mm, what is the power of the reduced eye? If the length of the eye is 31 mm, where must an object be placed in front of the eye to form the image on the retina (back of the eye)?

Known
Radius of surface: $r = 7.5\text{ mm} = 0.0075\text{ m}$
Index of object space: $n = 1.00$
Index of image space: $n' = 1.33$
Length of eye (image distance): $\ell' = 31\text{ mm} = 0.031\text{ m}$

Unknown
Power of cornea: $F = ?$
Object position: $\ell = ?$
Reduced vergence object/image

Equations/Concepts
(4-4) $F = (n'-n)/r$
(4-7) $L' = L + F$
(4-5), (4-6) Vergence

The first step is to diagram the eye and label the radius of the cornea and the indices (see Figure 4-10). The radius, as drawn in the diagram, is positive. Using Equation 4-4, the power is

$$F = \frac{n'-n}{r} = \frac{1.33-1.00}{0.0075} = +44.00\text{ D}$$

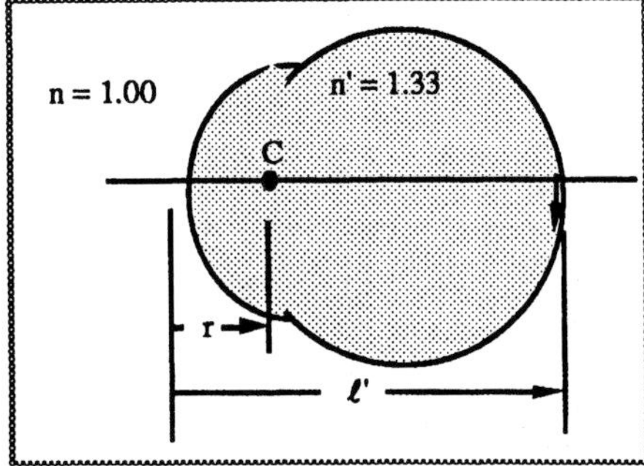

Figure 4-10. *Example 4-f.*

The image is located on the retina 31 mm from the refracting surface. The object position must be determined. At this point, it may be difficult to know if the object will be located to the left or to the right of the interface (i.e., real or virtual). Our sign convention will help. Start by calculating the reduced vergence of the image using Equation 4-6. Because the image is formed on the retina (screen), it is real and must be positioned to the right of the interface. As indicated by an arrow in the figure, the image distance is measured from the cornea to the image (left to right), the same direction that light travels. This makes the image distance a positive value according to the sign convention, with the emerging reduced vergence of:

$$L' = \frac{n'}{\ell'} = \frac{1.33}{+0.031} = +42.90\text{ D}$$

Thus an unknown incident vergence strikes the interface of + 44.00 D, and an emergent vergence of + 42.90 D results. The incident vergence is calculated by rearranging Equation 4-7 to solve for L:

$$L = L' - F = +42.90 - (+44.00) = -1.10\text{ D}$$

This result shows that the incident vergence is negative. Thus there is divergent light incident on the corneal refracting surface, and the object is real and positioned to the left of the interface. To solve for the object position, rearrange Equation 4-5 to solve for ℓ:

$$\ell = \frac{n}{L} = \frac{1.00}{-1.10} = -0.91\text{ m} = -91\text{ cm}$$

The object is therefore located 91 cm to the left of the interface.

Example 4-g

When an object is placed 25 cm in front of a glass rod, an image is formed within the rod 1.57 m from the front surface. If the surface of the rod is spherical with a power of + 5.00 D, what is the index of the glass rod ?

Known	***Unknown***	***Equations/Concepts***
Power of surface: $F = +5.00$ D	Index of the rod	(4-5) , (4-6) Reduced vergence
Index of object space: $n = 1.33$		(4-7) $L' = L + F$
Object distance: $\ell = 25$ cm $= 0.25$ m		
Image distance: $\ell' = 1.57$ m		

Because the object is in front of the glass rod, it is in object space and the object is real. The distance from the rod to the object is negative or –25 cm. Solve for the incident using Equation 4-5 and substitute the result into Equation 4-7 to determine the emerging vergence:

$$L = \frac{n}{\ell} = \frac{1.00}{-0.25\text{m}} = -4.00 \text{ D} \qquad L' = L + F = +5.00 \text{ D} + (-4.00 \text{ D}) = +1.00 \text{ D}$$

The image is formed to the right of the surface (within the rod), and therefore it is real and has a positive distance value. Using the emerging vergence and the image distance, use Equation 4-6 to calculate the index of the rod (image space):

$$L' = \frac{n'}{\ell'} \qquad n' = \ell' L' = (+1.57 \text{ m})(+1.00 \text{ D}) = 1.57 = \text{index of the rod}$$

Example 4-h

The pupil in the eye in *Example 4-f* is located 3.6 mm from the cornea. Where does the pupil appear to be?

Known	***Unknown***	***Equations/Concepts***
Power of surface: $F = +5.00$ D	Image position: ℓ'	(4-5), (4-6) Vergence
Index of object space: $n = 1.33$		(4-7) $L' = L + F$
Object distance: $\ell = 25$ cm $= 0.25$ m		
Image distance: $\ell' = 1.57$ m		

The question asks for the apparent position of the pupil, which is the image position of the pupil. The object is the pupil, which is located inside the eye in index 1.33. Because the object is real (the pupil exists), it must be positioned to the left of the interface, which means that Figure 4-10 must be reversed as shown in Figure 4-11, where n is the index of the eye (before refraction) and n' is the index of air (after refraction). Because the object is real and to the left of the interface, the value of the object distance is negative. The incident vergence and emerging vergence are calculated using Equations 4-5 and 4-7:

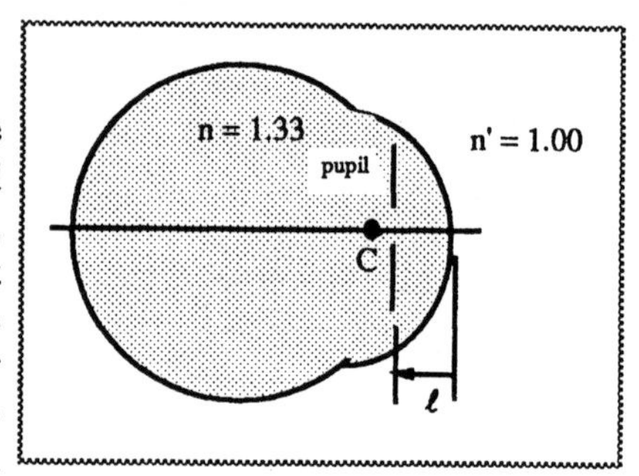

Figure 4-11. *Example 4-h.*

$$L = \frac{n}{\ell} = \frac{1.33}{-0.0036\text{m}} = -369.44 \text{ D}$$
$$L' = L + F = 44.00 \text{ D} + (-369.44 \text{ D}) = -325.44 \text{ D}$$

Solving for the image position using Equation 4-6:

$$\ell' = \frac{n'}{L'} = \frac{1.00}{-325.44 \text{ D}} = -0.003073 \text{ m} = -3.07 \text{ mm}$$

The image has a negative value and therefore according to our sign convention, is located to the left of the interface (in object space) and is virtual. The pupil appears to be 3.07 mm inside of the eye.

Image-Object Relationships: Focal Points

There are special image and object positions inherent in all optical systems called **focal points**. The **secondary focal point** is the image position that corresponds to an infinite object position. An object at an infinite distance from the refracting surface has zero vergence, as shown by Equation 4-5:

$$L = \lim_{\ell \to \infty} \frac{n}{\ell} = 0.00 \text{ D}$$

If you find the limit notation to be confusing think about what happens to a fraction as the denominator increases (i.e., the fraction decreases). Take the reciprocals of say 10, 100, and 10,000. When the value of ℓ approaches an extremely large number, the vergence then approaches zero.

Using zero incident vergence, the emergent vergence is calculated using Equation 4-7:

$$L' = L + F = 0 + F = F$$

For this special object position, the vergence leaving the refracting surface is the same as the power of the surface. This will be true for any optical system.

The resulting axial image position is the secondary focal point. The distance from the surface to this image position is the secondary focal length, which may be calculated using Equation 4-6:

$$\ell' = \frac{n'}{L'} \quad \text{or in this case} \quad f' = \frac{n'}{F} \tag{4-8}$$

where:

f' = the secondary focal length
F = the power of the surface
n' = the index of refraction of the image space

The value of f' is positive for a positive power surface and negative for a negative power surface. This means that the secondary focal point is located to the right of a positive refracting surface and to the left of a negative refracting surface, as illustrated in Figure 4-12.

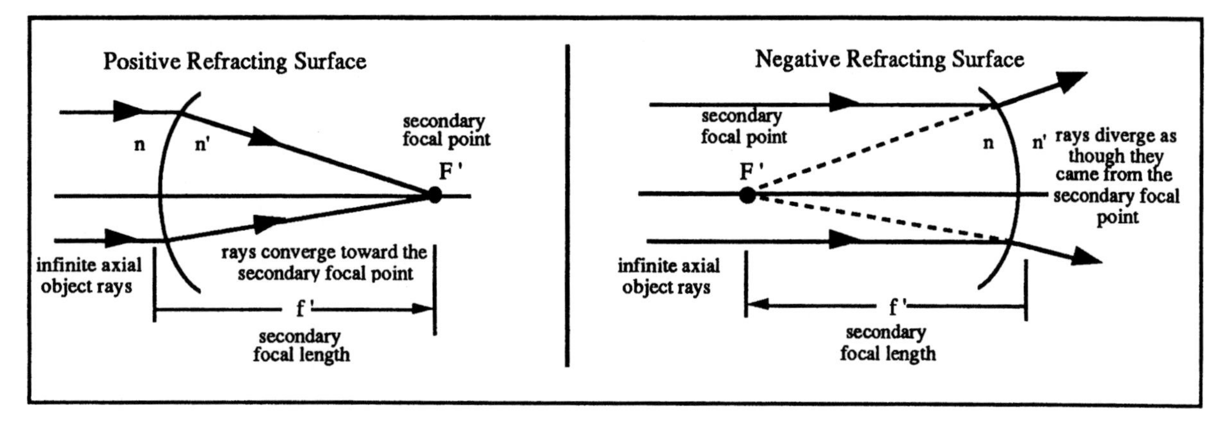

Figure 4-12. Positive and negative power surfaces and secondary focal point location. The incident rays are axial rays from an infinite object source with an axial image formed at the secondary focal point.

The **primary focal point** is another location that is extremely important in defining optical systems. To better understand this position, let's first explore how the vergence and image position change as an object approaches the refracting surface from infinity.

Example 4-i

Calculate the vergence incident on a + 5.00 D refracting surface for object positions of infinity, 1 m, 50 cm, 20 cm, and 10 cm. In each case calculate the emerging vergence and the image location. Assume an air-glass (n = 1.50) interface.

Known	***Unknown***	***Equations/Concepts***
Object locations:	Incident vergence for each object location	(4-5), (4-6), (4-7)
ℓ = infinity, 1 m, 50 cm, 20 cm, 10 cm	Emergent vergence for each image location	Sign conventions
Power of surface: F= + 5.00 D	Image locations	
Index of object space: n = 1.00		
Index of image space: n = 1.50		

The incident vergence (0.00 D), emergent vergence (L' = F) and image position (secondary focal length) have been calculated for the infinite object position in defining the secondary focal point. Let's work the complete problem for the 1 m object position. The objects are real and located in object space (to the left of the interface) (refer to Figures 4-8 and 4-9). The measurement from the surface to the object is against the direction of light travel and is therefore a negative number (i.e., ℓ = – 1.00 m). First calculate the reduced vergence of the object using Equation 4-5:

$$L = \frac{n}{\ell} = \frac{1.00}{-1.00} = -1.00 \text{ D}$$

Using this incident vergence and Equation 4-7, calculate the emergent vergence L':

$$L' = L + F = +5.00 \text{ D} + (-1.00 \text{ D}) = +4.00 \text{ D}$$

The resulting image location (ℓ') is calculated by using Equation 4-6. The index to the right of the interface (n' = 1.50) is used in the calculation.

$$\ell' = \frac{n'}{L'} = \frac{1.50}{+4.00 \text{ D}} = +0.375 \text{ m} = +37.50 \text{ cm}$$

The image is located to the right of the interface (a positive image location) at a distance of 37.50 cm. The calculations for the other object positions are shown in Figure 4-13. By reviewing this table, it should be clear that as a real object approaches a refracting surface from infinity, the incident vergence increases from zero and becomes more negative. For the same surface power, this results in a decrease in positive (or an increase in negative) vergence leaving the surface. For a positive refracting surface, the image moves from the secondary focal point (when the object is at infinity) to the right, away from the surface. This occurs as long as the incident divergence is less than the positive power of the refracting surface. At some object location, the magnitude of the incident divergence matches the positive power of the refracting surface, resulting in an emergent vergence of zero. In our example, this occurs when the object is located 20 cm in front of the surface. The image is then at an infinite position. *The object position that yields a reduced vergence with the same magnitude (opposite sign) as the power of the refracting surface (image is at an infinite position) is the primary focal point.*

As seen in this example, if the object moves closer to the refracting surface (inside the primary focal point), the magnitude of the negative incident divergence is greater than the power of the surface, resulting in a negative emergent vergence. This means that divergent rays leave the surface. This will result in a virtual image and will be explained further in the diagrams.

Object position	Incident vergence	Emergent vergence	Image position
ℓ	$L=\frac{n}{\ell}$	$L'=L+F$	$\ell'=\frac{n'}{L'}$
∞	$L=\lim_{\ell\to\infty}\frac{1.00}{\ell}=0.00\text{ D}$	$L'=0.00+5.00=+5.00\text{ D}$	$\ell'=\frac{1.50}{+5.00}=+0.30\text{m}=+30.00\text{ cm}$
-1.00 m	$L=\frac{1.00}{-1.00}=-1.00\text{ D}$	$L'=-1.00+5.00=+4.00\text{ D}$	$\ell'=\frac{1.50}{+4.00}=+0.375\text{m}=+37.50\text{ cm}$
-0.50 m	$L=\frac{1.00}{-0.50}=-2.00\text{ D}$	$L'=-2.00+5.00=+3.00\text{ D}$	$\ell'=\frac{1.50}{+3.00}=+0.50\text{m}=+50.00\text{ cm}$
-0.20 m	$L=\frac{1.00}{-0.20}=-5.00\text{ D}$	$L'=-5.00+5.00=+0.00\text{ D}$	$\ell'=\lim_{l'\to 0}\frac{1.50}{l'}=\infty$ infinite *
-0.10 m	$L=\frac{1.00}{-0.10}=-10.00\text{ D}$	$L'=-10.00+5.00=-5.00\text{ D}$	$\ell'=\frac{1.50}{-5.00}=-0.30\text{ m}=-30.00\text{ cm}$**

Figure 4-13. Solutions for *Example 4-i.* * infinite image position; ** virtual image position.

From this example, the definition of the primary focal point may be further developed. As stated, the primary focal point is the object position that yields an emergent vergence of zero (L' = 0.00 D) and thus an infinite image distance. Using Equations 4-7 and 4-5, this may be expressed as follows:

$$L = L' - F = 0.00\text{ D} - F = -F$$

$$\ell = \frac{n}{L} \quad \text{or in this case} \quad f = -\frac{n}{F} \tag{4-9}$$

where:

f = the primary focal length
F = the power of the surface
n = the index of refraction of object space

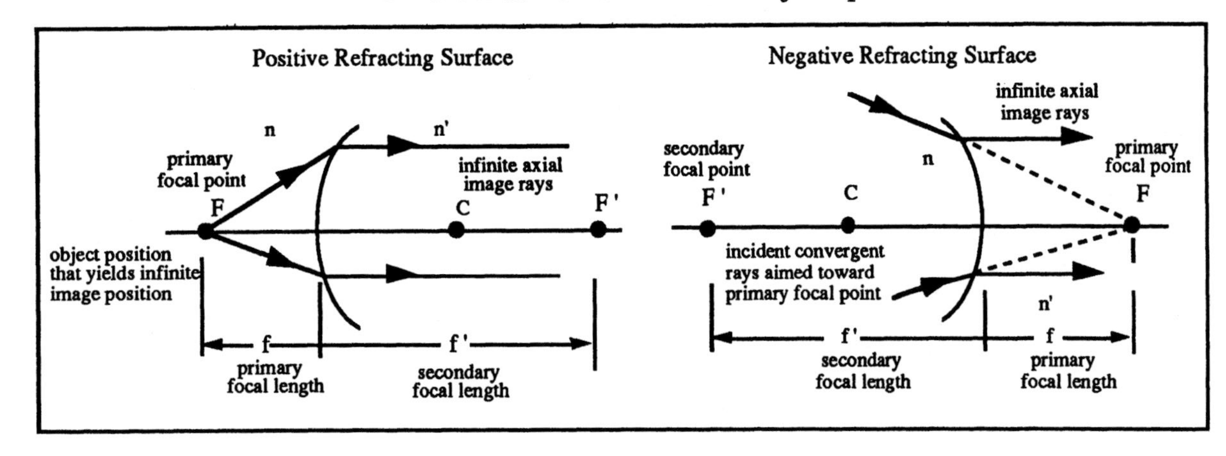

Figure 4-14. Positive and negative power surfaces and primary focal point location. The object is located at the primary focal point and the emergent rays are parallel to the axis and form an image at infinity.

The primary focal point location is illustrated in Figure 4-14 for positive and negative refracting surfaces. To have an incident vergence match the magnitude of the refracting surface (with opposite sign) for the negative power surface, convergent rays must be incident on the surface. These rays are aimed toward the primary focal point, but before arriving at that position, the rays are diverged by the surface and leave parallel toward an infinite image position.

Equations 4-8 and 4-9 may be solved for the power of the surface F and equated:

$$F = -\frac{n}{f} = \frac{n'}{f'} \qquad (4\text{-}10)$$

It is strongly suggested that at this point you review Chapter 9 on ray tracing through curved single refracting surfaces.

Example 4-j

A – 7.00 D single refracting surface has a primary focal length of 19 cm and a secondary focal length of 25 cm. What are the signs of these lengths? What is the index of refraction on either side of the interface?

Known	***Unknown***	***Equations/Concepts***
Primary focal length: $f = 19\text{ cm} = 0.19\text{ m}$	Signs of focal lengths	(4-10) Focal length - power
Secondary focal length: $f' = 25\text{ cm} = 0.25\text{ m}$	Index of object space: $n = ?$	Sign convention for focal length
Power of surface $F = -7.00\text{D}$	Index of image space: $n' = ?$	

The surface has negative power. Parallel incident rays will diverge (after refraction) as though they were coming from the secondary focal point. Thus the secondary focal point must be to the left of the interface and must have a negative value. The primary focal point must be to the right of the surface and must have a positive value. Solve for the indices by solving for n and n' in Equation 4-10:

$$F = -\frac{n}{f} \qquad n = -fF = -(+0.19\text{ m})(-7.00\text{ D}) = 1.33 = \text{ index before refraction}$$

$$F = \frac{n'}{f'} \qquad n' = f'F = (-0.25\text{ m})(-7.00\text{ D}) = 1.75 = \text{ index after refraction}$$

Example 4-k

How would the power and the focal lengths change if the water medium for *Example 4-j* were changed to air?

Known	***Unknown***	***Equations/Concepts***
From *Example 4-j*	Primary focal length: $f = ?$	(4-4) Power - radius
Primary focal length: $f = 19\text{ cm} = 0.19\text{ m}$	Secondary focal length: $f' = ?$	(4-10) Focal length - power
Secondary focal length: $f' = 25\text{ cm} = 0.25\text{ m}$	Power of surface : $F = ?$	
Power of surface $F = -7.00\text{D}$		
Index of object space: $n = 1.33$		
Index of image space: $n = 1.75$		
New index of object space : $n = 1.00$		

Solve for the radius of the surface using the old indices and Equation 4-4:

$$F = \frac{n' - n}{r} \qquad r = \frac{n' - n}{F} = \frac{1.75 - 1.33}{-7.00\text{ D}} = -0.06\text{ m} = -6.00\text{ cm}$$

Use the radius to solve for the new power (Equation 4-4) and the focal lengths (Equation 4-10) in air:

$$F = \frac{n'-n}{r} = \frac{1.75-1.00}{-0.06} = -12.50 \text{ D}$$

$$f = -\frac{n}{F} = -\frac{1.00}{-12.50 \text{ D}} = +0.08 \text{ m} = +8.00 \text{ cm} \qquad f' = \frac{n'}{F} = \frac{1.75}{-12.50 \text{ D}} = -0.14 \text{ m} = -14.00 \text{ cm}$$

Axial points - object or image points located on the optical axis.

Extra axial points - object or image points that are off the optical axis. In our diagrams, these positions will be above or below the axis.

Reversibility of light - light will travel the same path independent of direction of travel (i.e., light will follow the same path through a media whether traveling right to left or left to right).

Conjugate points - an image and object relationship such that if the object is placed in the image plane, an image will be formed in the original object plane. The primary and secondary focal points are conjugate to positive and negative infinity, respectively.

Paraxial region - the region about the optical axis where all rays focus to a single point. In this region, the sine of an angle is equal to the tangent of the angle and to the angle itself in radian measure ($\sin \theta = \tan \theta = \theta$ in radians). In paraxial, first order, or Gaussian optics, it is assumed that all rays fall within the paraxial region.

Extrafocal distance - the distance from the primary focal point to the object (x) or the distance from the secondary focal point to the image (x'). The sign convention is followed.

Chief ray - a ray that passes through the refracting surface undeviated. This ray passes through the center of curvature of the surface (nodal point).

Image-Object Relationships: Size and Orientation

Just as the positions of the images and objects are related, their relative sizes and orientation (right side up or upside down) are also related. These relationships may be determined using several methods that are illustrated in this section. Although all of the methods presented yield similar answers, it is suggested that you learn all of the relationships. This will allow you flexibility in problem solving.

Figure 4-15 illustrates a positive single refracting surface. The object and image are shown with respective sizes of h and h'. The relative size of the image and object can be expressed as a ratio called the **lateral magnification**:

$$\text{Lateral Magnification} = \text{LM} = \frac{\text{image size}}{\text{object size}} = \frac{h'}{h} \qquad (4\text{-}11)$$

This ratio is greater than unity if the image is larger than the object and it is less than unity when the image is smaller than the object. A special condition exists when the image and object are the same size. This will be discussed further in the next chapter.

The sign associated with the lateral magnification indicates the image-object orientation. If the image and object are on opposite sides of the axis (one above and the other below), the relationship is called **inverted**, and the lateral magnification is negative. If the image and object are on the same side of the axis (both above or both below), the relationship is called **erect**, and the lateral magnification is positive.

In Figure 4-15, a chief (undeviated) ray connects the top of the object to the top of the image. This ray passes through the center of curvature of the surface; it is therefore normal

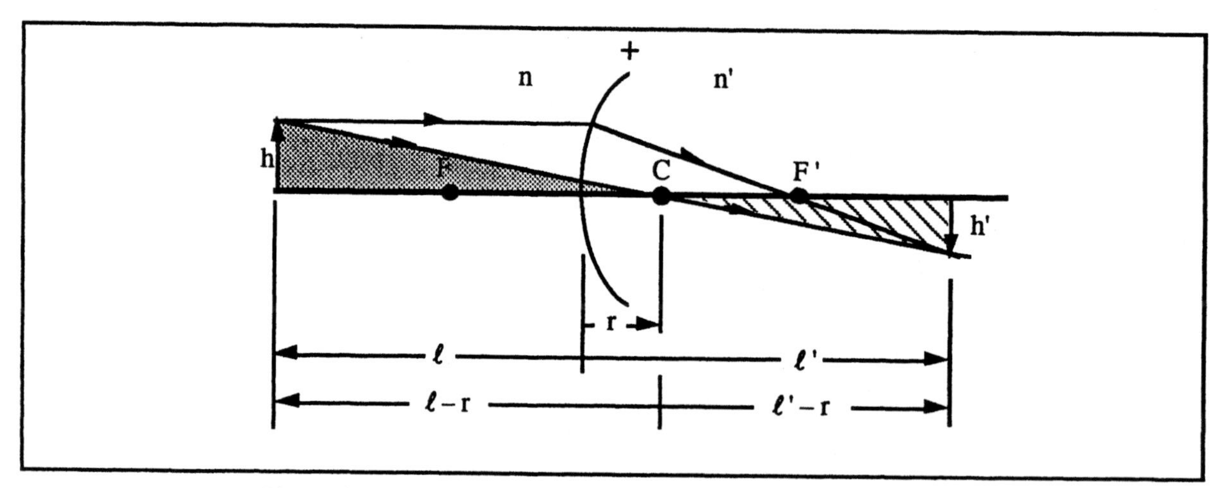

Figure 4-15. Radius method for determining lateral magnification.

to the surface and undeviated. Using the two shaded similar triangles shown in the figure, the following relationship may be developed:

$$\frac{h}{\ell - r} = \frac{h'}{\ell' - r}$$

Setting the equation equal to the lateral magnification (h'/h), this becomes

$$LM = \frac{h'}{h} = \frac{\ell' - r}{\ell - r} \tag{4-12}$$

The same positive refracting surface with image and object may be drawn as shown in Figure 4-16. Here a ray is drawn from the top of the object to the point where the surface intersects the optical axis. The ray is refracted at the surface following Snell's Law. Note that the axis is the normal in this case, and the incident and refracted angles are ω and ω', respectively. Using Snell's Law, this becomes

$$n \sin \omega = n' \sin \omega'$$

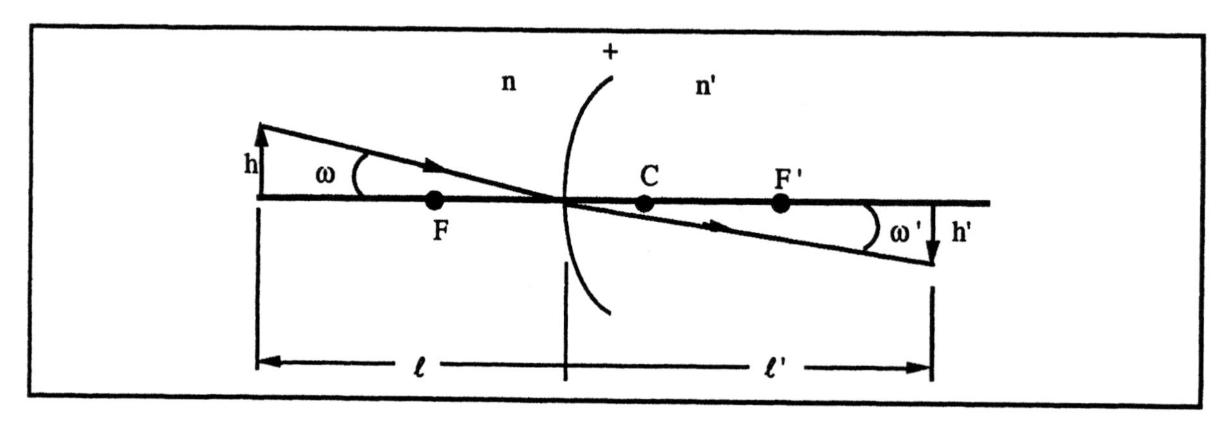

Figure 4-16. Illustration of Snell's Law to determine lateral magnification.

Because the region about the axis is considered to be paraxial, the sine of the angle is equal to the tangent of the angle, which is equal to the angle itself in radians. This may be expressed as:

$$n \tan \omega = n' \tan \omega' \quad (4\text{-}13)$$

Looking at the figure, it can be seen that tan ω and tan ω' are defined as

$$\tan \omega = \frac{h}{\ell} \qquad \tan \omega' = \frac{h'}{\ell'} \quad (4\text{-}14)$$

Substituting Equation 4-14 into 4-13 and solving for lateral magnification, this becomes

$$\frac{nh}{\ell} = \frac{n'h'}{\ell'}$$

$$LM = \frac{h'}{h} = \frac{n\ell'}{n'\ell} \quad (4\text{-}15)$$

The final method for determining lateral magnification uses a measurement called the **extrafocal distance** (Figure 4-17). The extrafocal object distance x is the distance from the primary focal point to the object. The extrafocal image distance x' is the distance from the secondary focal point to the image. The sign convention must be maintained. *These are the only distances that are not measured from the interface.* The relationship between the extrafocal distances, focal lengths, and image and object distances may be seen in Figure 4-17.

$$\ell = f + x \quad (4\text{-}16)$$

$$\ell' = f' + x' \quad (4\text{-}17)$$

Remember that when the extrafocal distances are measured from right to left, the values are negative; when measured from left to right, the values are positive.

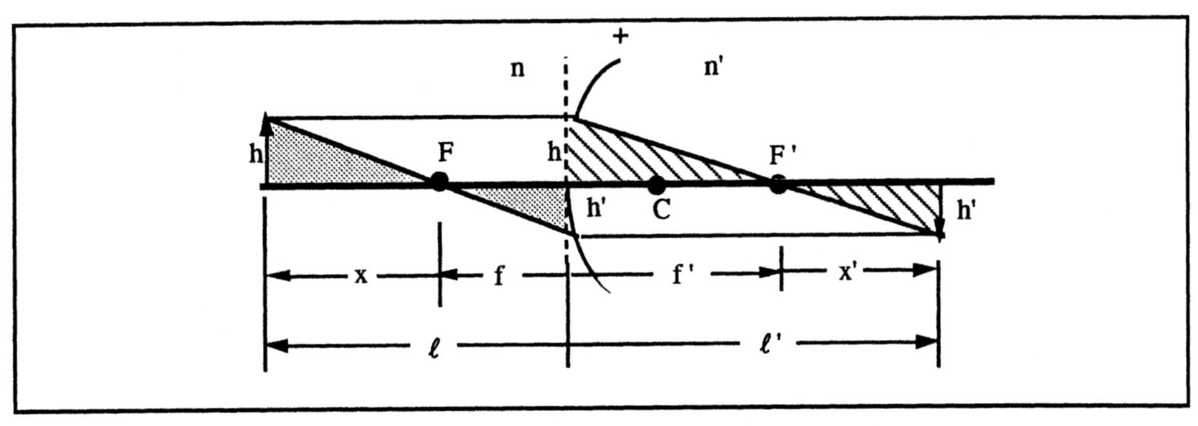

Figure 4-17. Extrafocal distance method for determining the lateral magnification.

The lateral magnification may be determined by using the shaded triangles. The shaded triangles are similar, and the relationship may be written as

$$\frac{h}{x}=\frac{h'}{f} \quad \text{or} \quad LM=\frac{h'}{h}=-\frac{f}{x} \qquad (4\text{-}18)$$

The minus sign is included in the equation so that the sign convention can be consistant.

Another lateral magnification relationship may be developed by using the similar striped triangles on the right:

$$\frac{h'}{x'}=\frac{h}{f'} \quad \text{or} \quad LM=\frac{h'}{h}=-\frac{x'}{f'} \qquad (4\text{-}19)$$

Equations 4-18 and 4-19 may be set equal to each other to yield what is called *Newton's Relationship:*

$$xx'=ff' \qquad (4\text{-}20)$$

This relation is useful in solving problems.

Note that the lateral magnification may also be determined by using the ray tracing procedure discussed in Chapter 9.

Example 4-l

In *Example 4-i*, an object was placed 1 m in front of a + 5.00 D single refracting surface, and a real image was located 37.50 cm from the surface. The index of object space was 1.00, and the index of image space was 1.50. The radius of the surface is 10 cm. Solve for the lateral magnification using each of the techniques discussed. Is the final image inverted or erect?

Known	***Unknown***	***Equations/Concepts***
Object distance: $\ell=-1.00$ m $=-100.00$ cm	Lateral magnification: LM = ?	(4-12), (4-15), (4-18), (4-19)
Image distance: $\ell'=+37.50$ cm $=+0.3740$ m		Magnification formula
Power of surface: F= + 5.00 D		(4-16), (4-17) Extrafocal distance
Index of object space: n = 1.00		(4-10) Focal length-power
Index of image space: n = 1.50		
Radius of surface: r = +10.00 cm		

Solve for the lateral magnification using Equations 4-12 and 4-15:

$$LM=\frac{\ell'-r}{\ell-r}=\frac{+37.50\text{ cm}-10.00\text{ cm}}{-100.00\text{ cm}-10.00\text{ cm}}=\frac{+27.5\text{ cm}}{-110.00}=-0.25$$

$$LM=\frac{n\ell'}{n'\ell}=\frac{(1.00)(+37.50\text{ cm})}{(1.50)(-100\text{ cm})}=\frac{37.50\text{ cm}}{-150\text{ cm}}=-0.25$$

Use Equations 4-10, 4-16 and 4-17 to calculate the focal lenghts and extrafocal distances, respectively:

$$f=-\frac{n}{F}=-\frac{1.00}{5.00\text{ D}}=-0.20\text{ m}=-20.00\text{ cm} \qquad f=\frac{n'}{F'}=\frac{1.50}{5.00\text{ D}}=+0.30\text{ m}=+30.00\text{ cm}$$

$$\ell=f+x \qquad x=\ell-f=-100\text{ cm}-(-20.00\text{ cm})=-80.00\text{ cm}$$

$$\ell'=f'+x' \qquad x'=\ell'-f'=+37.50\text{ cm}-(+30.00\text{ cm})=+7.50\text{ cm}$$

Solving for the lateral magnification using Equations 4-18 and 4-19:

$$LM=-\frac{f}{x}=-\frac{-20.00\text{cm}}{-80.00\text{cm}}=-0.25 \qquad LM=-\frac{x'}{f'}=-\frac{+7.50}{+30.00}=-0.25$$

Supplemental Problems

Sag Problems

4-1. The 2 mm center thickness of a plano-concave lens is one-half the edge thickness. If the diameter of the lens is 80 mm, what is the radius of the curved surface?
ANS. 0.40 m

4-2. What is the edge thickness of a −5.00 D biconcave glass lens (n = 1.523), if the center thickness is 10 mm, and the diameter is 60 mm?
ANS. Edge thickness = 14.3 mm

4-3. A meniscus lens has an edge thickness of 5 mm and a diamter of 40 mm. If the sag of the convex surface is 1.2 mm, and the sag of the other surface is 2.4 mm, what is the center thickness of the lens?
ANS. 3.8 mm

Power

4-4. A negative single refracting surface between air and glass (n = 1.75) has a radius of 25 cm. Calculate the power of the surface.
ANS. −3.00 D

4-5. A lens has a power of + 4.00 D in air. Under water (n = 1.33), the lens has a power of + 1.73 D. What is the index of refraction of the lens?
ANS. n = 1.58

Spherometer, Lens Clock

4-6. Using a spherometer, you find the sag of a +5.50 D convex surface to be 3.30 mm. Calculate the distance between the outer and middle legs (i.e., solve for y). Assume an index of 1.523.
ANS. y = 2.51 cm

4-7. A lens clock measures + 10.00 D when placed on a ball bearing. Find the diameter of the ball bearing.
ANS. 10.6 cm

4-8. Using a lens clock to measure the surface power of a plastic lens (n = 1.49), you find the power to be −3.75 D. What is the actual power of the surface?
ANS. −3.47 D

Object-Image Relationships

4-9. Where is the image formed when an object is placed 25 cm in front of a + 4.00 D single refracting surface? Assume air to be the index to the left of the interface.
ANS. at infinity

4-10. A long glass rod (n = 1.5) has a curved surface with a power of + 2.50 D. If an object located in air is placed an infinite distance in front of the rod, where is the image formed?
ANS. 60 cm behind the refracting surface

4-11. An object located 100 cm in front of a glass rod (n = 1.49) forms an image 30 cm behind the front surface. Calculate the radius of the end of the rod.
ANS. Radius = +8.21 cm

Reduced Vergence: Image-Object Relationships

4-12. Light converges toward a single refracting surface to a point located 20 cm behind the surface. After refraction, the light is focused 10 cm behind the surface. What is the power of the glass surface (n = 1.50)?
ANS. + 10.00 D

4-13. Light diverges from a real object located 15 cm in front of a plastic single refracting surface (n = 1.49) with a radius of + 9.8 cm. Where is the image formed, and what is the emergent vergence?
ANS. A virtual image is formed 89.4 cm in front of the surface; emergent vergence = – 1.67 D

4-14. A concave glass rod (n = 1.523) is held under water (n = 1.33). An object inside the rod appears to be 10 cm from the water-glass interface, but the actual distance is 7.50 cm. Find the power of the glass rod.
ANS. Power = – 2.50 D

Focal Lengths

4-15. What is the secondary focal length of a glass single refracting surface (n = 1.50) if the surface is convex, and the center of curvature is located 45 cm from the surface?
ANS. 1.35 m

4-16. For a biconvex lens with a front surface power of + 8.50 D, a back radius of 8.823 cm, and an index of 1.75, what is the power of the back surface of the lens and the focal length of the front surface?
ANS. Back surface power = +8.50D; front focal length = – 11.76 cm

4-17. A single refracting surface has a power of – 6.00 D with primary and secondary focal lengths of 22 cm and 25 cm, respectively. Calculate the index of refraction of object space and image space.
ANS. Index of object space = 1.32; index of image space = 1.50

Lateral Magnification

4-18. A virtual object is one meter from a + 9.00 D single refracting surface (n=1.50). What is the lateral magnification of the image formed?
ANS. LM = +0.10x

4-19. If a real object in air is 50 cm from a −3.00 D single refracting surface (n = 1.49), determine the image location, the type (virtual or real) and orientation (erect or inverted), and the lateral magnification.
ANS. Image is 29.8 cm in front of the surface; it is virtual and erect; LM=+0.40x

Newton's Relationship

4-20. An object in air is located 20 cm in front of a concave single refracting surface. The resultant image is formed 5 cm closer to the surface than the secondary focal point. The image is one-third as large as the object. What is the index of refraction of the single refracting surface?
ANS. n = 1.50

Chapter 5

Thin Lenses

The most common refracting element in ophthalmic use is the lens. A lens may be considered to be a transparent material bounded by two polished single refracting surfaces. The shapes of the refracting surfaces determine the type of lens (spherical, cylindrical, aspheric, etc.) and the dioptric power. In this chapter, spherical lenses (lenses with two spherical surfaces) are presented. Spherical lenses may have different shapes or forms, depending on the surface radii. Some of these forms are illustrated in Figure 5-1. A *convex lens* is a converging element and has *positive power*; a *concave lens* is a diverging element and has *negative power*. (See Figure 5-2.) The fundamental properties and definitions for curved refracting surfaces also apply to spherical lenses.

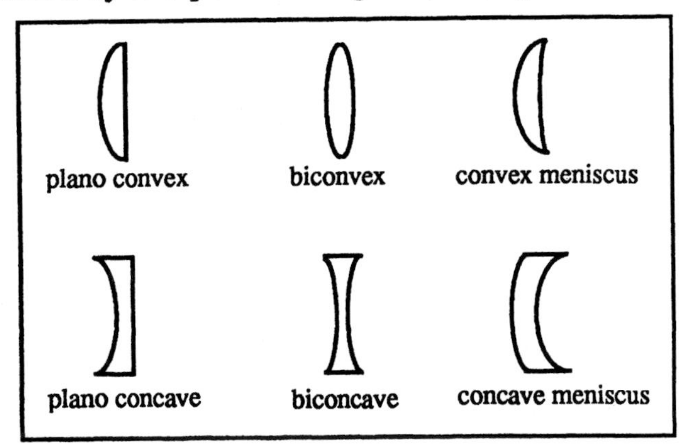

Figure 5-1. Shapes of converging and diverging lenses.

Bending a lens - changing the surface radii while maintaining the same total power.

Optical axis - a line connecting the two radii of curvature. In the case of a plano-concave or plano-convex lens, the optical axis is a line that runs through the center of curvature (of the curved surface) and is perpendicular to the flat surface.

Vertex (A) - the points where the lens surfaces and the optical axis intersect. A thin lens has one common vertex at the intersection of a plane through the center of the lens and the optical axis.

Optical center (OC) - the axial point through which the chief ray passes. For the single refracting surface, the optical center is the center of curvature; for a thin lens, it is the center of the lens (the common vertex).

Thin lens - a lens is considered to be thin when its thickness is small and has a negligible effect on the power. All refraction is considered to take place in one plane centered between the two surfaces.

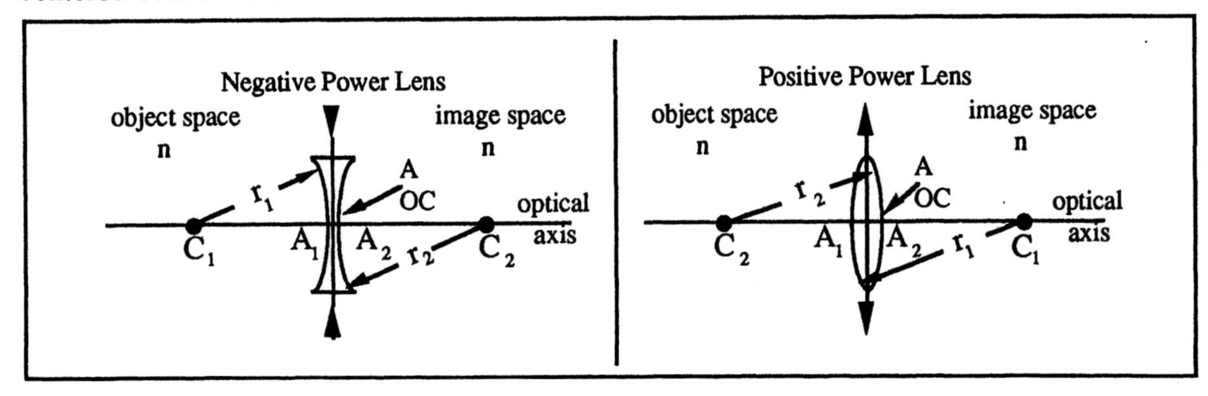

Figure 5-2. Definitions of thin lenses.

Object space - the region in front of (to the left of) the lens.*

Image space - the region behind (to the right of) the lens.*

Surrounding medium - the medium surrounding the lens. This medium fills both object space and image space (i.e., both regions have the same index of refraction, which is usually lower than the index of the lens). In most cases, the surrounding medium is air.

Geometric center - the physical center of the surface (i.e., if a lens has an overall size of 60 mm in diameter, the geometrical center is located 30 mm from the edge).

Chief ray - a ray that follows a path such that the incident ray and the emergent ray are parallel to each other (i.e., the rays make the same angle with the optical axis). This ray is considered to be the **undeviated ray**. For the single refracting surface, the chief ray passes through the center of curvature. For the thin lens, the ray passes through the center of the lens (common vertex and optical center).

Thin Lens Power

Each surface of a thin lens is a single refracting surface and therefore has **surface power**. The surface power may be determined using the techniques and equations described for curved single refracting surfaces. In the equations developed, n represents the index before refraction (left of the surface) and n' represents the index after refraction (right of the surface). To determine the thin lens power, we will treat each surface independently, paying attention to the index to the left (n) and right (n') of the interface. Be sure to follow the convention in determining the sign associated with the radius.

The power of the first refracting surface (F_1), shown in Figure 5-3 (top) is

$$F_1 = \frac{n'-n}{r} = \frac{n_L - n_o}{r_1} \tag{5-1}$$

where:

F_1 = power of the first surface

$n' = n_L$ = index of the lens

$n = n_o$ = index of object space

$r = r_1$ = radius of the first surface

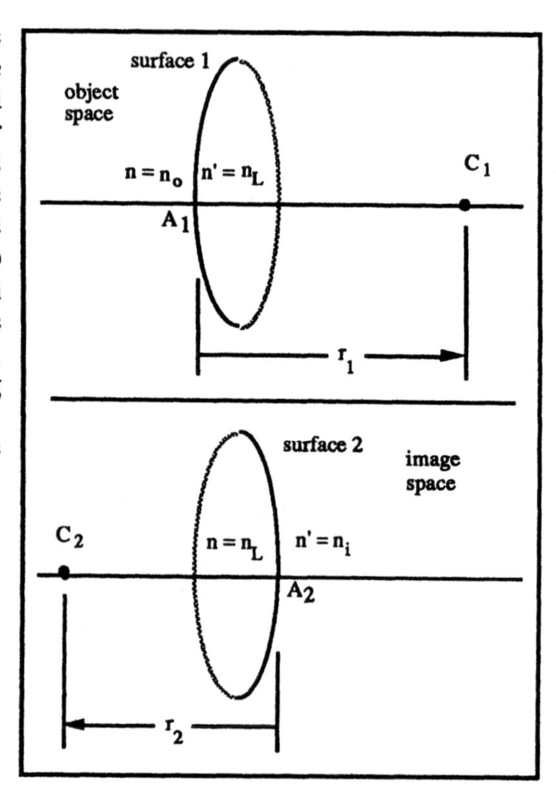

Figure 5-3. Power calculation of a thin lens. *(Top)* Front surface power. *(Bottom)* Back surface power.

The power of the second refracting surface (F_2) is calculated by redefining the indices and using the radius for the back surface (Figure 5-3, bottom):

$$F_2 = \frac{n'-n}{r} = \frac{n_i - n_L}{r_2} \tag{5-2}$$

*These definitions differ from the classic definitions in which object space is anywhere an object is located and image space is anywhere an image is located. In this workbook, the definitions are simplified to help you work problems using the elementary rules that are defined here. Consult your textbook for more details.

where:

F_2 = power of the second surface
$n = n_L$ = index of the lens
$n' = n_i$ = index of image space
$r = r_2$ = radius of the second surface

The dioptric power (F) of a thin lens is is simply the sum of the two surface powers:

$$F = F_1 + F_2 \qquad (5\text{-}3)$$

By substituting Equations 5-1 and 5-2 for F_1 and F_2, respectively, and assuming that the surrounding medium (n_s) is the same [i.e., the object space medium (n_o) is the same as the image space medium (n_i)], Equation 5-3 may be rewritten:

$$F = (n_L - n_s)\left(\frac{1}{r_1} - \frac{1}{r_2}\right) \qquad (5\text{-}4a)$$

In many texts, n' is used for the index of the lens, and n is used for the index of the surrounding medium when solving for lens power. Equation 5-4a may be written as:

$$F = (n' - n)\left(\frac{1}{r_1} - \frac{1}{r_2}\right) \qquad (5\text{-}4b)$$

Both Equations 5-4a and 5-4b are referred to as the Lens Makers Formula.

Example 5-a
Derive Equation 5-4a from Equation 5-3.

Substituting the equations for the front and back surface powers (Equations 5-1 and 5-2) gives

$$F = F_1 + F_2 = \frac{n_L - n_o}{r_1} + \frac{n_i - n_L}{r_2} \qquad (5\text{-}4c)$$

Rewriting F_2 with a negative sign in front and switching the order of the indices yield

$$F_2 = \frac{n_i - n_L}{r_2} = -\frac{n_L - n_i}{r_2} \qquad (5\text{-}4d)$$

Substituting the power of the second surface (Equation 5-4d) into Equation 5-4c gives

$$F = F_1 + F_2 = \frac{n_L - n_o}{r_1} - \frac{n_L - n_i}{r_2}$$

Factoring out the indices and substituting $n_s = n_o = n_i$, this becomes:

$$F = (n_L - n_s)\left(\frac{1}{r_1} - \frac{1}{r_2}\right) \qquad (5\text{-}4a)$$

Example 5-b
A plastic biconcave lens (n = 1.49) has surface radii of 40 cm and 20 cm. Calculate the surface and lens powers.

Known
Type of lens: biconcave
Radius of front surface: $r_1 = 40$ cm $= 0.40$ m
Radius of back surface: $r_2 = 20$ cm $= 0.20$m
Index of lens: n = 1.49
Index of surround assumed to be air: $n_s = 1.00$

Unknown
Surface powers of front and back of lens

Equations/Concepts
(5-1), (5-2), (5-4a)
Power formula

The first step in solving this problem is to diagram and label the lens (Figure 5-4). One difficulty students have with this problem is the orientation of the lens. Refracting surfaces have only one power, and therefore as long as you are consistent and remember the sign convention, the orientation is insignificant. Whenever the surrounding medium is not mentioned in a problem, assume that the medium is air (n = 1.00) and that the indices of object and image space are the same. The surface powers are found by substituting the values into Equations 5-1 and 5-2:

$$F_1 = \frac{n_L - n_i}{r_1} = \frac{n_L - n_s}{r_1} = \frac{1.49 - 1.00}{-0.40} = \frac{0.49}{-0.40} = -1.23 \text{ D}$$

$$F_2 = \frac{n_i - n_L}{r_2} = \frac{n_s - n_L}{r_2} = \frac{1.00 - 1.49}{+0.20} = \frac{-0.49}{+0.20} = -2.45 \text{ D}$$

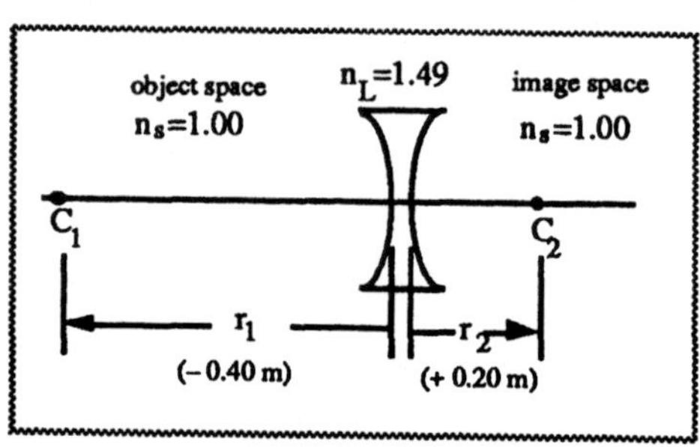

Figure 5-4. *Example 5b.*

The total lens power is the sum of the two surface powers:

$$F = F_1 + F_2 = -1.23 + (-2.45) = -3.68 \text{ D}$$

The power may also be calculated with the Lens Makers Formula (Equation 5-4a):

$$F = (n_L - n_s)\left(\frac{1}{r_1} - \frac{1}{r_2}\right) = (1.49 - 1.00)\left(\frac{1}{-0.40} - \frac{1}{+0.20}\right)$$
$$= (0.49)(-2.50 - 5.00) = (0.49)(-7.50) = -3.68 \text{ D}$$

Example 5-c
Show that there is only one lens power by switching the first and second radii and recalculating the lens power in *Example 5-b*.

Known
Information from *Example 5-b*
Radius of front surface: $r_1 = 20$ cm $= 0.20$ m
Radius of back surface: $r_2 = 40$ cm $= 0.40$ m

Unknown
Lens power: F = ?

Equations/Concepts
(5-1), (5-2), (5-3)
Power formula

Because the front and back radii are switched, the signs associated with these radii are also switched (see Figure 5-4). Making the appropriate changes in the signs and radii, solve for the power using Equations 5-1, 5-2, and 5-3. (Equation 5-4 could also be used.)

$$F_1 = \frac{n_L - n_i}{r_1} = \frac{n_L - n_s}{r_1} = \frac{1.49 - 1.00}{-0.20} = \frac{0.49}{-0.20} = -2.45 \text{ D}$$

$$F_2 = \frac{n_i - n_L}{r_2} = \frac{n_s - n_L}{r_2} = \frac{1.00 - 1.49}{+0.40} = \frac{-0.49}{+0.40} = -1.23 \text{ D}$$

$$F = F_1 + F_2 = -2.45\text{D} + (-1.23\text{D}) = -3.68 \text{ D}$$

Example 5-d
A plastic thin lens (n = 1.49) has a power of – 12.00 D in air. What would the power of the lens be if it were submerged in water?

Known	***Unknown***	***Equations/Concepts***
Lens power in air: F = – 12.00 D	Lens power under water	(5-1), (5-2), (5-3)
Index of water: n = 1.33		Power formula

Because the radii of the surface remain the same for all surrounding indices, use the Lens Makers Formula (Equation 5-4a) to solve for the difference in the reciprocal radii:

$$F = (n_L - n_s)\left(\frac{1}{r_1} - \frac{1}{r_2}\right) \qquad \left(\frac{1}{r_1} - \frac{1}{r_2}\right) = \frac{F}{(n_L - n_s)} = \frac{-12.00\text{ D}}{(1.49 - 1.00)} = -24.49$$

Now use this value and solve for the power of the lens with a surrounding medium of water:

$$F = (n_L - n_s)\left(\frac{1}{r_1} - \frac{1}{r_2}\right) = (1.49 - 1.33)(-24.49) = -3.92\text{ D}$$

This represents a significant change in the power of the lens. Note that as the index of the surrounding medium increases, the surface and total powers of the lens decrease.

Image-Object Relationships: Position

Once the power of the lens has been calculated, the surface curves and index of the lens are no longer required for imaging problems. The lens is reduced to a single plane perpendicular to the optical axis (see Figure 5-2) where all refraction is assumed to occur. Imaging through the lens is similar to imaging through a single refracting surface with one exception: Object space and image space have the same index of refraction. This surrounding medium is usually air. This means that the formulas developed for single refracting surfaces may be simplified. From Equations 4-5 and 4-6, the reduced (incident) object and (emergent) image vergences become

$$\text{Incident object vergence:} \quad L = \frac{n_o}{\ell} = \frac{n_s}{\ell} = \frac{1.00}{\ell} \quad \text{(in air)} \qquad (5\text{-}5)$$

$$\text{Emergent image vergence:} \quad L' = \frac{n_i}{\ell'} = \frac{n_s}{\ell'} = \frac{1.00}{\ell'} \quad \text{(in air)} \qquad (5\text{-}6)$$

where: ℓ = the object distance relative to the lens
ℓ' = the image distance relative to the lens

The Gaussian Imaging Formula (Equation 4-7) may be used to calculate the image-object-power relationship:

$$L' = L + F \qquad (5\text{-}7)$$

where: F = the total power of the lens

The types of images and objects and the sign convention are the same as those described for single refracting surfaces. They are summarized in Figure 5-5.

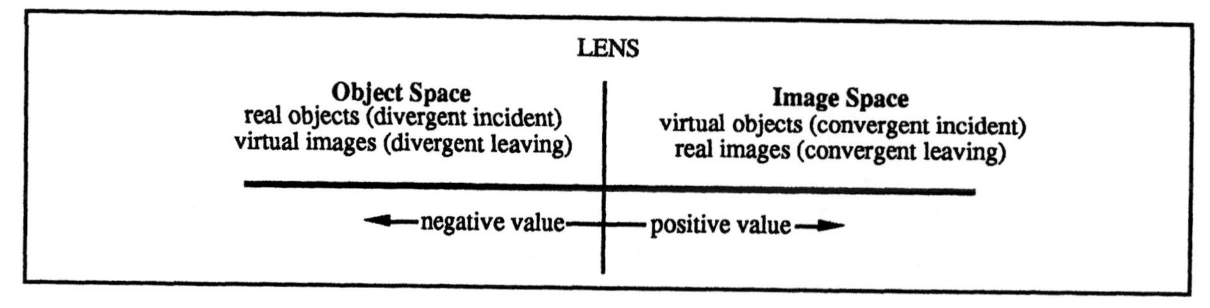

Figure 5-5. Location of types of images and objects, wavefront vergence, and the sign convention.

Example 5-e
A virtual image is formed 25 cm from a thin lens. If the object is real and positioned 50 cm from the lens, what is the power of the lens?

Known
Surrounding index is air: $n = 1.00$
Real object (left of lens): $\ell = -50$ cm $= -0.50$ m
Virtual image (left of lens): $\ell' = -25$ cm $= -0.25$ m

Unknown
Power of lens: $F = ?$

Equations/Concepts
(5-5), (5-6), (5-7)
Gaussian Imaging Formula

Refer to Figure 5-6, which shows the relationships between the types of objects and the types of images formed by positive and negative lenses.

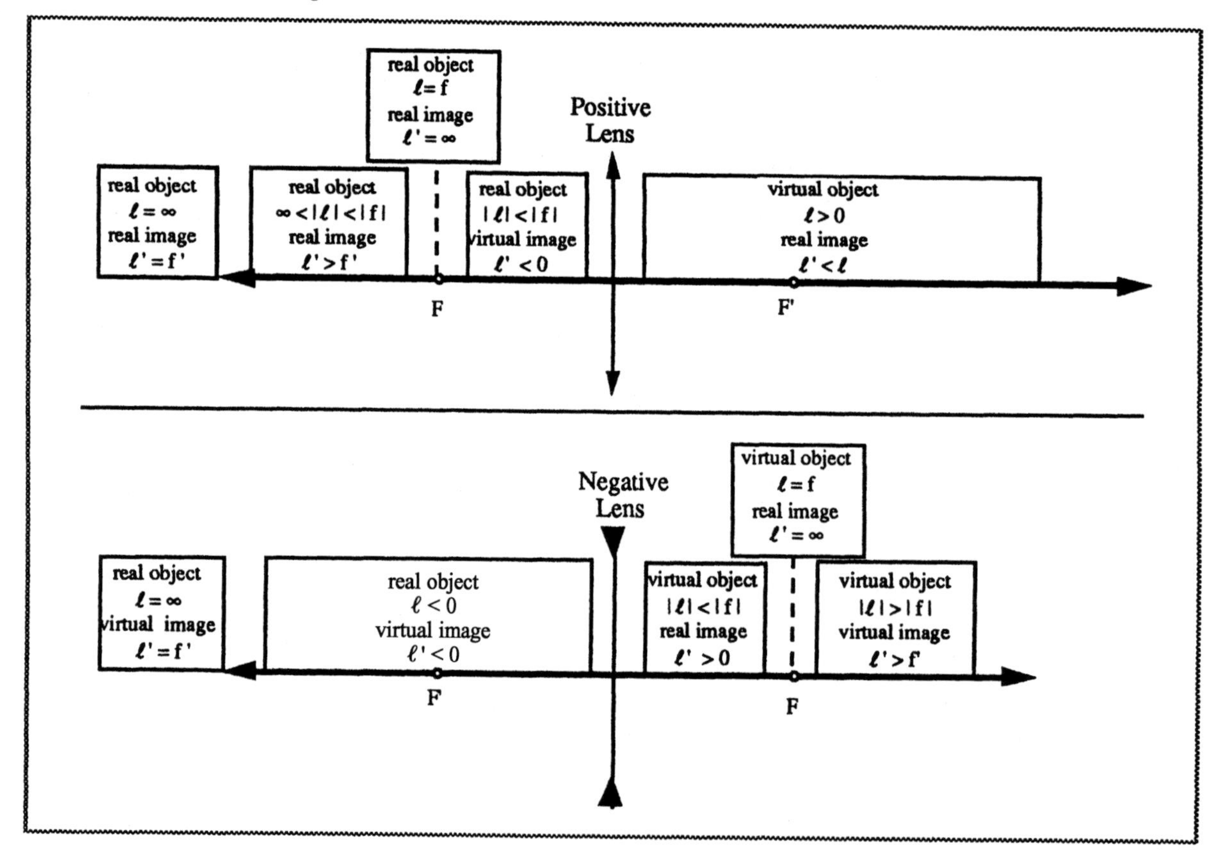

Figure 5-6. *Example 5-e.* Types of objects and resulting types of images for positive and negative lenses.

In this problem, there is a real object and a virtual image. Figure 5-6 shows that for a positive lens, there is only one real object region (between the primary focal point and the lens) that yields a virtual image. Again referring to Figure 5-6 for a negative lens, any real object yields a virtual image. Start by calculating the incident and emergent vergence for the object and the image using Equations 5-5 and 5-6:

$$L = \frac{1.00}{\ell} = \frac{1.00}{-0.50} = -2.00 \text{ D} \qquad L' = \frac{1.00}{\ell'} = \frac{1.00}{-0.25} = -4.00 \text{ D}$$

Using Equation 5-7, solve for the lens power:

$$F = L' - L = -4.00 - (-2.00) = -2.00 \text{ D}$$

The lens has negative power. It may have been obvious that the lens power was negative because a real object between a primary focal point and a positive lens yields a magnified, virtual image that is farther from the lens than the object.

Example 5-f

When an object is placed 25 cm from an equiconvex lens, a real image is formed 1 m from the lens. If the radii of the surfaces are 32 cm, what is the index of refraction of the lens?

Known	***Unknown***	***Equations/Concepts***
Assume object is real because it is "placed"	Index of lens material	(5-5), (5-6), (5-7)
Object distance: $\ell = -25$ cm $= -0.25$ m		(5-4a) Lens Makers Formula
Equi -convex lens: positive power		(5-3) Power of a lens
Radii $r_1 = -r_2 = 32$ cm $= 0.32$ m		
Real image: $\ell' = 1$ m		

The first step is to solve for the total power of the lens using the image and object distances and Equations (5-5), (5-6), and (5-7):

$$L = \frac{1.00}{\ell} = \frac{1.00}{-0.25} = -4.00 \text{ D} \qquad L' = \frac{1.00}{\ell'} = \frac{1.00}{+1.00} = +1.00 \text{ D}$$

$$F = L' - L = +1.00 - (-4.00) = +5.00 \text{ D}$$

The Lens Makers Formula (Equation 5-4a) may be rewritten for an equiconvex lens by substituting $r_2 = -r_1$:

$$F = (n_L - 1.00)\left(\frac{1}{r_1} - \frac{1}{r_2}\right) = (n_L - 1.00)\left(\frac{1}{r_1} - \frac{1}{-r_1}\right) = (n_L - 1.00)\left(\frac{1}{r_1} + \frac{1}{r_1}\right) = (n_L - 1.00)\left(\frac{2}{r_1}\right)$$

Substituting the values for the power and the radii, solve for the index of the lens:

$$+5.00 = (n_L - 1.00)\left(\frac{2}{0.32}\right) \qquad +5.00 = (n_L - 1.00)(6.25)$$

$$n_L = \left(\frac{5.00}{6.25}\right) + 1.00 = 1.80$$

Practice working thin lens problems. Review the section on ray tracing through thin lenses in Chapter 9.

Image-Object Relationships: Focal Points

The location of the primary and secondary focal points of a thin lens may be calculated using the relationships developed for single refracting surfaces (Equation 4-9). Because the surrounding medium is the same (usually air), the equation becomes

$$F = -\frac{n}{f} = \frac{n'}{f'} = -\frac{n_s}{f} = \frac{n_s}{f'} \qquad (5\text{-}8) \qquad\qquad F = -\frac{1.00}{f} = \frac{1.00}{f'} \quad \text{(in air)} \qquad (5\text{-}9)$$

These equations show that for thin lenses, *the magnitudes of the primary and secondary focal lengths are equal.* The focal lengths have opposite signs (they are on opposite sides of the lens).

Just as with single refracting surfaces, an infinite axial object forms an image at the secondary focal point; and an object placed at the primary focal point forms an image at infinity. These imaging situations represent conjugate image-object relationships (i.e., an infinite object is conjugate to the secondary focal point, and the primary focal point is conjugate to infinity). The conjugate relationship may also apply to real images and real objects and to virtual images and virtual objects. (Refer to Figure 5-6.)

Example 5-g

What are the primary and secondary focal lengths for the lens in *Example 5-d,* both in air and under water?

Known	***Unknown***	***Equations/Concepts***
Power in air: $F = -12.00$ D	Primary focal length: $f = ?$	(5-8), (5-9)
Power under water: $F = -3.92$ D	Secondary focal length: $f' = ?$	Focal length definitions
Index of air: $n = 1.00$		
Index of water: $n = 1.33$		

The primary and secondary focal lengths in air ($n = n' = 1.00$) are solved by using Equations 5-8 and 5-9, respectively:

$$F = -\frac{n}{f} \qquad f = -\frac{n}{F} = -\frac{1.00}{-12.00\ \text{D}} = +0.0833\ \text{m} = +8.33\ \text{cm}$$

$$F = \frac{n'}{f'} \qquad f' = \frac{n'}{F} = \frac{1.00}{-12.00\ \text{D}} = -0.0833\ \text{m} = -8.33\ \text{cm}$$

The primary and secondary focal lengths in water ($n = n' = 1.33$) are solved by using Equation 5-8:

$$F = -\frac{n}{f} \qquad f = -\frac{n}{F} = -\frac{1.33}{-3.92\ \text{D}} = +0.3393\ \text{m} = +33.93\ \text{cm}$$

$$F = \frac{n'}{f'} \qquad f' = \frac{n'}{F} = \frac{1.33}{-3.92\ \text{D}} = -0.3393\ \text{m} = -33.93\ \text{cm}$$

The higher the power, the shorter the focal length. Also, notice the signs of the focal lengths. For the negative lens, the primary focal length is positive, and therefore the primary focal point is located to the right of the lens in image space. The secondary focal length is negative, and the secondary focal point is located to the left of the lens in object space. This is illustrated in Figure 5-6 for the negative lens.

For a positive lens, the primary focal point is to the left of the lens (in object space), and the secondary focal point is to the right of the lens (in image space). The primary focal length is negative, and the secondary focal length is positive. See Figure 5-6 for the positive lens.

Image-Object Relationships: Size and Orientation

The lateral magnification of an image formed by a thin lens may be calculated using several of the equations developed for single refracting surfaces in Chapter 4. The definition of the lateral magnification ratio (image size/object size) and the sign (–, inverted; +, erect) holds for thin lenses. The single refracting surface equation that uses the object and image distance (Equation 4-11) may be modified for thin lenses by using the same surrounding medium:

$$LM = \frac{h'}{h} = \frac{n\ell'}{n'\ell} = \frac{n_s\ell'}{n_s\ell} = \frac{\ell'}{\ell} \qquad (5\text{-}10)$$

This relationship is shown by the similar triangles in Figure 5-7. (See the section on lateral magnification in Chapter 4 for more details.) This equation will be true for all surrounding media (not just air).

The extrafocal distances may also be used to calculate the lateral magnification of the thin lens.

$$LM = -\frac{f}{x} = -\frac{x'}{f'} \qquad (5\text{-}11)$$

where: x = the object extrafocal distance (from the primary focal point to the object)
x' = the image extrafocal distance (from the secondary focal point to the image)

Shown in Figure 5-7 are the similar triangles (dotted for x and f and striped for x' and f') that are used to derive this relationship (see Chapter 4). The relationship between the object and image distances and the respective extrafocal distances are shown in Figure 5-7 as

$$\ell = x + f \qquad (5\text{-}12) \qquad\qquad \ell' = x' + f' \qquad (5\text{-}13)$$

Newton's Relationship (Equation 4-20) is derived from the lateral magnification in Equation 5-10. For thin lenses, $f' = -f$ may be substituted and the relationship changed to

$$xx' = ff' \qquad xx' = (-f')(f') \qquad xx' = -(f')^2 \qquad (5\text{-}14)$$

This equation is fundamental for measuring the power of lenses with a lensometer.

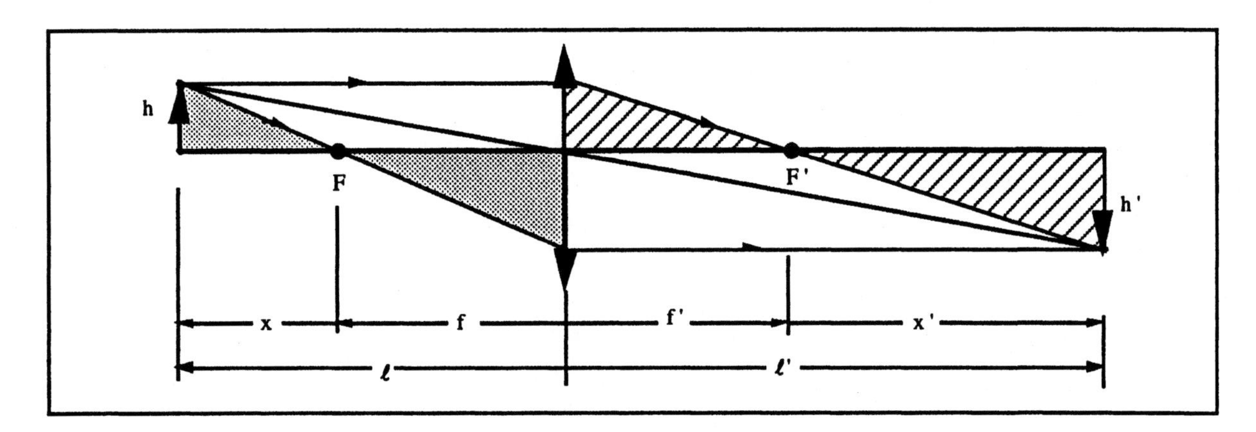

Figure 5-7. Similar triangles used in the derivation of the lateral magnification formulas.

Example 5-h

A real image moves to 5 cm away from a lens when an object is moved from infinity to a position 40 cm in front of the lens. What is the power of the lens?

Known	***Unknown***	***Equations/Concepts***
Object position 1: $\ell_1 = \infty$	Power of lens: F = ?	Infinite object distance; image at F'
Object position 2: $\ell_2 = -40\text{cm} = -\ 0.40\text{m}$		Extrafocal length definitions
Extrafocal length for image 2: x' = 5 cm = 0.05 m		(5-14) Newton's Formula
Real image formed		

Diagram the optical system(Figure 5-8). The lens must have positive power because a real image is formed with an object at an infinite distance. This indicates that the secondary focal plane is in image space. When the object moves from infinity, the image moves from the secondary focal point to the new image position. This distance is the same as the extrafocal distance x' or in this case + 5 cm. The object position ℓ is 40 cm in front of the lens. Because we know from Equation 5-12, $\ell = f + x$, the extrafocal distance x may be calculated in terms of ℓ:

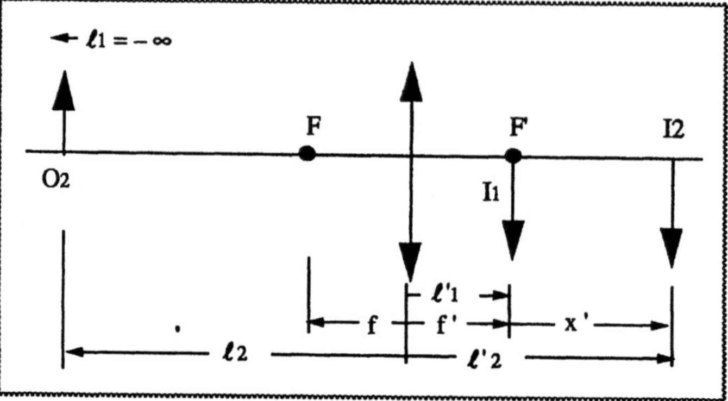

Figure 5-8. *Example 5-h*

$$\ell = f + x = -40 \text{ cm}$$
$$x = -40 - f = -40 + f' \quad \text{(because } f' = -f)$$

Substituting the values for x and x' into Newton's Formula, solve for the secondary focal length of the lens:

$$-(f')^2 = xx' = (-40 + f')(5) = 5f' - 200 \qquad \text{or} \qquad f'^2 + 5f' - 200 = 0$$

Using the quadratic equation of the form: $ax^2 + bx + c = 0 \quad a = 1, b = 5, c = -200$

$$f' = \frac{-b \pm \sqrt{b^2 - 4ac}}{2a} = \frac{-5 \pm \sqrt{5^2 - 4(-200)}}{2(1)} = \frac{-5 \pm \sqrt{825}}{2} = \frac{-5 \pm 28.73}{2}$$

For a positive lens, use the positive root:

$$f' = \frac{+23.73}{2} = +11.86 \text{ cm} = +0.1186 \text{ m}$$

$$F = \frac{n'}{f'} = \frac{1.00}{+0.1186} = +8.43 \text{ D}$$

To check the answer, let's solve for the image and extrafocal distance using the calculated power and the given object distance of 40 cm.

$$L = \frac{n}{\ell} = \frac{1.00}{-0.40} = -2.50 \text{ D}$$

$$L' = F + L = 8.43 + (-2.50) = +5.93 \text{ D}$$

$$\ell' = \frac{n'}{L'} = \frac{1.00}{+5.93 \text{ D}} = +0.1686 \text{ m} = +16.86 \text{ cm}$$

Solving for the extrafocal distance x'

$$x' = \ell' - f' = 16.86 - 11.86 = 5 \text{ cm} \qquad \text{check}$$

Example 5-i
A – 3.00 D lens forms an image that is erect and five times larger than the object. Where is the object located?

Known	***Unknown***	***Equations/Concepts***
Image 5x larger than object: LM = + 5	Object distance: $\ell = ?$	(5-10) Laterial magnification Positive lateral magnification (erect image) (5-5), (5-6), (5-7) Image - object

Set up the lateral magnification relationship using Equation 5-10:

$$LM = \frac{h'}{h} = \frac{\ell'}{\ell} = +5 \qquad \frac{\ell'}{\ell} = +5 \qquad \ell' = 5\ell$$

Substitute Equations 5-5 and 5-6 into 5-7 and substitute for ℓ':

$$L' - L = \frac{n'}{\ell'} - \frac{n}{\ell} = F \qquad \frac{1.00}{5\ell} - \frac{1.00}{\ell} = \frac{0.20}{\ell} - \frac{1.00}{\ell} = \frac{0.20 - 1.00}{\ell} = -3.00 \text{ D}$$

$$\frac{-0.80}{\ell} = -3.00 \qquad \ell = \frac{-0.80}{-3.00} = +0.2667 \text{ m} = 26.67 \text{ cm}$$

Check the answer by substituting the object distance and power into Equation 5-7 and solving for the image distance (Equation 5-6), then check the magnification:

$$L = \frac{n}{\ell} = \frac{1.00}{+0.2667} = 3.75 \text{ D}$$

$$L' = L + F = 3.75 + (-3.00) = +0.75$$

$$\ell' = \frac{n'}{L'} = \frac{1.00}{+0.75} = +1.33 \text{ m}$$

$$LM = \frac{\ell'}{\ell} = \frac{+1.33}{+0.267} = +5.00\text{x} \qquad \text{check}$$

The **symmetrical planes** are the image and object distances that yield a lateral magnification of – 1X. This means that the object and image are the same size but inverted. This relationship may be written as

$$LM = \frac{h'}{h} = \frac{\ell'}{\ell} = -1 \qquad \text{or} \qquad \ell' = -\ell \quad h' = -h \qquad (5\text{-}15)$$

Example 5-j
By using the technique described in *Example 5-i,* develop the relationship between the image and object distances and the focal lengths for the symmetrical points.

Use the technique described to solve for the object and image distances in terms of the focal lengths:

$$L' - L = F \qquad \frac{n_s}{\ell'} - \frac{n_s}{\ell} = \frac{n_s}{f'}$$

Substituting $\ell' = -\ell$

$$\frac{n_s}{-\ell} - \frac{n_s}{\ell} = \frac{n_s}{f'} \qquad -\frac{2n_s}{\ell} = \frac{n_s}{f'}$$

$$\ell = -2f \text{ and } \ell' = 2f' \qquad (5\text{-}16)$$

From this example, the positions of the symmetrical planes for any lens are easily located by simply doubling the secondary focal length. Note that the image and the object are on opposite sides of the lens (as indicated by the sign). For a positive lens, the symmetrical points are represented by a real object and a real image and for a negative lens, the symmetrical points are represented by a virtual object and a virtual image. In Figure 5-9 the object and image symmetrical point positions for positive and negative lenses are shown.

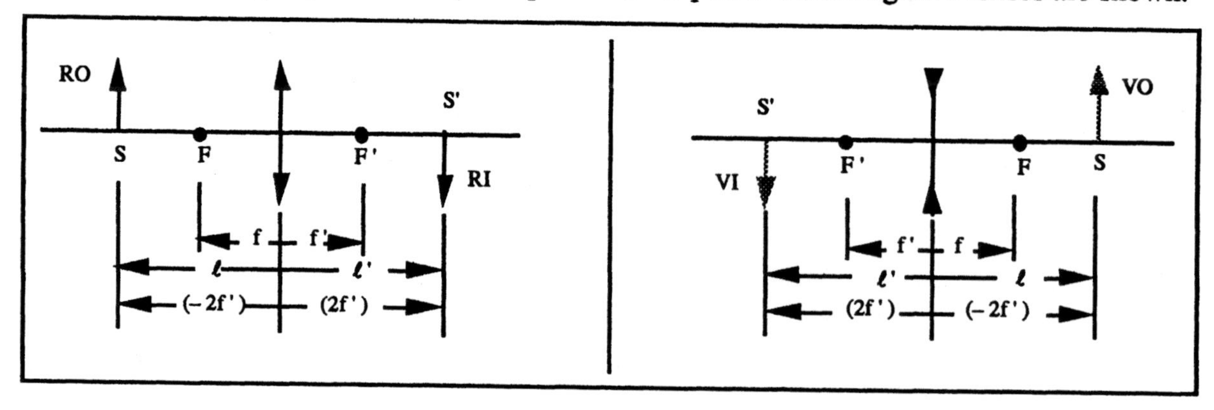

Figure 5-9. Location of the symmetrical points for positive and negative lenses.

The *minimum distance* between the symmetrical planes is the difference between the image and object positions. This may be written as:

$$\ell' - \ell = 2f' - (-2f') = 4f' \qquad (5\text{-}17)$$

Example 5-k
Locate the symmetrical planes for a + 5.00 D lens and a – 5.00 D lens.

Known
Power of lens: F = + 5.00 D
Power of lens: F = – 5.00 D

Unknown
Symmetrical planes

Equations/Concepts
(5-8), (5-9) Focal length-power
(5-16) Symmetrical points

This a fairly simple "plug in the numbers" type of problem. The symmetrical planes are shown in Figure 5-9 for positive and negative lenses.Solve for the focal length for the lenses using Equation 5-9, and then solve for the image and object distances using Equation 5-16:

$$f' = \frac{n'}{F} = \frac{1.00}{+5.00} = +0.20\text{m} = +20\text{cm}$$

$$\ell' = 2f' = 2(+20) = +40\text{cm}$$
$$\ell = -2f' = 2(-20) = -40\text{cm}$$

$$f' = \frac{n'}{F} = \frac{1.00}{-5.00} = -0.20\text{m} = -20\text{cm}$$

$$\ell' = 2f' = 2(-20) = -40\text{cm}$$
$$\ell = -2f' = -2(-20) = +40\text{cm}$$

Planes with positive unit magnification (LM = + 1) are called the **principal planes** (H and H'). These planes are used for solving problems involving thick lenses and lens systems (Chapter 7). For a thin lens, the principal planes coincide with the optical center of the lens.

Distant Object Size

When an object is at an infinite distance, the lateral magnification cannot be defined since ℓ is infinite (ℓ is in the denominator of Equation 5-10). As shown in Figure 5-10, the undeviated ray from a distant object subtends an angle ω at the lens. Another ray from the same object position may be drawn parallel to the undeviated ray and through the primary focal point. This ray leaves the lens parallel to the axis. The image in the secondary focal plane may be drawn with the proper height. The shaded triangle is used to calculate the height of the image:

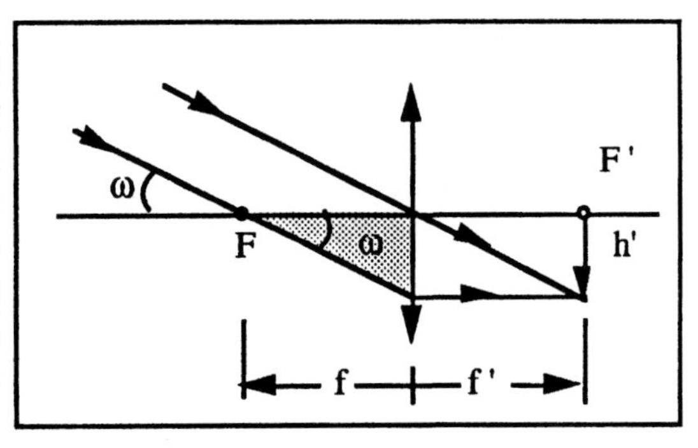

Figure 5-10. Image size with an infinite object.

$$\tan\omega = \frac{h'}{f} = -\frac{h'}{f'}$$
$$h' = -f'\tan\omega \qquad (5\text{-}18)$$

For small angles; this may be written as

$$h' = -f'\omega \quad (\omega \text{ in radians}) \qquad (5\text{-}19)$$

Example 5-1

The angular subtense of an object is often used in vision to describe the size of a distant object. For example, a 20/60 letter on an eye chart subtends a 15 minute arc at 20 feet (1 degree = 60 minutes). If a + 10.00 D lens is used to form a distant letter on a screen, what is the size of the image formed:

Known	***Unknown***	***Equations/Concepts***
Power of lens: F = + 10.00 D	Size of image	(5-18) Image size of infinite object
Object is at infinity		(5-9) Focal length-power

First, solve for the secondary focal length of the lens using Equation 5-9:

$$f' = \frac{n'}{F} = \frac{1.00}{+10.00\ \text{D}} = +0.10\ \text{m} = 10\ \text{cm}$$

This represents the image plane location. Next, convert the minutes of arc to degrees:

$$\omega = (15')\left(\frac{1^\circ}{60'}\right) = 0.25^\circ$$

Using this value, solve Equation 5-18 for the height of the image:

$$h' = -f'\tan\omega = 10\tan(0.25^\circ) = -0.0436\ \text{cm} = -0.436\ \text{mm}$$

The negative sign indicates that the image is inverted. A small image is formed by the lens using the 20/60 letter as the object. The eye actually resolves images that have critical dimensions (1/5 the size of the letter) on the order of 0.75 min (20/15 acuity).

Prismatic Effect of Thin Lenses

The amount of bending an incident ray undergoes after refraction through a thin lens is a function of the distance of the ray from the optical axis. As seen in Figure 5-11, an incident ray close to the axis bends less than a peripheral incident ray because both are refracted toward the same secondary focal point. This change in the direction of the incident ray is the **deviation**, and it may be considered to be the **prismatic effect** of the lens. Unlike prisms, which have one deviation for any position of the incident ray, a thin lens may be considered to be an infinite series of prisms with increasing deviation (and apical angles) toward the periphery.

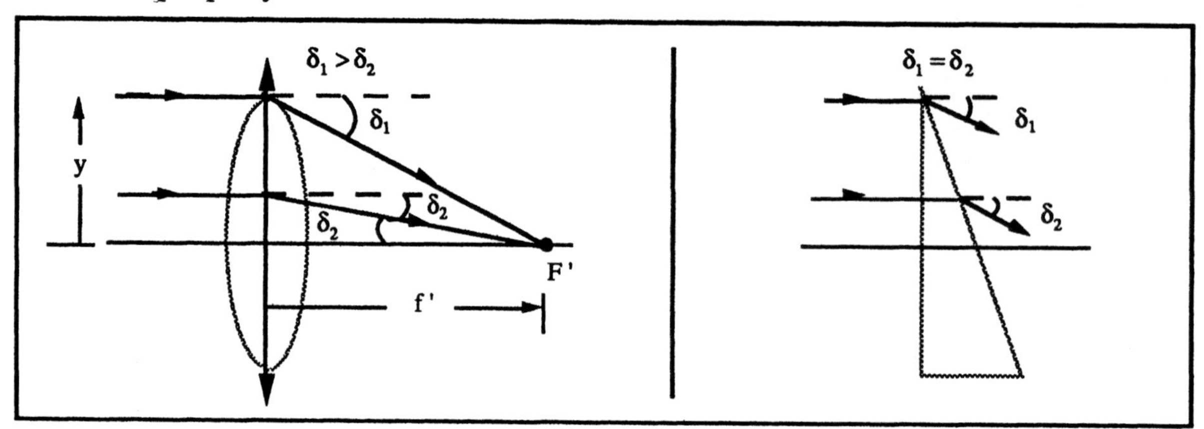

Figure 5-11. Prismatic effect of a thin lens and a prism. The deviation increases toward the periphery of the thin lens. The deviation of a prism is constant no matter where on the prism the ray is incident.

Independent of the surface curves, positive lenses may be considered to be prisms base-to-base at the optical axis, and negative lenses, apex-to-apex at the optical axis (Figure 5-12). The direction of the prismatic effect may be described in terms of the base. There is no prismatic effect at the optical center of the lens. However, if a lens is decentered, a prismatic effect is induced.

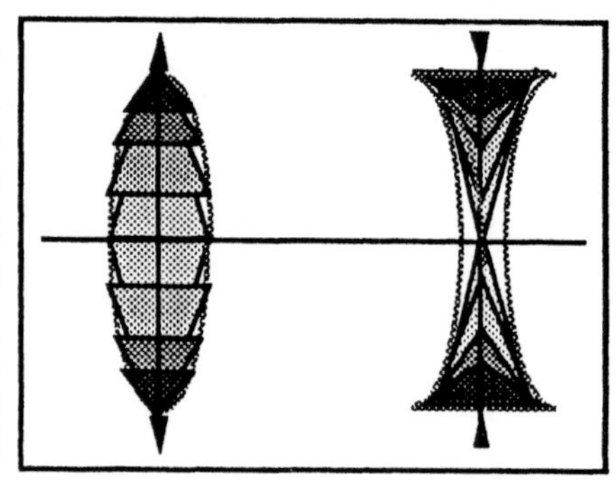

Figure 5-12. Prismatic effect lenses.

If the positive lens in Figure 5-12 is decentered downward so that one views through the top portion of the lens, the effect will be similar to a base-down prism, and the image will appear to move up (Figure 5-13). If the negative lens in Figure 5-12 is decentered downward so that one views through the upper portion of the lens, the effect will be similar to a base-up prism, and the object will appear to move down (Figure 5-13). In general, positive lenses will displace images

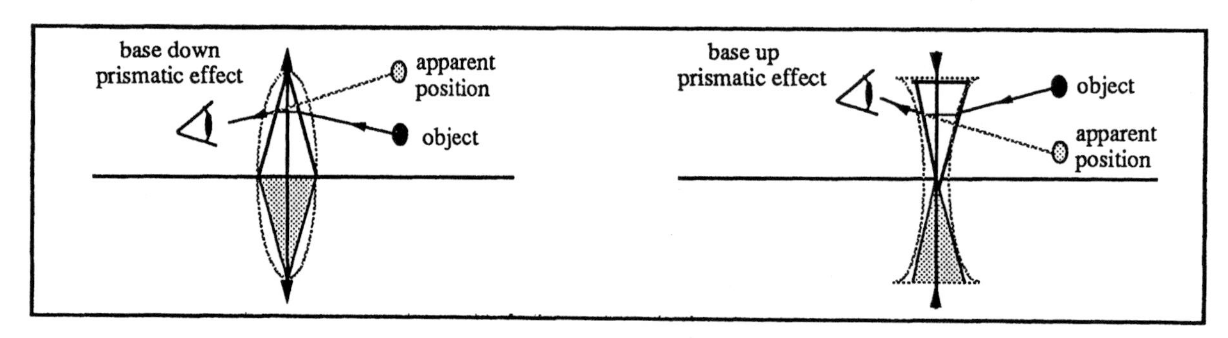

Figure 5-13. Location of image formed through decentered lens. Notice the base of the prism.

opposite the direction of decentration, and negative lenses will displace images in the same direction as decentration. Think about decentering the lens nasally (in) or temporally (out). This is desirable at times in prescribing glasses. The prismatic effect is also used in hand neutralization of lenses where the movement due to the prismatic effect is neutralized by another lens of the same power magnitude but opposite sign.

As seen in Figure 5-11, the magnitude of the prismatic deviation in prism diopters can be expressed in terms of the displacement from the optical center (y) in centimeters, and the secondary focal length of the lens in meters:

$$\delta^{\Delta} = \frac{y_{cm}}{f'_m} = y_{cm}F_D \qquad (5\text{-}20)$$

Note that the power was substituted for the reciprocal of the secondary focal length in Equation 5-20.

Example 5-m

A patient with an interpupillary distance (pd) of 60 mm is given a frame with a 70 mm pd. If the patient's Rx is -3.00 D in each eye, what is the direction and amount of induced prism?

Known	***Unknown***	***Equations/Concepts***
Patient's pd = 60 mm	Base of induced prism	(5-20) Prismatic effect of lenses
Frame pd = 70 mm	Magnitude of induced prism	Figure 5-12 and 5-13
Patient's Rx = -3.00 D each eye		

The first step is to diagram the problem (Figure 5-14). Note where the patient is viewing relative to the base of the prisms. It should be clear that the prismatic effect is base in (BI) for both eyes. The magnitude of the effect may be calculated using Equation 5-20. The total decentration is the difference between the frame pd and the patient's pd or

$$70 - 60 = 10 \text{ mm}$$

The displacement for each eye is equal to half the total difference (5mm). Using Equation 5-20, the prismatic effect is

$$\delta = (y)(F) = (0.50 \text{ cm})(3.00 \text{ D}) = 1.50^{\Delta}$$

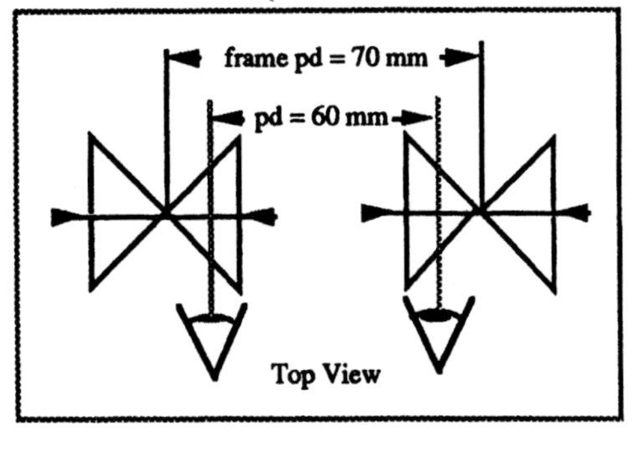

Figure 5-14. *Example 5-m.*

The total prismatic effect is the sum of both eyes or 3.00^{Δ} BI.

Example 5-n

A patient requires 5^{Δ} base up (BU) in front of the left eye. If the size of the largest lens blank is 70 mm, what is the minimum power of the lens required to produce this prism power?

Known	***Unknown***	***Equations/Concepts***
Patient's prism Rx: 5^{Δ} BU	Minimum power of lens for prism	(5-20) Prismatic effect of lenses
Diameter of lens: d = 70 mm		

The largest decentration for a 70 mm lens is 35 mm (which is not practical but is workable for this problem). The lens power is calculated using the known values and Equation 5-20:

$$\delta^{\Delta} = y_{cm}F_D \qquad F_D = \frac{\delta^{\Delta}}{y_{cm}} = \frac{5^{\Delta}}{3.5 \text{ cm}} = 1.42 \text{ D or approximately } \pm 1.50 \text{ D}$$

Effective Power

When parallel rays are incident on a thin lens, the rays leave the lens aimed toward (or away from) the secondary focal point. The vergence in any plane P behind the lens is easily calculated when the distance from the plane to the secondary focal point is known. As shown in Figure 5-15, the vergence leaving the lens is simply

$$L' = F = \frac{1.00}{f'}$$

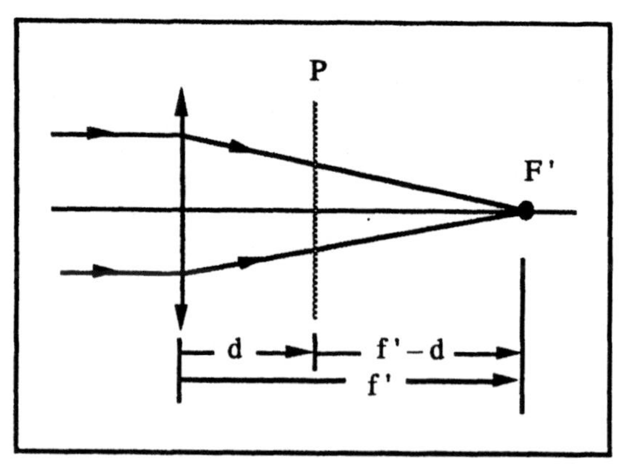

Figure 5-15. Effective power and vergence.

In the plane P at any distance d from the lens, the vergence may be calculated by

$$F_x = \frac{1.00}{f_x'} = \frac{1.00}{(f' - d)} = \frac{F}{1 - dF} \qquad (5\text{-}21)$$

This vergence is called the **effective power.** This is the power required for a new lens positioned at P to have the same effect on the incident rays (i.e.form an image at F').

Let's explore this concept further. Effective power becomes important if the distance between a lens and an image plane is changed (Figure 5-16). Because the incident rays are parallel, moving the lens does not change the object position (infinite), and the image will always be formed in the secondary focal plane, which coincides with the image screen. However, when the lens is moved (along with its secondary focal point), the vergence incident upon the screen will be changed. In Figure 5-16 (left), a positive lens forms an image on a screen positioned at the secondary focal point. The lens is moved distance d closer to the screen to position P_2. The vergence leaving the lens at P_2 is aimed toward its secondary focal point position, which was also moved distance d. It should be obvious that the lens in position P_2 has too little positive power to focus the image on the screen (the image is behind the screen). Therefore, the power of the lens at P_2 must be increased with its secondary focal length f'_x.

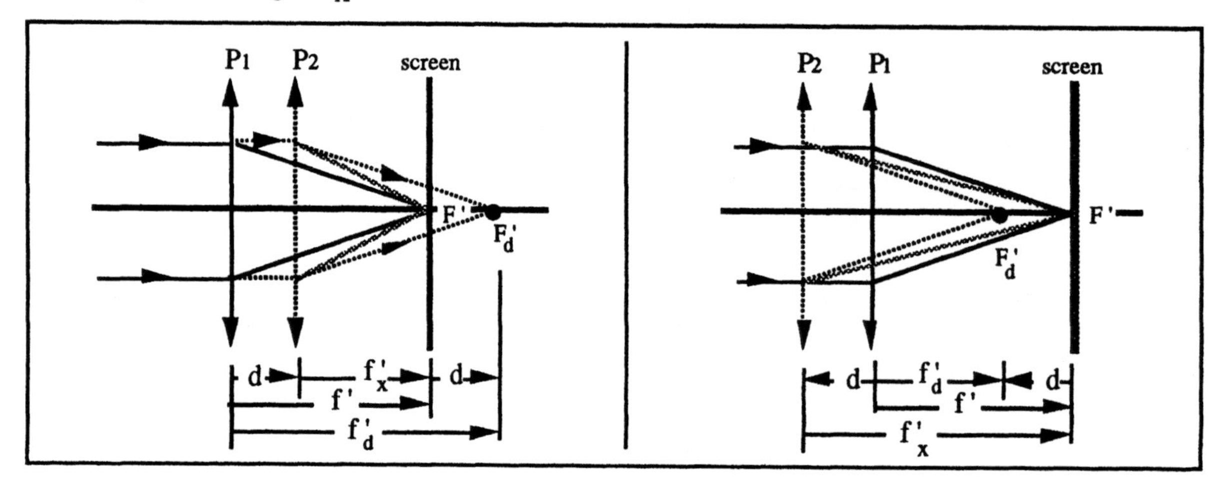

Figure 5-16. With parallel rays incident, a positive lens forms an image on a screen placed at its secondary focal plane. On the left, the lens is moved d closer to the screen, and on the right the lens is moved d farther from the screen. The arrow associated with d indicates the direction and sign. The secondary focal point of the lens is also moved d and is labeled F'_d. The dotted rays indicate the new image position. The dashed rays indicate the the path required to form an image on the screen. When the lens is moved closer to the screen, more positive power is required, and when the lens is moved farther from the screen, less power is required.

The new lens power would equal

$$L'_x = F_x = \frac{1.00}{(f'-d)} = \frac{1.00}{\left(1-\frac{d}{f'}\right)f'} = \frac{F}{1-dF} \qquad (5\text{-}21)$$

which has been defined as the effective power. The sign convention is maintained with the value d: if the lens is moved from left to right, d is a positive value, and if moved right to left, d is a negative value. Note the arrows on the diagrams.

It was shown that the lens at position P2 had too little positive power to focus the image on the screen. The effective reduced power would be equivalent to placing at P1 a lens that formed an image at F'_d. The power of the lens could be calculated by

$$F_d = \frac{1.00}{f'_d} = \frac{1.00}{f'+d} = \frac{F}{1+dF} \qquad (5\text{-}22)$$

This is the power error induced by moving the lens without changing the power.

Figure 5-16 (right) also shows what happens if the lens is moved farther from the screen. The lens in this position has too much positive power, the image is formed in front of the screen. Thus the power required would be less than in the original position. One may formulate a rule from this diagram:

Moving a lens away from the image plane effectively increases positive power (decreases negative power).

This rule may be used with spectacle corrections of the eye. By moving the glasses away from the eye, you effectively increase the positive power. By increasing the effective positive power (decreasing the negative power) you create a reading add for near tasks. You may have noticed individuals who wear their glasses at the end of their nose. They usually require a new bifocal with more positive power.

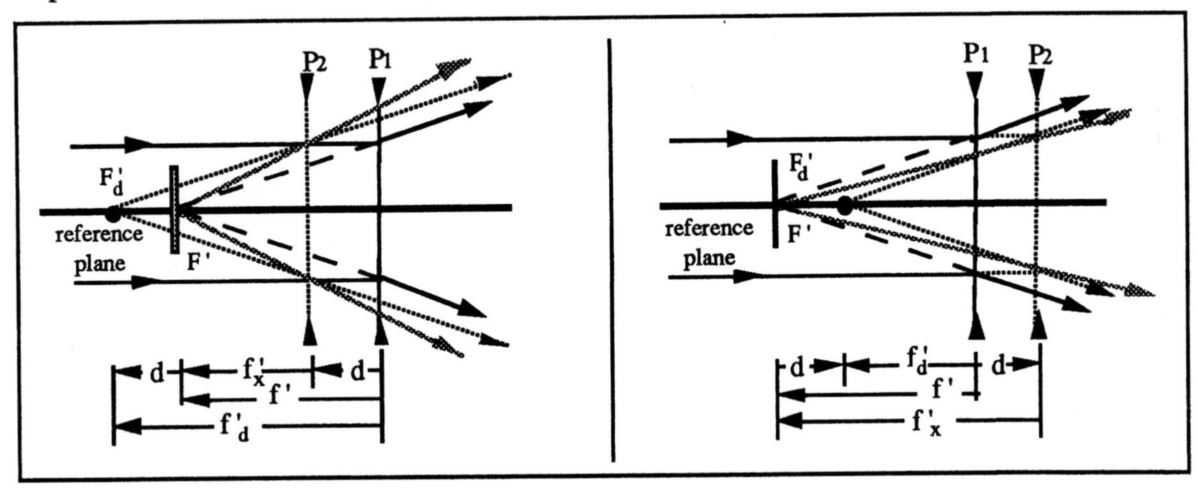

Figure 5-17. With parallel rays incident, a negative lens forms an image in the reference plane at its secondary focal plane. On the left, the lens is moved d closer to the reference plane, and on the right, the lens is moved d farther from the reference plane. The arrow associated with d indicates the direction and sign. The secondary focal point of the lens is also moved d and is labeled F'_d. The dotted rays indicate the new image position. The dashed rays indicate the path required to form an image on the screen. When the lens is moved closer to the reference plane, more negative power is required; when it is moved farther from the reference plane, less negative power is required.

Figure 5-17 shows the effective power of a negative lens when the lens is moved closer and farther from the image plane. Notice that the image is virtual and formed in front of the lens. The effective power equations are demonstrated in the following examples.

Example 5-o

A patient wears spectacle lenses with a power of – 5.00 D in the right eye and + 10.00 D in the left eye (this is not realistic but demonstrates the problem). The vertex distance of the glasses is 15 mm. What would be the undercorrection if the spectacle lens power were prescribed as contact lenses? What would be the patient's proper contact lens correction?

Known	***Unknown***	***Equations/Concepts***
Spectacle lens power: OD: – 5.00 D	Effective power by prescribing the	(5-21), (5-22) Effective power
OS: + 10.00 D	spectacle lens power in the contacts	Figures 5-16 and 5-17
Vertex distance: 15 mm = 0.015 m	Proper contact lens correction	

The right eye has a correction of – 5.00 D. In Figure 5-17, assume the eye is looking through the lens toward the reference plane. The right figure shows the lens moved closer to the eye and farther from the reference plane, a positive distance d. With the lens at P2, the focal point is moved closer to the original lens position P1, and thus f'_d is less than f' and has a greater negative power. Solve for the secondary focal length of the spectacle, the focal length lens f'_d, and the reduced effective power as follows:

$$f' = \frac{n'}{F} = \frac{1.00}{-5.00} = -0.20 \text{ m} = -20 \text{ cm}$$

$$f'_d = f' + d = -20 + (+1.5) = -18.5 \text{ cm} = -0.185 \text{ m}$$

$$F_d = \frac{n'}{f_d'} = \frac{1.00}{(-0.185)} = -5.41 \text{ D}$$

A similar answer can be obtained with Equation 5-22:

$$F_d = \frac{F}{1+dF} = \frac{-5.00}{1+(+0.015)(-5.00)} = \frac{-5.00}{1+(-0.075)} = \frac{-5.00}{0.925} = -5.41 \text{ D}$$

If the same spectacle lens power were prescribed for contact lenses, the patient would effectively be almost 0.50 over minused (i.e., the patient requires –5.00 D and the calculated effective power is –5.40 D).

When the lens is moved toward the eye, the distance from the new lens position to the reference plane is increased, and thus the lens has a longer focal length and lower power. This may be calculated by

$$f' = \frac{n'}{F} = \frac{1.00}{-5.00} = -0.20 \text{ m} = -20 \text{ cm}$$

$$f'_x = f' - d = -20 - (+1.5) = -21.5 \text{cm} = -0.215 \text{ m}$$

$$F_x = \frac{n'}{f'_x} = \frac{1.00}{(-0.215)} = -4.65 \text{ D}$$

Or, by using the Equation 5-21

$$F_x = \frac{F}{1-dF} = \frac{-5.00}{1-(+0.015)(-5.00)} = \frac{-5.00}{1-(-0.075)} = \frac{-5.00}{1.075} = -4.65 \text{D}$$

You should prescribe – 4.65 D (or – 4.62 D in eighth diopter steps) for the right contact lens.

For the left eye (+ 10.00 D spectacle correction), look at the left side of Figure 5-16. In this case, d is also a positive value. The power error may be calculated using the same approach:

$$f' = \frac{n'}{F} = \frac{1.00}{+10.00} = +0.10\ \text{m} = +10\ \text{cm}$$

$$f'_d = f' + d = 10 + (+1.5) = +11.5\ \text{cm} = +0.115\ \text{m}$$

$$F_d = \frac{n'}{f'_d} = \frac{1.00}{+0.115} = +8.70\ \text{D}$$

or

$$F_d = \frac{F}{1 + dF} = \frac{+10.00}{1 + (+0.015)(+10.00)} = \frac{+10.00}{1 + 0.15} = \frac{+10.00}{1.15} = +8.70\ \text{D}$$

By prescribing the spectacle power for a contact lens in the left eye, the eye would effectively be undercorrected by approximately 1.30 D. (+ 10.00 – 8.70 = 1.30 D).

The correct contact lens power is

$$f' = \frac{n'}{F} = \frac{1.00}{+10.00} = +0.10\ \text{m} = +10\ \text{cm}$$

$$f'_x = f' - d = +10 - (+1.5) = +8.5\ \text{cm} = +0.85\ \text{m}$$

$$F_x = \frac{n'}{f'_x} = \frac{1.00}{(+0.85)} = +11.76\ \text{D}$$

or

$$F_x = \frac{F}{1 - dF} = \frac{+10.00}{1 - (+0.015)(+10.00)} = \frac{+10.00}{1 - (0.15)} = \frac{+10.00}{+0.85} = +11.76\ \text{D}$$

Contact lens power is usually prescribed in 0.25 D steps. Thus contact lens power prescribed for the left eye would be + 11.75 D.

You should be able to work the problem using both methods described. Using the focal lengths to determine the power yields more insight than simply plugging values into a memorized formula. Read the problem carefully, diagram the problem with d labeled, then try to determine if the effective power should be greater or smaller than the starting lens power.

Supplemental Problems

Power of Thin Lenses

5-1. A thin meniscus lens (n = 1.5) has front and back radii of 20 cm each. If the front surface is in contact with water, what is the power of the lens?
ANS. F = −1.65 D

5-2. A thin biconcave lens (n = 1.70) has a front surface power of −6.54 D and a back radius of 45 cm. What is the power of the lens?
ANS. −8.10 D

5-3. A plastic thin biconvex lens (n = 1.49) has front and back surface radii of 20 cm and 35 cm, respectively. What is the power of the lens?
ANS. + 3.85 D

5-4. A thin lens has a front surface power of + 3.00 D, a back surface power of + 2.00 D, an overall diameter of 60 mm, and a center thickness of 10 mm. Find the total power of the lens and the edge thickness of the thickest edge.
ANS. Total power = + 5.00 D; thickest edge = 5.7 mm

Object-Image Relationships

5-5. A + 15.00 D lens is placed 20 cm behind a + 5.00 D lens. If a luminous-point object is placed 1 m in front of the + 5.00 D lens, what is the vergence striking the second thin lens?
ANS. + 20 D

Focal Points

5-6. An astonomical telescope consists of two positive power lenses separated by the sum of the secondary focal lengths. If the first lens has a power of + 10.00 D and the second lens has a power of + 20.00 D, what is the distance between the two lenses?
ANS. 15 cm

5-7. Alone, the primary focal point of a thin lens is 10 cm behind the lens, but when held next to another thin lens, the primary focal point of the system is 1 m in front of the lens combination. Where is the secondary focal point of the second thin lens?
ANS: 9.09 cm behind the lens

5-8. A real object is placed 45 cm from a − 5.00 D thin lens. What are the extrafocal distances x and x'?
ANS. +1.50 D; LM = +1.60x

Lateral Magnification

5-9. A virtual object is one meter from a + 9.00 D lens. What is the vergence in a plane 50 cm behind the lens, and what is the lateral magnification of the image formed?
ANS. − 2.50 D; lateral magnification = + 0.1 x

5-10. A 5 cm tall object is located 45 cm in front of a + 5.00 D thin lens. What is the size and type (real or virtual) of the image?
ANS. 4 cm; real, inverted

5-11. The minimum separation of a real object and its image for a given convex lens is 40 cm. How far from the lens must the object be placed so as to give a lateral magnification of + 2.0 x?
ANS. – 5.0 cm

5-12. A virtual image is formed by a + 5.00 D thin lens. If the lateral magnification is + 3 x, where is the image located?
ANS: 40 cm in front of the lens

5-13. An object is placed 25 cm in front of a lens. An erect image is formed 40 cm from the lens. What is the power of the lens and what is the lateral magnification of the image formed?
ANS. +1.50 D; LM = +1.60x

5-14. As an object is moved from an infinite position to a position 25 cm in front of the primary focal point of a positive lens, the image moves 10 cm. Determine the power of the lens and the lateral magnification of the image.
ANS. Power = + 6.32 D; lateral magnification = – 0.632 x

Newton's Relationship

5-15. When an object is placed 20 cm to the left of the primary focal point, the image formed is 40 cm to the right of the secondary focal point. What is the power of the lens?
ANS. +3.54 D

5-16. As an object is moved from an infinite position to a position 25 cm in front of the primary focal point of a positive lens, the image moves 10 cm. Find the power of the lens and the lateral magnification of the image.
ANS. + 6.32 D; LM = – 0.632 x

Symmetrical Planes

5-17. Locate the symmetrical planes for a thin lens that forms an inverted image 8 times the size of an object located 50 cm in front of the lens.
ANS. ℓ' =+ 89 cm; ℓ = – 89 cm

5-18. When an object is placed to the right of a thin lens, an inverted virtual image is formed that is equal in size to the virtual object. If the extrafocal distance (x) is + 10 cm, find the power of the lens.
ANS. power = – 10.0 D

Apparent Size

5-19. An infinite object subtends an angle of 1 minute at a +10 D lens (n = 1.5). What is the size of the image?
ANS. 0.029 mm

5-20. If a 20/20 letter forms a retinal image of 0.0264 mm in an axial hyperope (power of the eye = 55 D), what is the angular subtense of the letter?
ANS. 0.083 degrees or 5 minutes of arc

Prismatic Effects

5-21. A 5 D myope requires 2 pd base up and out. By how much must the lens be decentered?
ANS. 4 mm

5-22. A patient wearing + 4.00 D OU requires 2^{Δ} base up. What is the amount and direction the lens must be decentered?
ANS. 0.50 cm up

5-23. A patient requires the following Rx: OD: +12.00 D 4^{Δ} BI; OS: − 10.00 D 5^{Δ} BI. How would the lenses be decentered for this correction?
ANS. OD: 0.33 cm in; OS: 0.50 cm out

Effective Power

5-24. What spectacle Rx would correct a myope wearing − 2.75 D contact lenses if the glasses were to be worn 15 mm from the cornea?
ANS. − 2.87 D

5-25. If, by mistake, a patient were given + 4.00 D contact lenses when corrected with + 4.00 D spectacle lenses (vertex distance = 14 mm), by how much would the patient be undercorrected or overcorrected?
ANS. undercorrected by 0.24 D

Laboratory Measurements

5-26. Using a lens clock, you find the power of the front surface of a thin lens to be + 2.25 D and the back surface to be + 3.00 D. If the index of refraction of the lens is 1.49, what is the power of the lens?
ANS. F_1 = + 2.08 D; F_2 = + 2.77 D; total power = + 4.85 D

5-27. A real image is formed 20 cm behind a + 5.00 D thin lens. A negative thin lens of unknown power is placed 10 cm behind the + 5.00 D lens, and the real image is now formed 3.33 cm behind (to the right of) the original image. What is the power of the negative thin lens?
ANS. − 2.50 D

Chapter 6

Cylindrical and Spherocylindrical Lenses

The spherical surface is the most common form of imaging element. The simplest nonspherical form is a cylindrical surface. As shown in Figure 6-1, a cylindrical surface may be formed by cutting a glass rod along it length. If a lens clock is rotated about its middle leg on a cylindrical surface, the power readings will vary from a minimum (zero power in this case) to a maximum. In one direction, the radius is finite, and the surface has a maximum power. In Figure 6-1, this is represented by the horizontal merdian. The vertical meridian in Figure 6-1 (90° away) has an infinite radius (plane surface) and a power of zero.

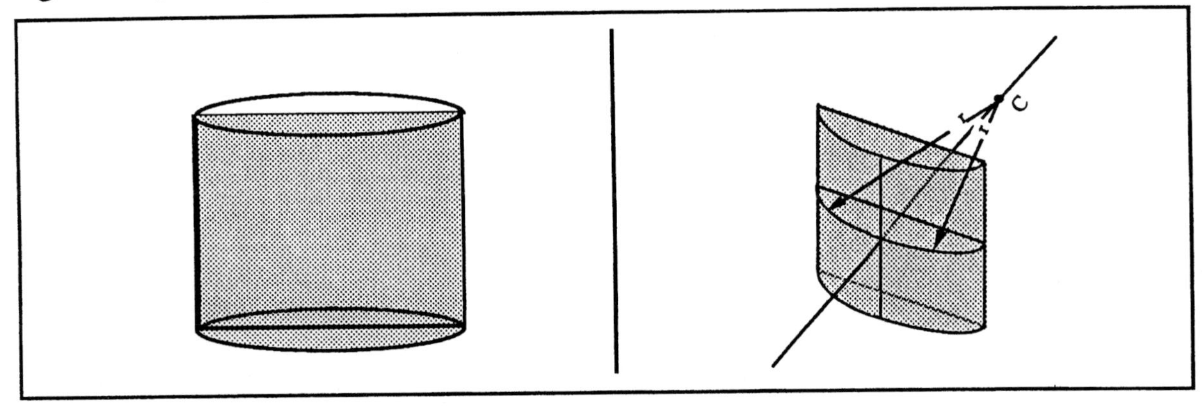

Figure 6-1. Cylindrical lens with an infinite radius (zero power) in the vertical and a finite radius (maximum) power in the horizontal. These meridians are called the *principal meridians.*

The two meridians that contain the *maximum* and *minimum powers* (and radii) represent the *principal meridians* of the lens. Cylindrical surfaces are described by the power and orientation of the principal meridians. Because the principal meridians are always 90° apart, the orientation of only one of the meridians needs to be specified. The orientation is usually described in terms of the angular location of the meridian with no power, with angles measured counterclockwise from the horizontal. Use the smallest angle to represent the meridian (with the exception being the 0° meridian - use 180°). Usually the degree symbol is not used, and three digits are always written. It is a good idea to draw a diagram much like that in Figure 6-2, with the power and meridians labeled. This diagram is referred to as an *optical cross.*

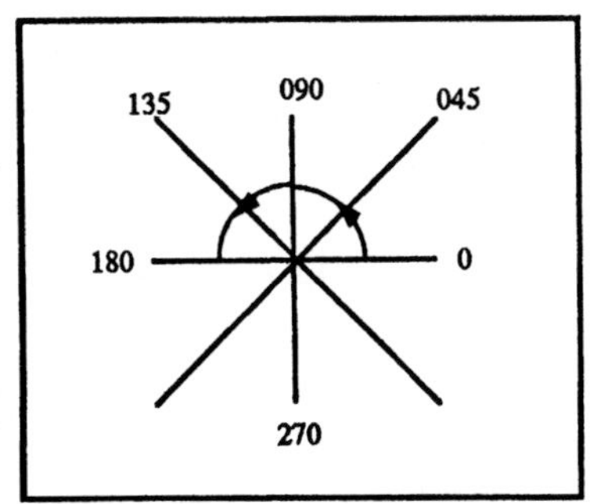

Figure 6-2. An optical cross.

Principal meridians - the two meridians of a cylindrical or spherocylindrical lens that represent the minimum and maximum powers. The two principal meridians of a lens are always 90° apart.

Axis meridian - the meridian of a cylindrical or spherocylindrical lens that contains the minimum or zero power.

Power meridian - the meridian of a cylindrical or spherocylindrical lens that contains the maximum power.

If the radii of the principal meridians of the cylindrical front surface are known, the powers in these meridians may be calculated using this power formula:

$$F = \frac{n_L - n_s}{r} \qquad (6\text{-}1)$$

where:

F = power of the meridian in diopters
n_L = index of the lens material
n_s = index of the surrounding medium
r = radius of curvature of the meridian in meters

Example 6-a
The cylindrical lens in Figure 6-1 has an index of refraction of 1.50, a positive radius of 50 cm in the horizontal meridian, and an infinite radius in the vertical meridian. Calculate the power in each meridian and draw and label an optical cross.

Known	***Unknown***	***Equations/Concepts***
Radius in 180 meridian: r = + 50 cm = + 0.50 m	Power in 180 meridian	(6-1) Power
Radius in 090 meridian: $r = \infty$	Power in 090 meridian	Figure 6-1
Index of surround (air): $n_s = 1.00$		
Index of lens: $n_L = 1.00$		

Use Equation 6-1 to calculate the power in the principal meridians, and label the optical cross accordingly:

$$F_{180} = \frac{n_L - n_s}{r_{180}} = \frac{1.50 - 1.00}{+0.50} = +1.00 \text{ D}$$

$$F_{090} = \lim_{r \to \infty} \frac{n_L - n_s}{r_{090}} = 0.00 \text{ D or plano}$$

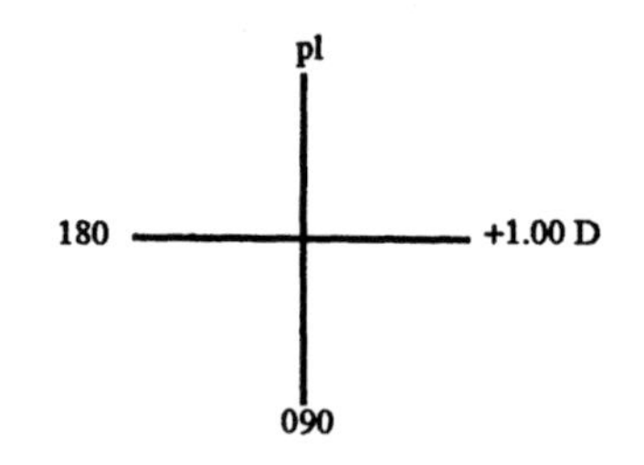

The lens has its maximum power in the 180 meridian (power meridian) and its minimum power in the 090 meridian (axis meridian).

The axis meridian (minimum or plano power) is specified in writing a description of the cylindrical lens. The format is shown in Figure 6-3: the power in the power meridian, followed by an x (which means axis), followed by the orientation of the axis meridian. In this example, the + 1.00 D power is in the 180 degree meridian, and the axis meridian is 090. Remember that the power in this notation is in the power meridian. Be careful in transcribing from a written description of the surface to the optical cross. In Figure 6-3, the written description is said, "plus one axis ninety."

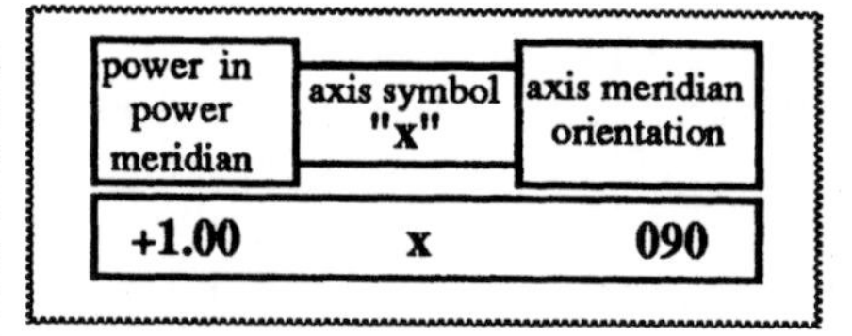

power in power meridian	axis symbol "x"	axis meridian orientation
+1.00	x	090

Figure 6-3. *Example 6-a.* The written description of a cylindrical lens.

Example 6-b
Represent the cylinder – 3.00 x 120 on an optical cross.

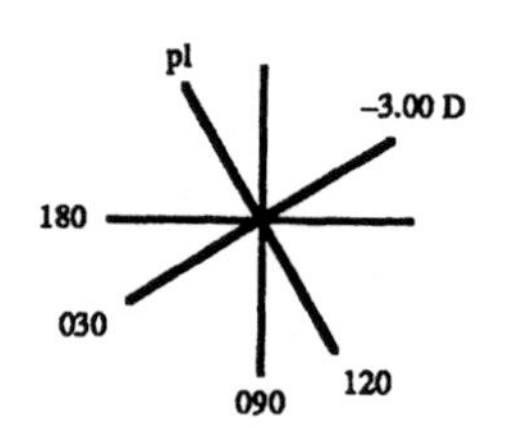

The way I remember how to write this on the optical cross is to first label the axis meridian by writing plano in the meridian given. Then I label the power in the meridian 90° away (in this case the 030 meridian). The optical cross diagrams help to minimize mistakes. Practice writing cylindrical lenses on the optical cross.

Power between Principal Meridians *

The power between the two principal meridians of the cylindrical surface varies between the maximum and minimum powers. In the case of a cylinder, the power varies between zero and the power in the power meridian. The power in any meridian θ degrees from the axis meridian may be calculated by

$$F_\theta = F_p \sin^2\theta \qquad (6\text{-}2)$$

where: F_θ = power in any meridian θ degrees from the axis meridian
F_p = power in the power meridian (maximum power)
θ = angle from the axis meridian to meridian of power calculation

Example 6-c
A cylindrical surface has a power of +5.00 x 030. What is the power in the 030, 090, 120, and 180 meridians ?

Known	***Unknown***	***Equations/Concepts***
Surface power: + 5.00 x 030	Power in 030, 090, 120, and 180	Cylinder power and axis (6-2) Power in meridians

Draw an optical cross of the surface powers.

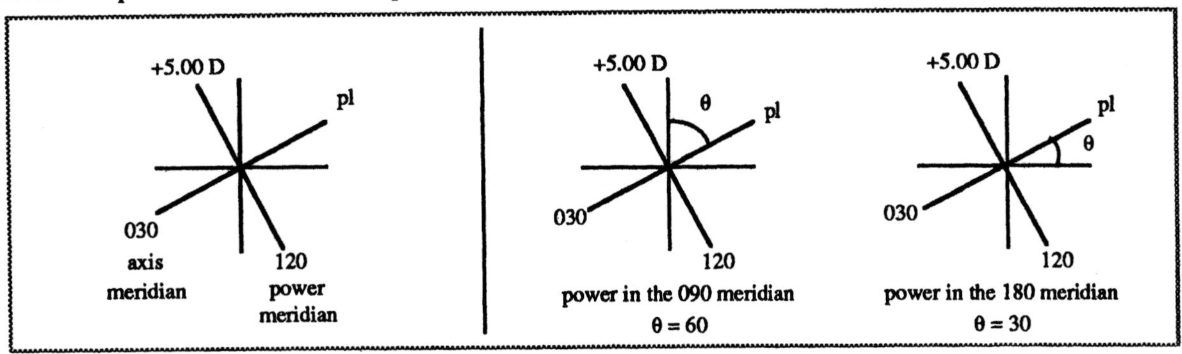

Figure 6-4. *Example 6-c. Left*: The optical cross representation of the lens. *Right*: Optical crosses for the power in the 090 and 180 meridians.

The power in the 030 and 120 meridians are determined directly from the lens powers; 030 is the axis meridian and has plano power, and 120 is the power meridian, which has + 5.00 D power (Figure 6-4 left). For the power in the 090 and 180 meridians, the angle between the axis meridian and these meridians must be determined from the optical crosses in Figure 6-4 (i.e., 60° and 30°, respectively). Substitute these values into Equation 6-2 to calculate the power:

$$F_{090} = +5.00\sin^2(60) = +5.00(0.75) = +3.75\text{ D}$$

$$F_{180} = +5.00\sin^2(30) = +5.00(0.25) = +1.25\text{ D}$$

Because sin(θ + 90) = cosθ, for meridians that are 090° apart, Equation 6-2 may be written

$$F_{\theta+090} = F_p \cos^2\theta \qquad (6\text{-}3)$$

*There has been recent interest in the actual power between the two principal meridians. This workbook addresses the classical definitions and equations. Interested students should consult their textbook or the current literature for more details.

In this example, let's use Equation 6-3 to calculate the power in the 180 meridian (using the angle θ from the power calculation in the 090 meridian, or 60 degrees):

$$F_{\theta+090} = +5.00\cos^2(60) = +5.00(0.25) = +1.25\ \text{D}$$

From equation 6-2, it can also be shown that the sum of the power in two meridians that are 90° apart is equal to the maximum power (i.e., the power in the power meridian).

$$F_\theta + F_{\theta+90} = F_p \sin^2\theta + F_p\cos^2\theta = F_p\overbrace{(\sin^2\theta + \cos^2\theta)}^{=1} = F_p \qquad (6\text{-}4)$$

To test the equation, substitute in the powers in the 090 and 180 meridians:

$$F_{90} + F_{180} = +3.75 + 1.25 = +5.00\ \text{D}$$

Optical axis - for a cylindrical lens, the optical axis is contained in a plane that passes through the center of curvature and is perpendicular to the back surface. There are an infinite number of axes in this plane, but one in the center of the lens is usually chosen to represent the axis.

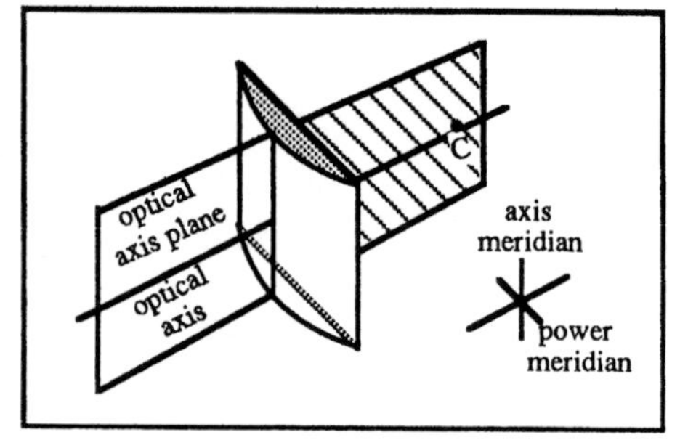

Figure 6-4. Optical axis for a cylindrical lens.

Image-Object Relationships: Position

As with spherical lenses, the relationship between an image, an object, and the power of a cylindrical lens may be calculated using the incident (L) and emergent (L') vergence and Gaussian Imaging Formula:

$$L = \frac{n_s}{\ell} \quad (6\text{-}5) \qquad L' = \frac{n_s}{\ell'} \quad (6\text{-}6) \qquad L' = F + L \quad (6\text{-}7)$$

where:

ℓ = object distance relative to the lens
ℓ' = image distance relative to the lens
n_s = index of the surrounding medium
F = power of the lens

With cylindrical lenses, both of the principal meridians must be considered in determining the position and orientation of an image. In Figure 6-6, a cylindrical lens is shown with power in the horizontal meridian and the axis in the vertical meridian. For a point object at an infinite distance, incident rays will be parallel to the optical axis. Many incident rays all lying in a horizontal plane, which contains the optical axis, are incident. Because there is power only along the horizontal, the rays can only bend (be refracted) in the horizontal direction toward the plane containing the optical axis (Figure 6-4). These rays, after refraction, form a point image. If another group of rays is incident in another horizontal

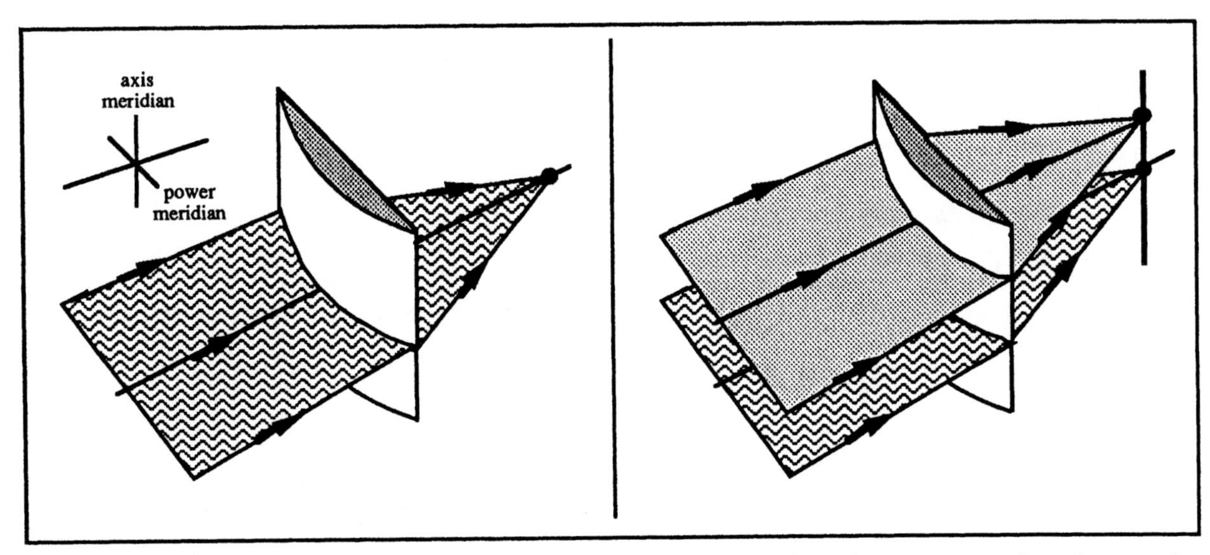

Figure 6-6. Image through a cylindrical lens. *Left*: Rays from and infinite point source in a plane that contains the optical axis form a point on the axis. *Right*: Rays from the same object in another plane form an image above the former image. A line image results.

plane, another point image is formed above the former, as shown Figure 6-6 (right). If an infinite number of horizontal planes containing rays were imaged, an infinite number of point images would be formed all lying directly above or below each other. These images form a line image parallel to the axis meridian of the lens. The distance from the lens to this line image is a function of the power of the cylindrical lens in the power meridian. Equations 6-5, 6-6 and 6-7 are used to calculate the line image position.

Example 6-d

A point object is located 60 cm in front of a + 10.00 x 090 cylindrical lens. Determine the position and orientation of the image formed.

Known	***Unknown***	***Equations/Concepts***
Lens power: + 10.00 x 090	Position of image	Power - axis concepts
Object in front of lens (object space)	Orientation of image	(6-5), (6-6), (6-7) Imaging formula
Object position: $\ell = -$ 60 cm = $-$ 0.60 m		Imaging through a cylinder
		Line image parallel to axis meridian

Start by drawing an optical cross and identifying the axis and power meridians. Remember that the axis meridian has plano power and is the 090 meridian in this case. The power meridian is 90° away or in the 180 meridian. Just consider the power meridian for the location of the image. The object is 60 cm in front of the lens or at – 0.60 m. Using Equations 6-5, 6-7, and 6-6, calculate the emerging vergence and the resulting image location:

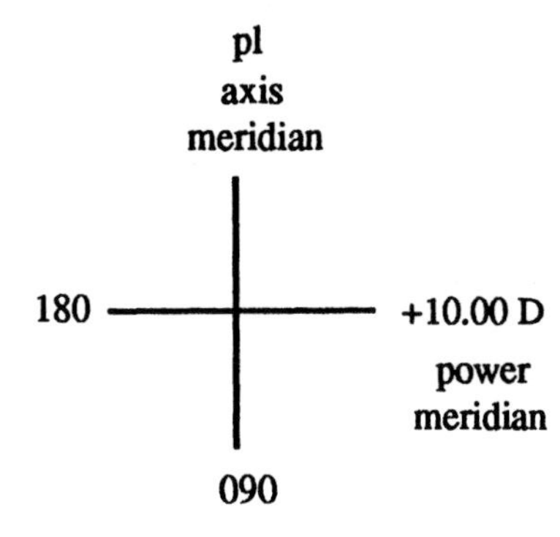

$$L = \frac{n}{\ell} = \frac{1.00}{-0.60} = -1.67 \text{ D}$$

$$L' = F + L = +10.00 + (-1.67) = +8.33 \text{ D}$$

$$\ell' = \frac{n'}{L'} = \frac{1.00}{+8.33} = +0.12 \text{ m} = +12 \text{ cm}$$

A line image is formed 12 cm behind the lens. The orientation of the lens may be determined by looking at the optical cross. The power is in the horizontal meridian, and therefore all the bending of the incident rays is in the horizontal meridian. The resulting image is a vertical line (parallel to the axis meridian) as in Figure 6-6.

Spherocylindrical Lenses

As we learned in Chapter 5, the power of a thin lens is the sum of the front and back surface powers. In this chapter, one surface is a spherical or plano, and the other is cylindrical. (Note that when both surfaces are cylindrical, the total lens power is not a simple addition unless the axes are aligned.) See Obliquely Crossed Cylinders at the end of this chapter and refer to other ophthalmic optics texts for more details. The total power of the lens will be the sum of both surfaces, added meridian-by-meridian. Because the spherical surface has the same power in all orientations, this power is added to both principal meridians. In the cylindrical examples discussed thus far, the back surface power was plano, and therefore the lens had the same powers as the front surface. A lens that has one spherical and one cylindrical surface is called a *spherocylinder*. For a spherocylinder, the principal meridians contain the maximum and minimum (not plano) powers.

Example 6-e

A thin lens has the following specifications: $F_1 = -3.00 \times 120$; $F_2 = +3.00$ D.
What is the power of the lens? Draw an optical cross to specify the powers of each surface.

Known	***Unknown***	***Equations/Concepts***
Surface powers F1 = – 3.00 x 120	Power of lens	Power thin lens $F = F_1 + F_2$
F2 = + 3.00 D		Add surface powers meridian by meridian
Front surface cylindrical		
030 meridian front surface power: – 3.00 D		
120 meridian front surface power: plano		

Draw an optical cross of the front and back surfaces, and label the powers and meridians (Figure 6-7).

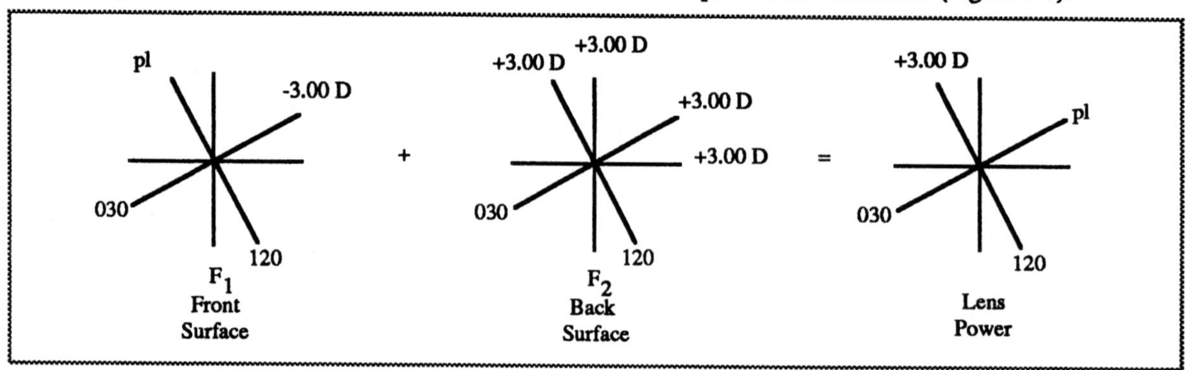

Figure 6-7. *Example 6e.* Optical crosses for front, back, and total powers of a cylindrical lens.

The cylindrical front surface has a power of – 3.00 D in the 030 meridian and plano in the 120 meridian; the spherical back surface has the same power (+ 3.00 D) in all the meridians. For the total lens power, add the powers in the two principal meridians to the sphere power:

030 meridian: $-3.00 + 3.00 = 0.00$ D

120 meridian: pl $+3.00 = +3.00$ D

The resulting powers is shown in the optical cross and may be written as + 3.00 x 030. Note that the front surface axis is changed by 90° in the total lens power. Combining two lenses in contact will yield the same result (i.e., adding a minus cylinder to a positive spherical lens with the same power will change a minus cylinder lens into a plus cylinder lens with the axis rotated by 90°).

Example 6-f

If the lens in the above example had a spherical back surface power of + 5.00 D, what would be the power of the resulting lens?

Using the information from above, draw the optical crosses (Figure 6-8).

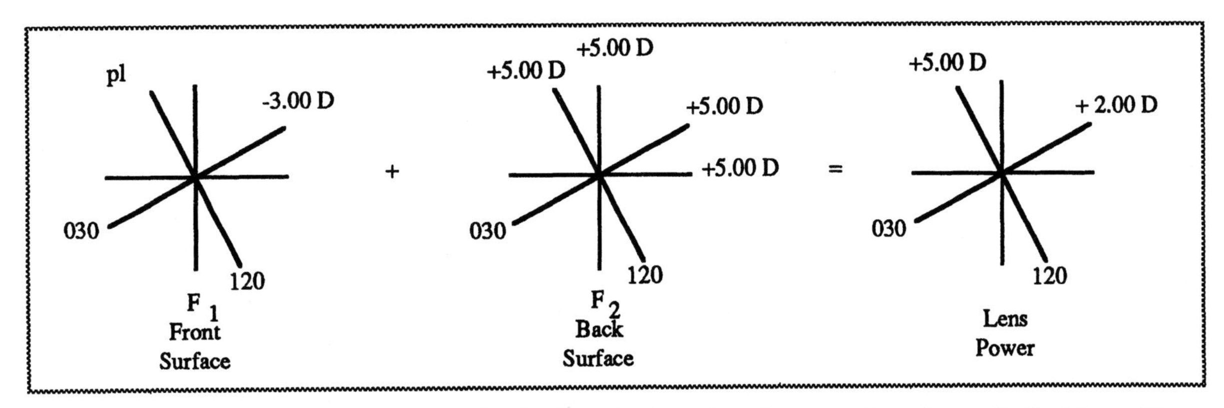

Figure 6-8. *Example 6-f.* Optical crosses for the front, back, and total powers of a spherocylindical lens. The total lens power is expressed on the cross in meridian form.

The cylindrical powers in the 030 and 120 meridians are added to the spherical power (+ 5.00 D), and the resulting lens power is shown on the optical cross. This may be solved by:

030 meridian: – 3.00 + 5.00 = + 2.00 D
120 meridian: pl + 5.00 = + 5.00D

The powers in the principal meridians represent the maximum and minimum powers in the spherocylindrical lens.

Forms of Spherocylindrical Lenses

The power of a spherocylindrical lens may be written in several ways, all of which represent the same lens properties. One may express the lens in terms of the powers in the principal meridians by using the @ sign. The lens in *Example 6-f* in **meridian form** is written as:

+ 2.00 D @ 030 and + 5.00 D @ 120

The total lens power on the optical cross in Figure 6-8 represent the meridian form.

The same lens may also be expressed in terms of a sphere and a cylinder. In general, this has the form

sphere power ⊃ cylinder power x meridian

sphere power *combined with* cylinder power *axis* meridian

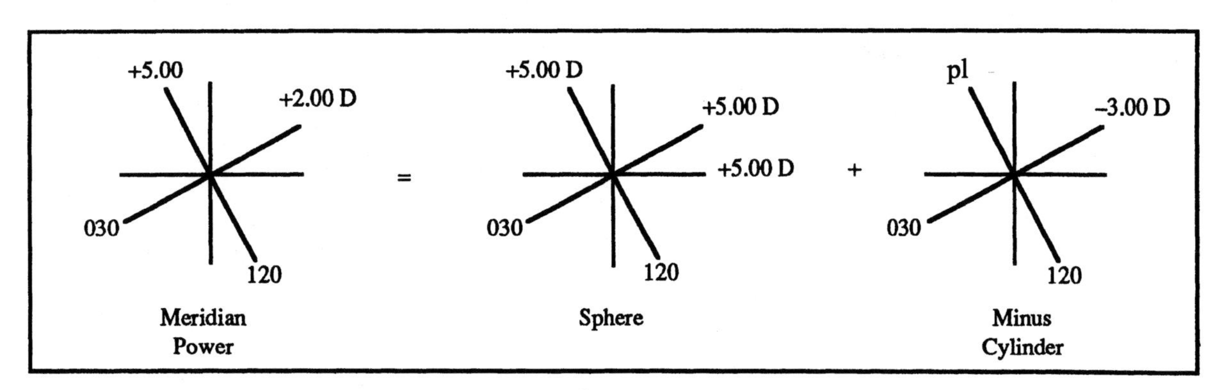

Figure 6-9. Minus cylinder form of a spherocylindrical lens.

For a negative cylinder power, this form is called **minus cylinder** and for a postive cylinder, the form is called **plus cylinder**. The "combined with" symbol is usually dropped (as it will be in this workbook).

An optical cross with the total power from *Example 6-f* is shown in the Figure 6-8 (left). To express this lens as a minus cylinder, take the meridian with the most plus (or least minus) and make it a sphere. In this case, the meridian with + 5.00 D is the most plus, and an optical cross sphere is drawn with this power. A cylinder must be combined with the sphere to arrive at the meridian powers. Draw the optical cross with the meridians labeled. In the 120 meridian, the lens has + 5.00 D, thus the cylinder added to the + 5.00 D sphere must be plano. In the 030 meridian, the lens has + 2.00 D. The cylinder power that must be added to a + 5.00 D sphere to yield + 2.00 D is found by

$$+5.00\text{ D} + C = +2.00\text{ D} \quad \text{or} \quad C = -3.00\text{ D}$$

Thus the power in the 030 meridian of the cylinder is – 3.00 D. To express the power in minus cylinder form, write the powers of the sphere and cylinder from the optical crosses:

+ 5.00 – 3.00 x 120

A lens written in plus cylinder form follows the same procedure (Figure 6-10). For plus cylinder form, the meridian with the least plus (or most minus) is used as the sphere, and the cylinder power is determined to yield the proper meridian powers. Using the optical crosses, the plus cylinder form of the lens in *Example 6-f* is:

+ 2.00 + 3.00 x 030

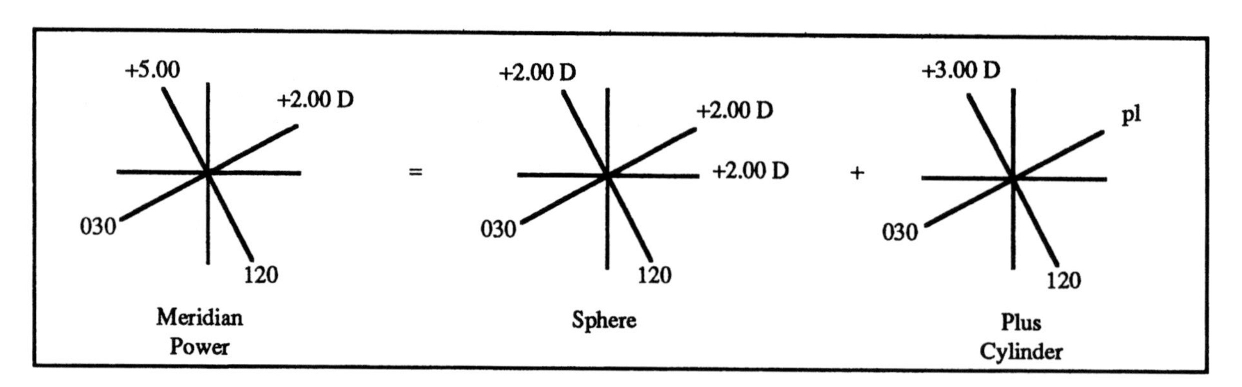

Figure 6-10. Plus cylinder form of a spherocylindrical lens.

Although an understanding of the plus and minus forms of the lenses is essential, one may use some simple rules in converting from one form to another without drawing the optical crosses:

1. For the new sphere, add the cylinder power to the sphere (use the proper sign)
2. Change the sign of the cylinder.
3. Rotate the axis of the cylinder by 90 degrees.

Change from the minus cylinder form of the lens (+ 5.00 – 3.00 x 120) to the plus cylinder form by using these rules:

1. + 5.00 + (– 3.00) = + 2.00 D sphere.
2. – 3.00 D cylinder becomes a + 3.00 D cylinder.
3. 120 axis meridian rotated by 90° is 030 axis meridian.

Using the rules, the spherocylindrical lens in *Example 6-f* is written in plus cylinder form as

+ 2.00 + 3.00 x 030

The spherocylindrical lens may also be written in **crossed cylinder form**, which is a combination of two cylindrical lenses with axes representing the principal meridians. Starting with the meridian power, simply transpose one of the meridian powers onto an optical cross, and make plano the meridian 90° away. On another optical cross, make another cylinder with the power of the other meridian. Adding the two cylinders meridian-by-meridian should yield the total lens power. This is shown in Figure 6-11 for the lens in *Example 6-f* and is written as

+ 2.00 x 120 combined with + 5.00 x 030

The crossed cylinder form of the lens is used in imaging through spherocylindrical lenses.

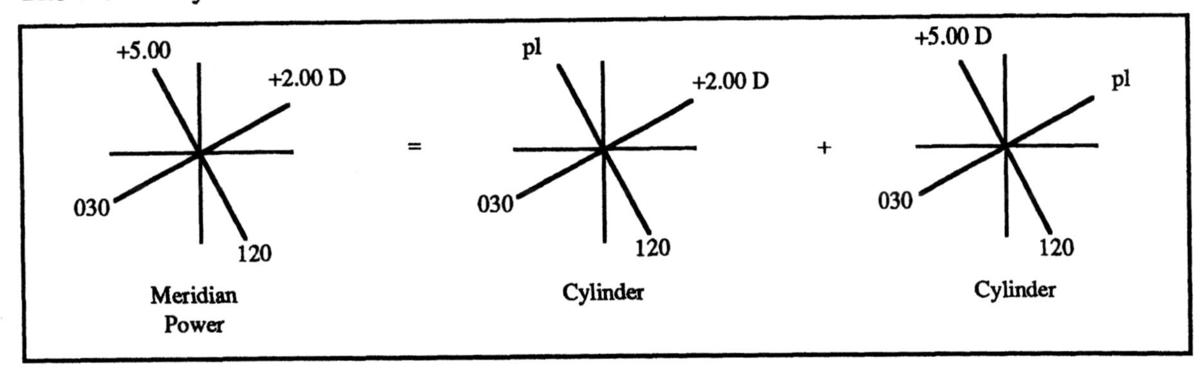

Figure 6-11. Cross cylinder form of a spherocylindrical lens.

It is recommended that you practice to develop your ability to convert from one form of the lens to another.

Example 6-g
A patient brings in the following presciption:

OD: + 4.00 – 3.00 x 045 OS: – 6.50 + 1.50 x 180

How would the prescription be written in plus cylinder, crossed cylinder and meridian forms?

Using the rules provided, the forms for the right eyewould be written as:

+ 4.00 – 3.00 x 045	minus cylinder
+ 1.00 + 3.00 x 135	plus cylinder
+ 4.00 x 135 combined with + 1.00 x 045	crossed cylinder
+ 4.00 @ 045 and + 1.00 @ 135	meridian form

For the left eye, the forms would be:

−5.00 − 1.50 x 090	minus cylinder
−6.50 + 1.50 x 180	plus cylinder
−6.50 x 090 combined with −5.00 x 180	crossed cylinder
−6.50 @ 180 and −5.00 @ 090	meridian form

Power between Principal Meridians of a Spherocylindrical Lens

The power between the principal meridians of a spherocylindrical lens falls between the maximum and minimum powers of the lens (powers in the principal meridans). By modifying Equation 6-2, the power in meridians between the principal meridans may be calculated. If the lens is converted to plus or minus cylinder form, the result is a spherical component and a cylindrical component. The power in any meridian θ degrees from the axis meridian of the cylinder component is given by

$$F_\alpha = F_s + F_c \sin^2\theta \qquad (6\text{-}8)$$

where:

F_α = the power in meridian α
F_s = the power of the sphere
F_c = the power of the cylinder
θ = the angle from the axis meridian to meridian α

The power between the principal meridians of the cylinder is added to the sphere power.

Example 6-h
What is the power in the 150 meridian of a + 5.00 @ 045 and – 7.00 @ 135 spherocylindrical lens?

Known	***Unknown***	***Equations/Concepts***
Power in the 045 meridian: + 5.00 D	Power in the 150 meridian	Transposing spherocylindrical to plus or minus cyl form
Power in the 135 meridian: – 7.00 D	θ angle	(6-8) Power between meridians

Use the optical cross and draw the meridian powers of the lens. Convert the lens to plus or minus cylinder form

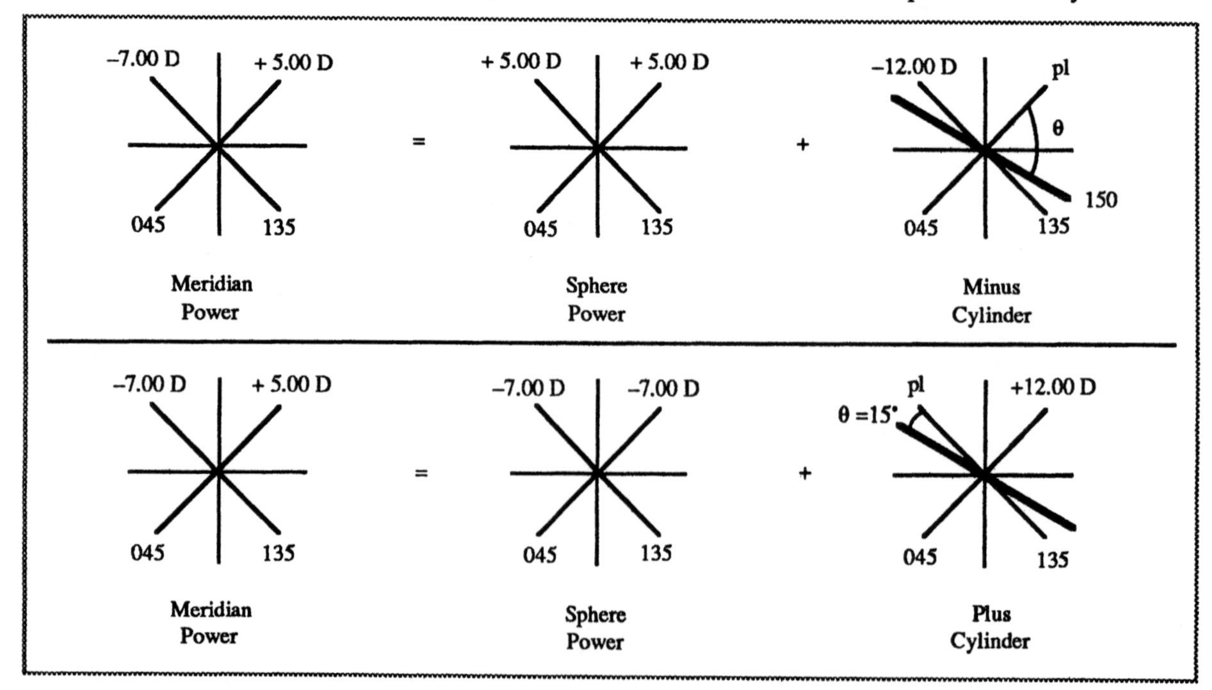

Figure 6-12a. *Example 6-h.* Plus and minus cylinder forms for power between meridians.

and determine the angle θ (from axis meridian of cylinder to 150 meridian). This is labeled on the cylindrical optical cross. For the minus cylinder, θ = 75°, the cylinder power is – 12.00 D and the sphere power is + 5.00 D. The calculation from Equation 6-8 is

$$F_{150} = F_s + F_c \sin^2 \theta = +5.00 + (-12.00)\sin^2(75) = -6.20 \text{ D}$$

For the plus cylinder, θ = 15°, the cylinder power is + 12.00 D and the sphere power is – 7.00 D. The calculation from Equation 6-8 is

$$F_{150} = F_s + F_c \sin^2 \theta = -7.00 + (+12.00)\sin^2(15) = -6.20 \text{ D}$$

One may also solve the problem as two crossed cylinders (– 7.00 x 045 and + 5.00 x 135, as shown on the optical crosses in Figure 6-12b) by calculating the power in the 150 meridian for each cylinder using Equation 6-2 and then summing.

$$F_{150c1} = F_{c1} \sin^2 \theta = (-7.00)\sin^2(75) = -6.53 \text{ D}$$
$$F_{150c2} = F_{c2} \sin^2 \theta = (+5.00)\sin^2(15) = +0.33 \text{ D}$$
$$F_{150} = F_{150c1} + F_{150c2} = -6.53 + 0.33 = -6.20 \text{ D}$$

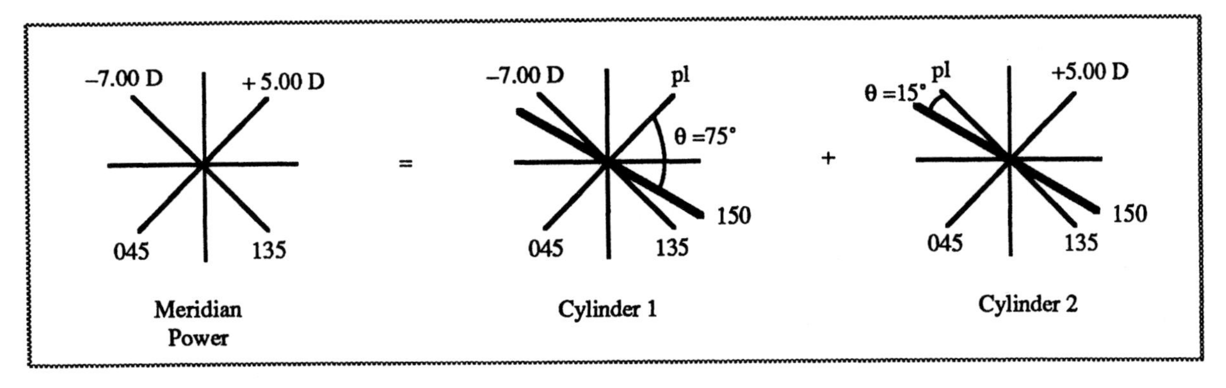

Figure 6-12b. *Example 6-h.* Crossed cylinder form for power between meridians.

Spherocylindrical Lenses and Image-Object Relationships: Position

In general, the form that specifies the power of a spherocylindrical lens does not provide information about the surfaces of the lens itself. The imaging properties of the lens, however, are the same no matter how the lens power is expressed. For a simple cylindrical lens, the line image formed may be located with the Gaussian Imaging Formula (Equation 6-7) with the orientation of the line image parallel to the axis meridian. Because a spherocylindrical lens may be expressed as two cylinders, the same procedure is used twice, once for each cylinder. The result is two line images (at right angles to each other) that are formed in different image planes. This is shown in Figure 6-13 for a spherocylindrical lens with the principal meridians in the 180 and 090 meridians. The distance between the two line images is called the **Interval of Sturm**. In the interval, the first line image changes to a series of ellipses with the major axes parallel to the line image. Farther along the optical axis, the major axes of these ellipses become shorter. The position where the major and

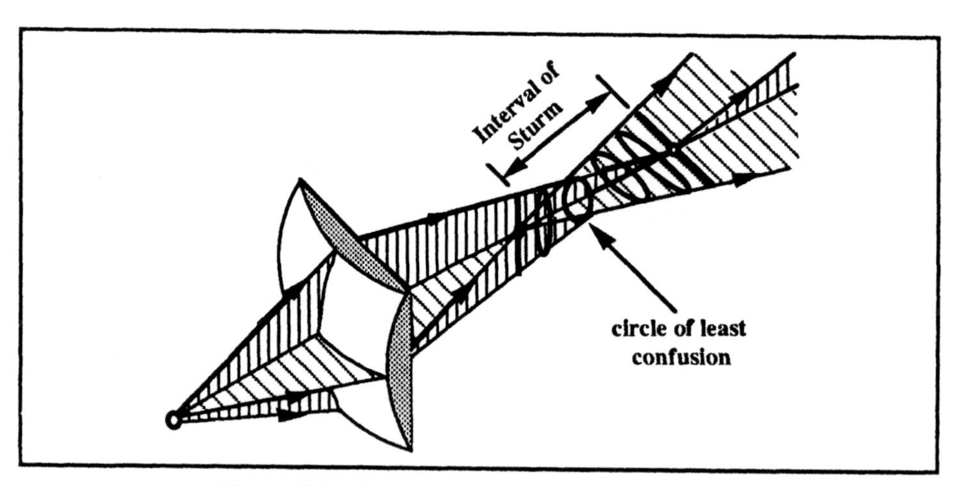

Figure 6-12. Images formed by a spherocylindrical lens.

minor axes are equal is called the **circle of least confusion.** The image plane that contains the circle of least confusion has equal blur in all directions, and thus vergence in this plane is the average emergent vergence leaving the lens.

The location of the circle may be calculated by averaging the vergence emerging from the lens:

$$L'_c = \frac{L'_{p1} + L'_{p2}}{2} \tag{6-9}$$

where:
L'_c = the average emerging vergence
L'_{p1} = the emergent vergence from one principal meridian
L'_{p2} = the emergent vergence from the other principal meridian

The reciprocal of the average vergence yields the location of the circle of least confusion:

$$\ell'_c = \frac{1}{L'_c} = \frac{2}{L'_{p1} + L'_{p2}} \tag{6-10}$$

When parallel rays are incident upon the lens, the average vergence leaving the lens is the average power of the lens. This is referred to as the **equivalent sphere:**

$$F_{es} = \frac{F_{p1} + F_{p2}}{2} \tag{6-11}$$

where:
F_{es} = the equivalent sphere
F_{p1} = the power in one principal meridian
F_{p2} = the power in the other principal meridian

With the lens in plus or minus cylinder form, the equivalent sphere may also be calculated by adding one-half the cylinder to the sphere:

$$F_{es} = F_s + \frac{F_c}{2} \tag{6-12}$$

Example 6-i
An object is placed 1 m in front of a spherocylindrical lens (+ 8.00 + 2.00 x 180). Calculate the location and orientation of the line images formed. Locate the circle of least confusion for this object. What is the equivalent sphere for the lens?

Known	***Unknown***	***Equations/Concepts***
Power of the lens: + 8.00 + 2.00 x 180	Location of line images and circle of least confusion	Imaging through cylinder
Real object position: $\ell = -1$ m	Orientation of line images	Axis cylinder notation
	Equivalent sphere	(6-5), (6-6), (6-7) Imaging formula
		(6-10) Circle of least confusion
		(6-11) Equivalent sphere

Express the spherocylindrical lens in cross cylinder form on optical crosses:

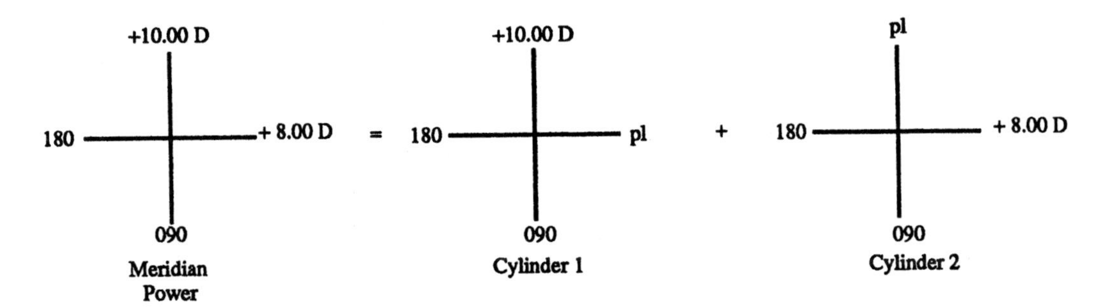

Image through cylinders 1 and 2 using the imaging Equations 6-5, 6-6, and 6-7. The incident vergence is −1.00 D.

$$L'_{c1} = F_{090} + L = +10.00 + (-1.00) = +9.00 \text{ D}$$

$$\ell'_{c1} = \frac{1.00}{L_{c1}} = \frac{1.00}{+9.00} = +0.1111 \text{ m} = +11.11 \text{ cm}$$

$$L'_{c2} = F_{180} + L = +8.00 + (-1.00) = +7.00 \text{ D}$$

$$\ell'_{c1} = \frac{1.00}{L_{c1}} = \frac{1.00}{+7.00} = +0.1428 \text{ m} = +14.28 \text{ cm}$$

The line image located 11.11 cm behind the lens is horizontal (parallel to the axis meridian of cylinder 1), and the line image 14.28 cm behind the lens formed by cylinder 2 is vertical (parallel to the axis meridian of cylinder 2).
The location of the circle of least confusion is calculated with Equation 6-10:

$$\ell'_{c} = \frac{2}{L'_{p1} + L'_{p2}} = \frac{2}{+9.00 + 7.00} = \frac{2}{+16.00} = +.125 \text{ m} = +12.50 \text{ cm}$$

The equivalent sphere may be calculated with Equation 6-11:

$$F_{es} = \frac{F_{p1} + F_{p2}}{2} = \frac{+10.00 + 8.00}{2} = \frac{+18.00}{2} = +9.00 \text{ D}$$

plus cylinder $F_{es} = F_s + \frac{F_c}{2} = +8.00 + \frac{2.00}{2} = +8.00 + 1.00 = +9.00 \text{ D}$

minus cylinder $F_{es} = F_s + \frac{F_c}{2} = +10.00 + \frac{(-2.00)}{2} = +10.00 - 1.00 = +9.00 \text{ D}$

Images Formed by Spherocylindrical Lenses

The images formed by a spherocylindrical lens of an object that has finite size also consists of line images (i.e., each point on the object forms a line image parallel to the axis meridian of a cylindrical lens). If the power is in the vertical meridian, the lines are horizontal, and if the power is in the horizontal meridian, the lines are vertical. This is shown in Figure 6-14.

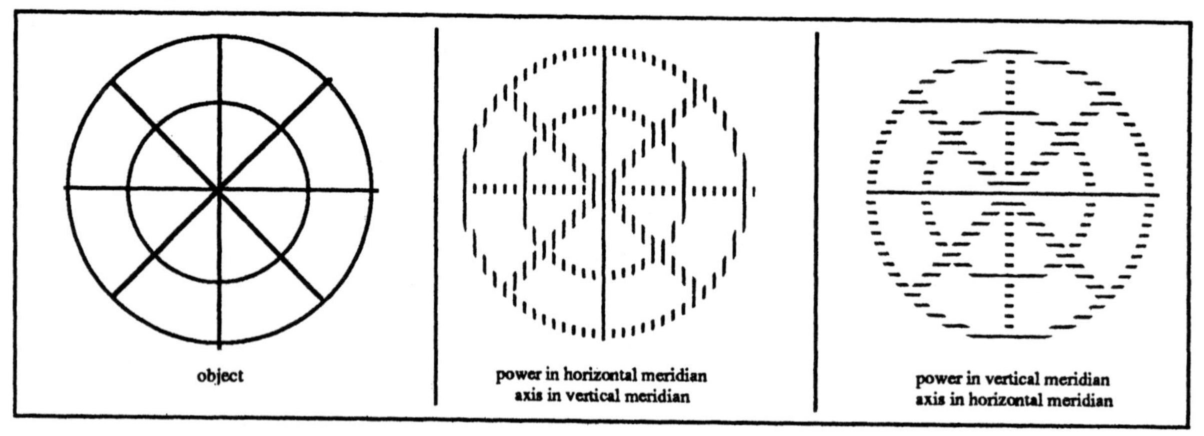

Figure 6-14. Object of a finite size imaged through a spherocylindrical lens with the principal meridians in the vertical and horizontal meridians. Line images are formed for each point in the object.

Obliquely Crossed Cylinders

The combination of two cylinders or spherocylinders that do not have the same principal meridians is not a simple addition. The resultant spherocylinder has sphere and cylinder powers and an axis that is not obvious from inspection. In practice, the simplest way to determine the resultant spherocylinder is to place the lens combination in a lensometer and read the powers and axis directly. There are two procedures for calculating the combination of obliquely crossed cylinders; one is graphical and the other involves mathematical calculations. In general, this topic is usually covered in ophthalmic optics courses. Only the graphical technique will be covered in this book. See Fannin and Grovsner, Keating, and other texts for the mathematical procedures.

The graphical procedure for adding obliquely crossed cylinders follows several rules that are outlined below:

1. Transpose the cylinders into plus cylinder form.

2. Using the cylinder with the smaller axis, draw a horizontal line with its length scaled to the dioptric power of the cylinder.

3. Determine the angular difference between the two cylinder axes. Double this difference.

4. Starting at the right end of the horizontal vector, lay out the cylinder with the larger axis at twice the angular difference (determined in Step 3) with a line scaled to the dioptric power of the cylinder.

5. Complete the triangle. The length of the resultant line is the dioptric power of the resulting cylinder. The counterclockwise angle from the horizontal cylinder vector to the resultant vector is twice the angular difference between the horizontal cylinder axis and the resultant cylinder axis. Therefore the resultant axis may be calculated using

$$A_{cr} = \frac{A_m}{2} + A_{c1} \qquad (6\text{-}13)$$

where: A_{cr} = resultant axis
A_m = measured axis from graph
A_{c1} = axis of smallest cylinder

6. Determine the spherical component of the two cylinders with this formula:

$$F_{sc} = \frac{F_{c1} + F_{c2} - F_{cr}}{2} \qquad (6\text{-}14)$$

where: F_{sc} = resultant sphere induced by the cylinders
F_{c1} = power of cylinder 1
F_{c2} = power of cylinder 2
F_{cr} = power of resultant cylinder (from graph)

7. Add all the spherical components.

$$F_s = F_{s1} + F_{s2} + F_{sc} \qquad (6\text{-}15)$$

where: F_s = power of resultant sphere
F_{s1} = power of sphere of first lens
F_{s2} = power of sphere of second lens

8. Write the resultant spherocylinder.

This procedure is illustrated for several cylinder combinations in the examples that follow.

Example 6-j
Determine the resultant spherocylinder for these cylinder combinations:

a) + 2.00 x 030 and + 1.50 x 050

b) − 2.00 − 1.00 x 030 and + 1.00 + 1.00 x 005

The solutions for each cylinder combination follow the steps outlined above.

For a)

Step 1. The cylinders are in plus cylinder form.

Step 2. The + 2.00 x 030 has the smallest axis and it is laid out on the graph with a length equal to two units. (See Figure 6-15, left.)

Step 3. The angular difference between the two cylinders is

050 − 030 = 020

Double difference = 2(020) = 040

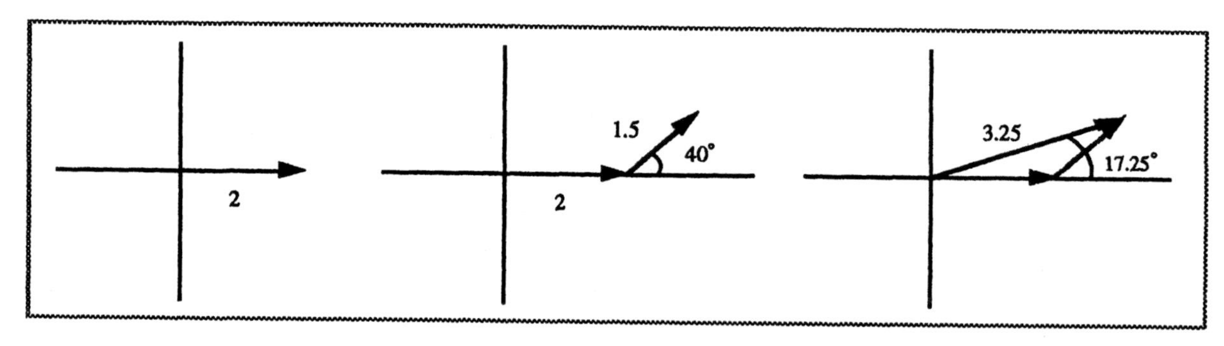

Figure 6-15. *Example 6-j. Left*: Vector that represents the cylinder with smaller axis. *Center*: Vector that represents the cylinder with the larger axis. *Right*: Resultant vector that represents the resultant cylinder magnitude and axis.

Step 4. Lay out the cylinder at the end of vector 1 at an angle equal to 040 and a length scaled to 1.50. (See Figure 6-15, center.)

Step 5. Complete the triangle and measure the resultant magnitude and angle. (See Figure 6-15, right.) The measured magnitude is 3.25, and the angle is 17.25°. The angle represents two times the difference between the horizontal cylinder axis, and the resultant cylinder axis and therefore the resultant cylinder axis is calculated using Equation 6-13:

$$A_{cr} = \frac{A_m}{2} + A_{c1} = \frac{17}{2} + 30 = 8.5 + 30 = 38.5°$$

Step 6. Calculate the spherical component of the cylinders using Equation 6-14:

$$F_{sc} = \frac{F_{c1} + F_{c2} - F_{cr}}{2} = \frac{+2.00 + 1.50 - 3.25}{2} = 0.125 \text{ D}$$

Step 7. The total sphere is calculated using Equation 6-15. There are no other spherical components in this problem. The total sphere is

$$F_s = F_{s1} + F_{s2} + F_{sc} = 0 + 0 + 0.125 = 0.125 \text{ D}$$

Step 8. The final resultant spherocylinder is written as a plus cylinder:

+ 0.125 + 3.25 x 038.5

For b) – 2.00 – 1.00 x 030 and + 1.00 + 1.00 x 005

Step 1. The first cylinder is written in plus cylinder form: – 3.00 + 1.00 x 120. This represents the cylinder with the largest axis.

Step 2. The + 1.00 + 1.00 x 005 lens has the smallest axis, and it is laided out on the graph with a length equal to three units. (See Figure 6-16, left.)

Step 3. The angular difference between the two cylinders is:

120 – 005 = 115

Double difference = 2(115) = 230

Step 4. Lay out cylinder at the end of vector 1 at an angle equal to 230 and a length scaled to 1.00. (See Figure 6-16, center.)

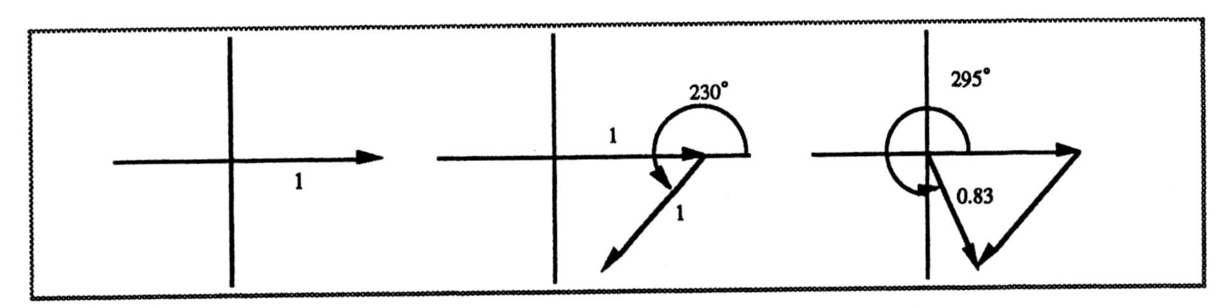

Figure 6-16. *Example 6-j. Left*: Vector that represents the cylinder with the smaller axis. *Center*: Vector that represents the cylinder with the larger axis. *Right*: Resultant vector that represents the resultant cylinder magnitude and axis.

Step 5. Complete the triangle and measure the resultant magnitude and angle. (See Figure 6-16, right.) The measured magnitude is + 0.83 D, and the angle is 295°. The angle represents two times the difference between the horizontal cylinder axis, and the resultant cylinder axis and therefore the resultant cylinder axis is calculated using Equation 6-13:

$$A_{cr} = \frac{A_m}{2} + A_{c1} = \frac{295}{2} + 005 = 147.5 + 005 = 152.5°$$

Step 6. Calculate the spherical component of the cylinders using Equation 6-14:

$$F_{sc} = \frac{F_{c1} + F_{c2} - F_{cr}}{2} = \frac{+1.00 + 1.00 - 0.83}{2} = +0.58 \text{ D}$$

Step 7. The total sphere is calculated using Equation 6-15:

$$F_s = F_{s1} + F_{s2} + F_{sc} = +1.00 + (-3.00) + 0.58 = -1.42 \text{ D}$$

Step 8. The final resultant spherocylinder is written as a plus cylinder:

– 1.42 + 0.83 x 152.5

Remember that when the principal meridians of the two spherocylinders align, simple addition meridian by meridian will result in the final powers. See ophthalmic texts for more details on adding obliquely crossed cylinders.

Supplemental Problems

Power of Cylindrical Lenses

6-1. What is the power in the 045 and 100 meridians of a plano-cylindrical lens with a front surface power of + 3.75 x 090?
ANS. F_{45} =+ 1.88 D; F_{100} =+ 0.11 D

6-2. Find the power in the 090 and 180 degree meridians of a plano-cylindrical lens with a power of + 4.00 x 120.
ANS. Power @ 090 = + 1.00 D; power @ 180 = + 3.00 D

Image-Object Relationships

6-3. A point object is placed 40 cm in front of a lens (in object space). A real, vertical line image is formed 25 cm from the lens, and a real, horizontal line image is formed 50 cm from the lens. What is the power of the lens?
ANS. + 6.50 – 2.00 x 180

6-4. How far from a + 7.00 – 2.00 x 090 lens will the horizontal line focus of an infinite object be found?
ANS. 14.3 cm

6-5. If a point source is placed 50 cm in front of a + 4.00 D.S. combined with a – 1.00 x 135, where would the images be located and oriented?
ANS. A line image in the 045 meridian is located 50 cm behind the lens, and a line image in the 135 meridian is located 1 m behind the lens

6-6. Locate the vertical line focus of an object 1 m in front of a lens with a power of + 2.00 – 1.00 x 090.
ANS. The image is at infinity

6-7. If a point object at infinity imaged through a lens forms a point image at + 40 cm, and the same object imaged through another lens forms two line foci, one at + 25 cm (the vertical image) and one at +33.3 cm (the horizontal image), locate the images formed if the two lenses are combined.
ANS. Vertical image = + 15.4 cm; horizontal image = + 18.2 cm

Spherocylindrical Lenses

6-8. Transcribe the following lens in plus, minus, and cross cylinder form:
+ 1.00 @ 090 and – 2.00 @ 180
ANS. – 2.00 + 3.00 x 180 plus cylinder; + 1.00 – 3.00 x 090 minus cylinder; + 1.00 x 180 combined with – 2.00 x 090 crossed cylinder form.

6-9. Which lens does not have a spherical equivalent of – 4.00 D? a) – 1.00 – 6.00 x 045 b) – 2.00 – 4.00 x 098 c) – 5.00 + 2.00 x 124 d) – 4.00 D.S. e) – 1.00 – 5.00 x 180
ANS. All have a spherical equivalent of – 4.00 D except (e) which has a spherical equivalent of – 3.50D

Power of Spherocylindrical Lenses

6-10. A + 5.00 – 7.00 x 180 lens is combined with a – 7.00 + 7.00 x 090 lens. In what meridian of the combination of the lenses is the power equal to – 3.50 D?
ANS. 051 meridian

6-11. What is the power of a + 5.00 – 2.00 x 090 lens in the 030 meridian?
ANS. + 3.50 D

6-12. The power of a cylindrical lens is + 2.00 + 3.00 x 180. If the power at angle θ measures + 2.75 D, then what is the power at θ + 90°?
ANS. + 4.25 D

The Circle of Least Confusion

6-13. What is the average vergence leaving a + 5.00 – 2.00 x 090 lens for an infinite object. Determine the location of the circle of least confusion for a point object located 1 m in front of the lens.
ANS. Average vergence = + 4.00 D; Location of circle of least confusion = + 33.33 cm

6-14. A spherocylindrical lens forms the circle of least confusion of an infinite object 25 cm from the lens. If the horizontal line focus is located 1 m from the lens, what is the power of the lens?
ANS. + 7.00 – 6.00 x 180

6-15. The circle of least confusion is formed 2 m behind a spherocylindrical lens of unknown power. When a point source (object) is located 25 cm in front of the lens, the vertical meridian of the lens forms a line image at infinity. What is the power of the lens?
ANS. +5.00 – 1.00 x 180

6-16. Find the length of the Interval of Sturm for a spherocylindrical lens with a power of – 1.00 – 1.50 x 170 if an object is located 30 cm in front of the lens.
ANS. 6 cm

Obliquely Crossed Cylinders

6-17. Calculate the combination of the following lenses using the graphical technique described.

a. – 1.50 – 1.00 x 170 and – 1.25 – 0.75 x 155
b. – 5.00 – 2.00 x 180 and + 4.00 – 3.00 x 015

ANS. a. – 4.46 + 1.68 x 073.5 b. – 5.95 + 4.90 x 099

Chapter 7

Thick Lenses and Multiple Lens Systems

Optical systems often consist of two or more imaging elements. In addition, a lens that cannot be considered thin (i.e., the center thickness changes the power), acts as a lens system with two refracting surfaces separated by the thickness of the lens. There are two methods in which image-object relationships may be determined with these systems. The first approach is to image through each element or surface separately. This technique is useful for simple systems. The other approach is to combine the elements or surfaces into a unique system by redefining reference planes and power. This method involves more calculations in setting up the new system. However, once established, the imaging calculations need only be performed one time for any object position. Both approaches are discussed in this chapter and are shown in Chapter 9 using ray tracing.

Element-by-Element Imaging: Location

The element-by-element method for determining image-object relationships uses previously developed single refracting surface and thin lens imaging formula (summarized in Figure 7-1) for each element or surface. The process involves locating an image and using this image as an object for the next element. This procedure is repeated until the final image is found. You must have an understanding of image-object relationships for both single refracting surfaces and thin lenses to master this method. If you feel deficient in these areas, it is recommended that you review Chapters 4 and 5.

$L' = F + L$ (7-1)

Incident vergence: $L = \frac{n}{\ell}$ (7-2)

Emergent vergence: $L' = \frac{n'}{\ell'}$ (7-3)

Surface power: $F = \frac{n'-n}{r}$ (7-4)

Thin lens power: $F = (n'-n)\left(\frac{1}{r_1} - \frac{1}{r_2}\right)$ (7-5)

where:
- n = index to left of interface
- n' = index to right of interface
- F = power (diopters)
- ℓ = object distance in meters
- ℓ' = image distance in meters
- r = radius of surface in meters

Figure 7-1. Summary of imaging and power formulas.

The imaging process is explained with the aid of Figure 7-2, which illustrates a two-element optical system. The first step in this method involves locating an image formed by the first element or surface. Isolate the first element (E1), and calculate its power (if necessary) using Equations 7-4 or 7-5. Equations 7-1, 7-2, and 7-3 are used to calculate the image location formed by the first element. Remember that if the index to the left of the surface differs from the index to the right (i.e., n and n' are not the same), the single refracting surface formulas are to be used. If the surrounding media are the same (n = n'), then treat the surface or element as a thin lens. Because there may be several indices represented in multiple lens systems, the index in each successive space will be assigned a new subscript number, and therefore the primed notation will correspond to the higher value subscript. For example, for the first element, n corresponds to n_1 and n' corresponds to n_2.

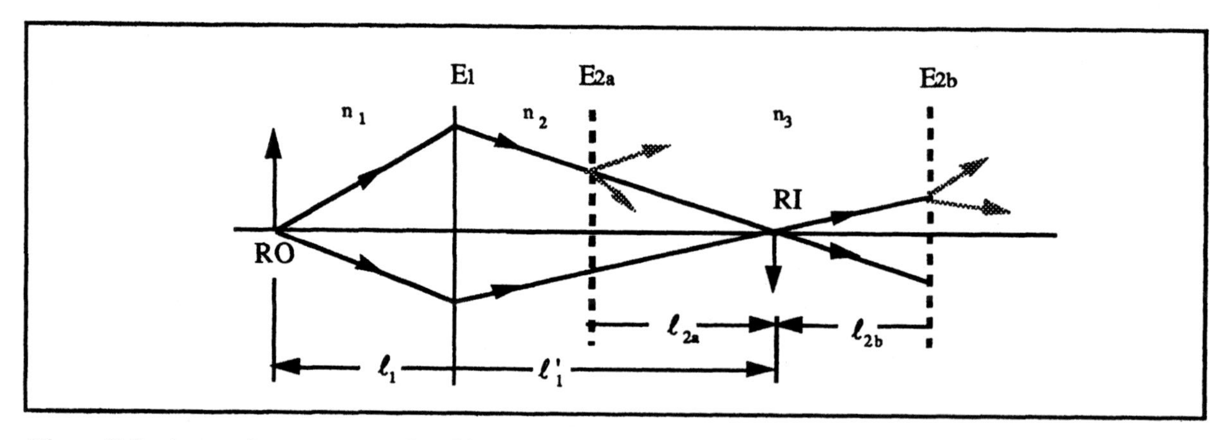

Figure 7-2. A two element system in which the first element forms a real image at position RI. There are two general positions for the next element relative to the image. If the element is to the left (E2a), convergent light will be incident on E2, and therefore the image acts as a virtual object. If the element is to the right of the image (E2b), divergent light is incident upon E2, and the image acts as a real object.

In this example, the first element (E1) forms a real image, which is labeled RI. This image will act as the object for the next element. The type of object will depend on the vergence incident on to the next element. Relative to this image, there are two general positions for the next element (E2); to the left (E2a) or to the right (E2b) of the image. When the element is in position E2a, convergent light is incident on it. This means that the image acts like a virtual object for E2. In fact, this is the way virtual objects are formed with a supplemental lens. (A detailed discussion of virtual objects is presented in Chapter 4.) If the second element is located at E2b, the incident light is divergent. The image, in this case, acts as a real object for the second element. The distance from the second element E2 to the image is used to calculate the incident vergence (Equation 7-2). Use the sign convention established, and the vergence will have the proper sign. For the system above, the distance ℓ_{2a} is positive, and the distance ℓ_{2b} is negative. The imaging formula (Equation 7-1) is used to determine emergent vergence and the next image location (Equation 7-3).

This process is repeated over and over again for each successive element until the final image is found. Using this process, the final image position is found relative to the last element or surface.

In the previous explanation, the first element formed a real image. If an element forms a virtual image, the next element will have divergent incident light for any position. This is illustrated in Figure 7-3. The incident vergence upon E2 is calculated using the distance ℓ_2, which has a negative value.

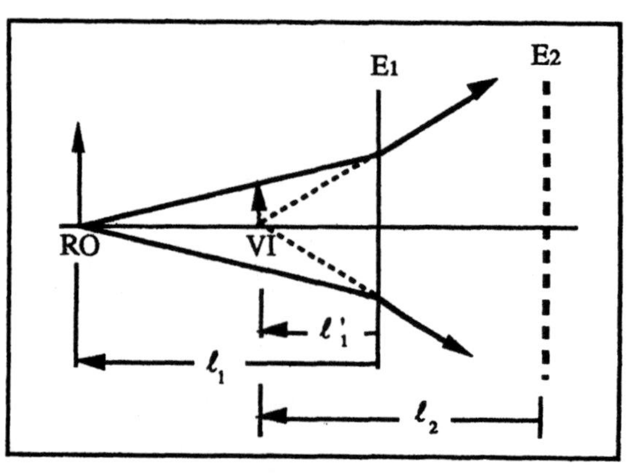

Figure 7-3. If the first element forms a virtual image, divergent light is incident upon the second element no matter where it is placed.

The same process can be used for a thick lens in which each refracting surface is an element of the system. The medium between the elements has the index of refraction of the lens. Because for each surface the image and object are formed in different media, single refracting surface relationships are used.

The process is demonstrated by several examples. Consult Chapter 9 for the graphical technique.

Example 7-a.
A glass rod (n = 1.50) has polished, convex, spherical ends such that the front has a radius of 5 cm and the back has a surface power of +5.00 D. The glass rod is 15 cm long. If an object is placed 50 cm in front of the front surface, where is the final image formed?

Known	***Unknown***	***Equations/Concepts***
Front surface radius: $r = 5$ cm	Final image location	Element by element imaging
Back surface power: $F = +5.00$ D		(7-2), (7-3) Vergence at each surface
Thickness of rod: $t = 15$ cm		(7-1) Imaging formula
Index of rod: $n = 1.50$		(7-4) Single refracting surface power
Index of surround: $n = 1.00$		
Object location: $\ell = -50$ cm $= -0.50$ m		

You may wish to draw a diagram of the entire system with a relative scale (Figure 7-4a). This helps to see the relationships before starting the problem. Label the information given, and determine positive or negative distances with arrows. Note that the ends of the rod are convex and the object distance is negative (real object to left of interface). Calculate the power using (Equation 7-4) the single refracting surface power formula. Be sure to note that the radius is positive, the medium to the right of the interface (n_2 = glass) corresponds to n', and the medium to the left of the interface (n_1 = air) corresponds to n. Check your answer for the proper sign (i.e., a convex surface should have positive power).

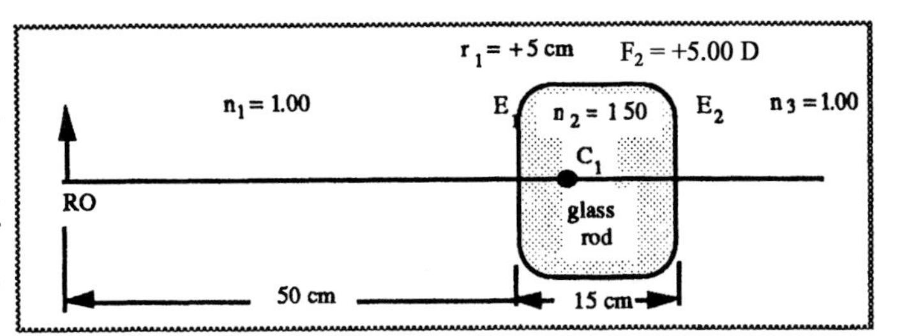

Figure 7-4a. *Example 7-a.* Diagram of optical system.

$$F_1 = \frac{n_2 - n_1}{r_1} = \frac{1.50 - 1.00}{+0.05\text{ m}} = +10.00\text{ D}$$

Isolate the front surface of the glass rod in a diagram (Figure 7-4b). In this case, assume that the medium to the right of the interface is the same (glass). Thus the element behaves as a single refracting surface. The next step is to locate the image formed by the first element. There is a real object to the left of the element, and therefore the distance is negative according to our sign convention. The incident vergence, emergent vergence, and image position are found using Equations 7-2, 7-1, and 7-3, respectively:

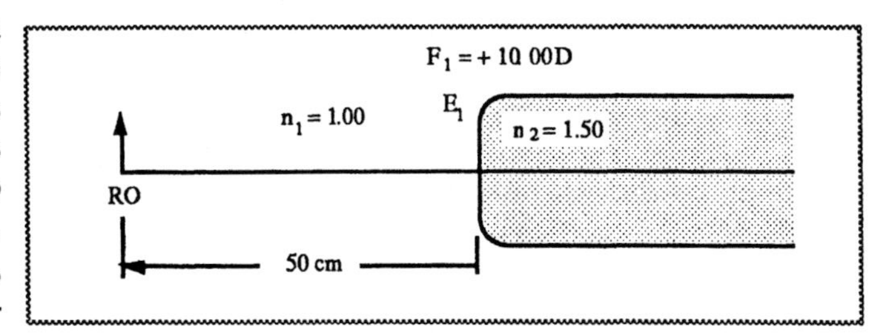

Figure 7-4b. *Example 7-a.* Isolated front element.

$$L_1 = \frac{n_1}{\ell_1} = \frac{1.00}{-0.50} = -2.00\text{ D}$$

$$L_1' = L_1 + F_1 = -2.00 + 10.00 = +8.00\text{ D}$$

$$\ell' = \frac{n_2}{L_1'} = \frac{1.50}{+8.00} = +0.1875\text{ m} = +18.75\text{ cm}$$

Thus a real image is formed 18.75 cm from the first element in the glass rod. At this point it is not important that the rod is shorter than this distance . Figure 7-4c illustrates the position of this image with respect to the second element. Note that the vergence incident on the second surface is convergent (the image is to the right of the surface). Thus the first image acts as a virtual object for the second surface. Because the incident rays are traveling in glass, this index must be considered when calculating the new vergence. Also note that the rays never actually reach the image position. They are intercepted by the second surface. From Figure 7-4c, the distance from the second element to its object (image 1) may be calculated

$$\ell_2 = 18.75 \text{ cm} - 15 \text{ cm} = +3.75 \text{ cm}$$

Using this value and the given power for the second element, the final image location may be calculated using Equations 7-2, 7-1, and 7-3:

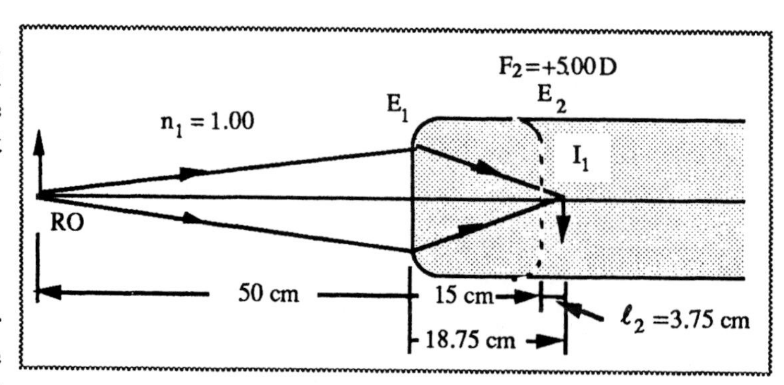

Figure 7-4c . *Example 7-a.* Image formed by first surface.

$$L_2 = \frac{n_2}{\ell_2} = \frac{1.50}{+0.0375 \text{ m}} = +40.00 \text{ D}$$

$$L'_2 = L_2 + F_2 = +40.00 + 5.00 = +45.00 \text{ D}$$

$$\ell'_2 = \frac{n_3}{L'_2} = \frac{1.00}{+45.00} = +0.0222 \text{ m} = +2.22 \text{ cm}$$

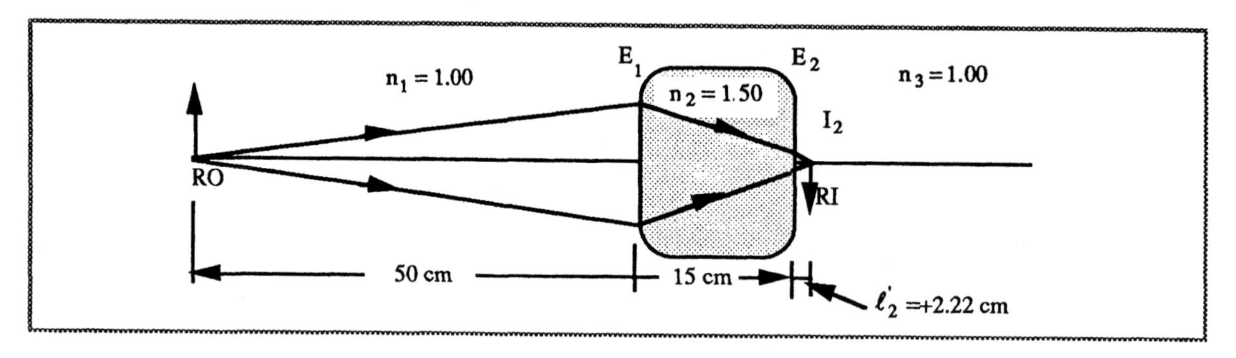

Figure 7-4d. *Example 7-a.* Final image formed by the glass rod.

Thus the final image is located 2.22 cm from the second surface. Because the emergent vergence is positive, the image is real and located in the medium to the right of the surface (Figure 7-4d).

Example 7-b

A lens system consists of two thin lenses, a +15.00 D and a – 3.00 D separated by 17 cm. An object is placed 1 m in front of the first lens. Where is the final image located?

Known	***Unknown***	***Equations/Concepts***
First lens power: $F_1 = +15.00$ D	Final image location	Element by element imaging
Second lens power: $F_2 = -3.00$ D		(7-2), (7-3) Vergence at each surface
Distance between lenses: $t = 17$ cm		(7-1) Imaging formula
Object location: $\ell_1 = -1.00$ m		

Diagram the entire lens system, as shown in Figure 7-5a. Solve for the location of the image formed by the first lens by substituting the given object distance and powers into Equations 7-2, 7-1, 7-3.

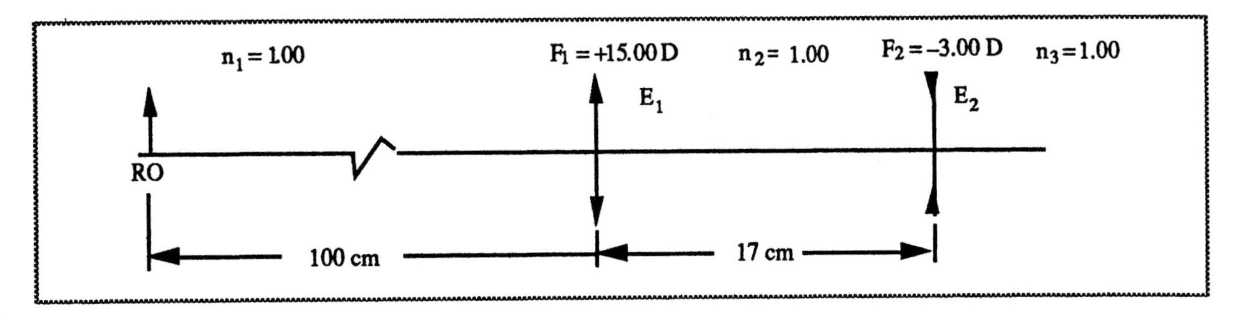

Figure 7-5a. *Example 7-b.* Entire lens system.

$$L_1 = \frac{n_1}{\ell_1} = \frac{1.00}{-1.00} = -1.00 \text{ D}$$

$$L'_1 = L_1 + F_1 = -1.00 + 15.00 = +14.00 \text{ D}$$

$$\ell'_1 = \frac{n_2}{L'_1} = \frac{1.00}{+14.00} = +0.0714 \text{ m} = +7.14 \text{cm}$$

The image formed by the first lens is 7.14 cm to the right of the lens. Next find the position of this image relative to the second lens (Figure 7-5b). This is easily calculated to be – 9.86 cm (17 cm – 7.14 cm = 9.86 cm). The image is formed to the left of the second lens (negative value).

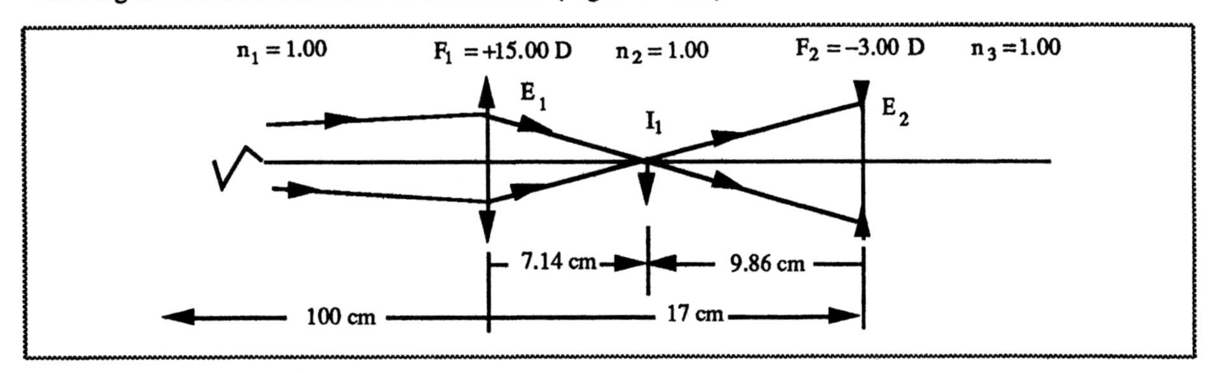

Figure 7-5b. *Example 7-b.* The image formed by the first lens acts as a real object for the second lens.

Because the object for the second lens is to the left, the distance is negative (from lens to object is against the direction of light travel), and negative vergence is incident on the lens. The final image location is calculated using the second lens power and the incident vergence:

$$L_2 = \frac{n_2}{\ell_2} = \frac{1.00}{-0.0986 \text{ m}} = -10.14 \text{ D}$$

$$L'_2 = L_2 + F_2 = -10.14\text{D} + (-3.00\text{D}) = -13.14 \text{ D}$$

$$\ell'_2 = \frac{n_3}{L'_2} = \frac{1.00}{-13.14 \text{ D}} = -0.0761 \text{ m} = -7.61 \text{ cm}$$

The final image is located to the left of the second lens and is thus a virtual image. Divergent light leaves the second lens. This is illustrated in Figure 7-5c.

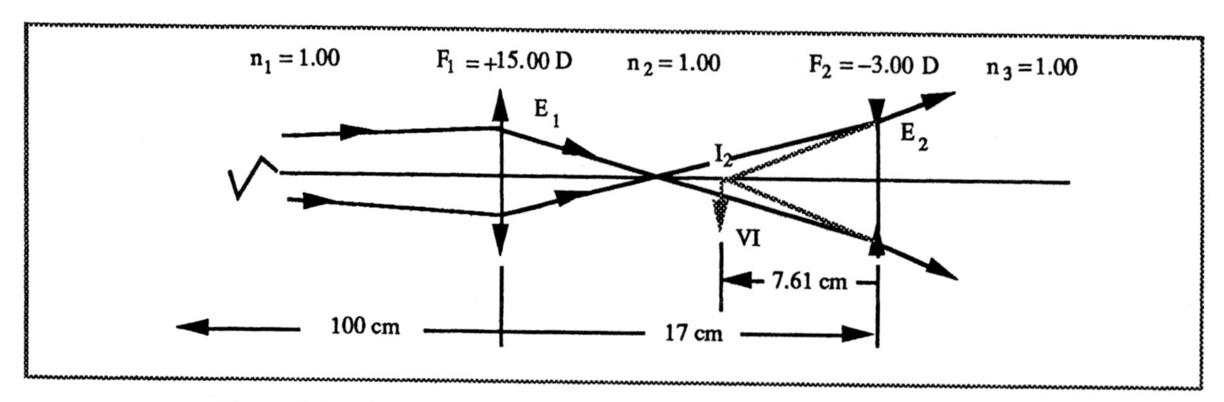

Figure 7-5c. *Example 7-b.* Final image location relative to the last element.

Element-by-Element Imaging: Size and Orientation

In the previous examples, the final image position is determined by using successive images as objects for each element. The image size may also be calculated with the same procedure (i.e., the product of the lateral magnification for each image-object will result in the final magnification of the system).

$$LM_{system} = (LM_1)(LM_2)(LM_3)\cdots \qquad (7\text{-}6)$$

The equations for lateral magnification previously presented for single refracting surfaces and thin lenses are summarized in Figure 7-6. The signs associated with each calculation are important because the sign of the final lateral magnification will reveal the final orientation of the image (+, erect, −, inverted) relative to the original object.

$$LM = \frac{h'}{h} \qquad (7\text{-}7) \qquad\qquad LM = \frac{n\ell'}{n'\ell} \qquad (7\text{-}8) \qquad\qquad LM = -\frac{f}{x} = -\frac{x'}{f'} \qquad (7\text{-}9)$$

where:

n = index to left of interface	ℓ = object distance
n' = index to right of interface	ℓ' = image distance
f = primary focal length	f' = secondary focal length
x = object extrafocal distance	x' = image extrafocal distance

Figure 7-6. Summary of lateral magnification equations.

If the object size and the final image size are known, Equation 7-7 may be used. However, the sign will not indicate whether the final image is inverted or erect relative to the object.

Note that the type of object or image cannot be used to predict the orientation of the final image, as previously established for single refracting surfaces and thin lenses.

Example 7-c
Calculate the lateral magnification of the systems in *Examples 7a* and *7b*. Indicate the orientation of the image-object in each case.

Example 7-a

Known	***Unknown***	***Concepts/Equations***
Index of rod: $n = 1.50$	System lateral magnification	(7-8) Lateral magnification
Index of surround: $n = 1.00$		(7-6) System lateral magnification
Object location: $\ell_1 = -50.00$ cm		
First image location: $\ell'_1 = +18.75$ cm		
Second object location: $\ell_2 = +3.75$ cm		
Final image location: $\ell'_2 = +2.22$ cm		

Calculate the lateral magnification for each image-object relationship using Equation 7-8 and then determine system magnification using Equation 7-6:

$$LM_1 = \frac{n\ell'}{n'\ell} = \frac{n_1\ell'_1}{n_2\ell_1} = \frac{(1.00)(18.75\text{ cm})}{(1.50)(-50.00\text{ cm})} = -0.25 \quad \text{inverted image relative to original object}$$

$$LM_2 = \frac{n\ell'}{n'\ell} = \frac{n_2\ell'_2}{n_3\ell_2} = \frac{(1.50)(2.22\text{ cm})}{(1.00)(3.75\text{ cm})} = +0.89 \quad \text{erect image relative to } I_1$$

$$LM_s = (LM_1)(LM_2) = (-0.25)(+0.89) = -0.22 \quad \text{inverted image relative to original object}$$

The system lateral magnification indicates that the final image is minified and inverted relative to the original object.

Example 7-b

Known	***Unknown***	***Concepts/Equations***
Index of between lenses: $n = 1.00$	System lateral magnification	(7-8) Lateral magnification
Index of surround: $n = 1.00$		(7-6) System lateral magnification
Object location: $\ell_1 = -100.00$ cm		
First image location: $\ell'_1 = +7.14$ cm		
Second object location: $\ell_2 = -9.86$ cm		
Final image location: $\ell'_2 = -7.61$ cm		

$$LM_1 = \frac{n\ell'}{n'\ell} = \frac{n_1\ell'_1}{n_2\ell_1} = \frac{(1.00)(7.14\text{cm})}{(1.00)((-100.00\text{cm}))} = -0.0714 \quad \text{inverted image relative to original object}$$

$$LM_2 = \frac{n\ell'}{n'\ell} = \frac{n_2\ell'_2}{n_3\ell_2} = \frac{(1.00)(-7.61\text{ cm})}{(1.00)(-9.86\text{ cm})} = +0.7718 \quad \text{erect image relative to } I_1$$

$$LM_s = (LM_1)(LM_2) = (-0.0714)(0.7718) = -0.055 \quad \text{inverted image relative to original object}$$

The system lateral magnification indicates that the final image is minified and inverted relative to the original object.

Cardinal Points Method

Another procedure for calculating image-object relationships involves the calculation of several special planes and points. Other definitions for power and focal length must also be established. We have assumed to this point that object and image distances and focal lengths are measured from the center of the lens. These measurements are not valid when the thickness of a system becomes significant. Therefore, new reference planes for distance measurements must be established so that the image formulas may be used.

Thick Lens Power

It is possible to establish the location of the focal points of a thick lens or lens system by using the definition of primary and secondary focal points. The primary focal point represents the object position that yields an image at infinity. The secondary focal point coincides with the image plane when an object is at infinity. These definitions remain the same, no matter how many lenses are contained in a system. Let's use these definitions to locate the primary and secondary focal points of a thick lens relative to these new reference planes.

Example 7-d

Using the element-by-element method, locate the primary and secondary focal points for a 6 cm thick glass lens (n =1.50) with surface powers of + 10.00 D.

Known	***Unknown***	***Equations/Concepts***
Front surface power: F_1 =+10.00D	Final image location	Element-by-element imaging
Back surface power: F_2 = +10.00D		(7-2), (7-3) Vergence at each surface
Distance between lenses: t = 6.00 cm = 0.06 m		(7-1) Imaging formula
		Definition of primary and secondary focal points

To determine the secondary focal point, locate the final image of an object positioned at infinity using the procedure described in *Examples 7a* and *7b*. The location of the secondary focal point of the thick lens coincides with the final image, as shown in Figure 7-7a.

$$L_1 = \lim_{\ell_1 \to \infty}\left(\frac{n_1}{\ell_1}\right) = 0.00 \text{ D}$$

$$L'_1 = L_1 + F_1 = 0.00 + 10.00 = +10.00 \text{ D}$$

$$\ell'_1 = \frac{n_2}{L_1'} = \frac{1.50}{+10.00} = +0.15 \text{ m} = +15.00 \text{ cm}$$

$$\ell_2 = \ell'_1 - t = 15.00 \text{ cm} - 6.00 \text{ cm} = +9.00 \text{ cm} = +0.09 \text{ m}$$

$$L_2 = \frac{n_2}{\ell_2} = \frac{1.50}{0.09 \text{ m}} = +16.67 \text{ D}$$

$$L'_2 = L_2 + F_2 = +16.67 \text{ D} + 10.00 \text{ D} = +26.67 \text{ D}$$

$$\ell'_2 = \frac{n_3}{L'_2} = \frac{1.00}{+26.67 \text{ D}} = +0.0375 \text{ m} = +3.75 \text{ cm}$$

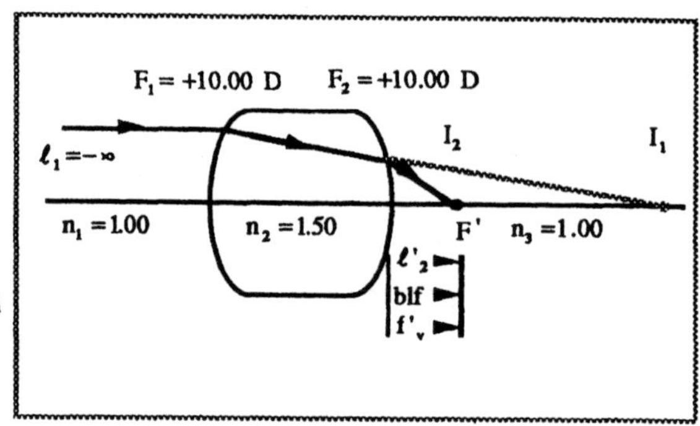

Figure 7-7a. *Example 7-d.* Secondary focal point location relative to the back surface of the lens.

This distance is relative to the back surface of the lens. The distance from the back surface of a lens or the last element of an optical system to the secondary focal point is called the **back focal length.** This distance is labeled **bfl** or $\mathbf{f'_v}$. Note that the secondary focal point is labeled **F '**, which follows our standard notation.

To locate the primary focal point, this problem must be worked backward (i.e., use the image position (+ ∞) to calculate the object position (which is the primary focal point). Use Equations 7-3, 7-1, and 7-2. Be sure to

$$L'_2 = \lim_{\ell'_2 \to \infty} \left(\frac{n_3}{\ell'_2} \right) = 0.00 \text{ D}$$

$$L_2 = L'_2 - F_2 = 0.00 - 10.00 = -10.00 \text{ D}$$

$$\ell_2 = \frac{n_2}{L_2} = \frac{1.50}{-10.00} = -0.15 \text{ m} = -15.00 \text{ cm}$$

$$\ell'_1 = \ell_2 - t = -15.00\text{cm} + 6.00 \text{ cm} = -9.00 \text{ cm} = -0.09 \text{ m}$$

$$L'_1 = \frac{n_2}{\ell'_1} = \frac{1.50}{-0.09 \text{ m}} = -16.67 \text{ D}$$

$$L_1 = L'_1 - F_1 = -16.67 \text{ D} - 10.00 \text{ D} = -26.67 \text{ D}$$

$$\ell_1 = \frac{n_3}{L'_1} = \frac{1.00}{-26.67 \text{ D}} = -0.0375 \text{ m} = -3.75 \text{ cm}$$

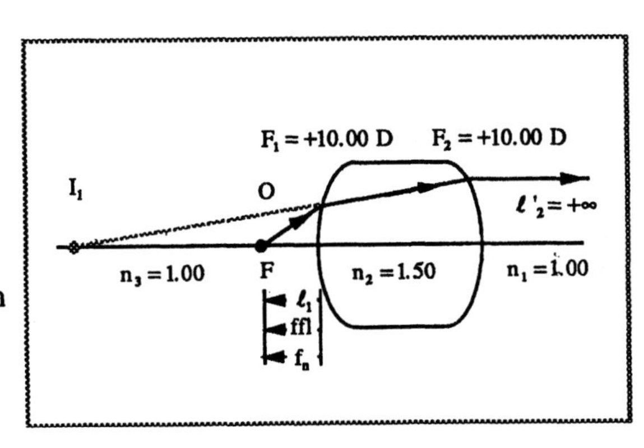

Figure 7-7b. *Example 7-d.* Primary focal point location relative to the front surface of the lens.

The object position that yields an image at infinity is the primary focal point of the system and is labeled **F**. The distance from the front surface of the lens (or the first element in a system) to the primary focal point is called the **front focal length** and is labeled **ffl** or $\mathbf{f_n}$.

Front focal length (ffl, f_n) - the distance from the front surface of a lens or the first element of an optical system to the primary focal point of the system.

Back focal length (bfl, f'_v) - the distance from the back surface of a lens or the last element of an optical system to the secondary focal point of the system.

Neutralizing power (F_n) - the incident vergence on the front surface of a lens or the first element of an optical system that yields a final image at infinity. The neutralizing power is related to the front focal length by

$$F_n = -\frac{n_1}{\text{ffl}} = -\frac{n_1}{f_n} \tag{7-10}$$

Back vertex power (F_v) - the emergent vergence from the back surface of a lens or the last element of an optical system for an object placed at an infinite position. The back vertex power is related to the back focal length by

$$F_v = \frac{n_3}{\text{bfl}} = \frac{n_3}{f'_v} \tag{7-11}$$

where: n_3 = the index after the last element or surface.

The back vertex power is the measure used in refraction and corrective lens prescriptions.

The back vertex power and the neutralizing power can also be calculated using the principle of effective power. The effective power of the front surface at the back surface plus the power of the back surface yields the back vertex power. The effective power of the back surface at the front plus the power of the front surface yields the neutralizing power.

$$F_v = \frac{F_1}{1 - \left(\frac{t}{n_2} \right) F_1} + F_2 \tag{7-12}$$

$$F_n = \frac{F_2}{1 - \left(\frac{t}{n_2} \right) F_2} + F_1 \tag{7-13}$$

Example 7-e

Using the data from *Example 7-d*, calculate the back vertex power and the neutralizing power using both the focal lengths and effective power formulas.

Known

Front surface power: $F_1 = +10.00$ D

Back surface power: $F_2 = +10.00$ D

Distance between lenses: $t = 6.00$ cm $= 0.06$ m

$n_1 = n_3 = 1.00$

$n_2 = 1.50$

bfl $= +3.75$ cm $= +0.0375$ m

ffl $= -3.75$ cm $= -0.0375$ m

Unknown

Back vertex power

Neutralizing power

Concepts/Equations

(7-10), (7-11) Front and back focal length method

(7-12), (7-13) Effective power method

Substitute the values into Equations 7-11 and 7-12 to calculate the back vertex power:

$$F_v = \frac{n_3}{\text{bfl}} = \frac{1.00}{+0.0375\text{ m}} = +26.67\text{ D}$$

$$F_v = \frac{F_1}{1-\left(\frac{t}{n_2}\right)F_1} + F_2 = \frac{+10.00\text{D}}{1-\left(\frac{0.06\text{ m}}{1.50}\right)(+10.00\text{ D})} + 10.00\text{D} = +26.67\text{ D}$$

Follow the same procedure with Equations 7-10 and 7-13 for the neutralizing power:

$$F_n = -\frac{n_1}{\text{ffl}} = -\frac{1.00}{-0.0375\text{ m}} = +26.67\text{ D}$$

$$F_n = \frac{F_2}{1-\left(\frac{t}{n_2}\right)F_2} + F_1 = \frac{+10.00\text{D}}{1-\left(\frac{0.06\text{m}}{1.50}\right)(+10.00\text{ D})} + 10.00\text{ D} = +26.67\text{ D}$$

Example 7-f

If the glass rod in *Example 7-a* were 35 cm long, What would be the neutralizing and back vertex powers? Locate the primary and secondary focal points relative to the front and back surfaces, respectively.

Known

Front surface power: $F_1 = +10.00$ D

Back surface power: $F_2 = +5.00$ D

Thickness of rod: $t = 35$ cm $= 0.35$ m

Index of rod: $n = 1.50$

Index of surround: $n = 1.00$

Unknown

Back vertex power: F_v

Neutralizing power: F_n

Front focal length: ffl

Back focal length: bfl

Equations/Concepts

(7-12), (7-13) Power formula

(7-10), (7-11) Focal lengths

Use the appropriate equations and solve for the powers (Equations 7-12 and 7-13):

$$F_v = \frac{F_1}{1-\frac{t}{n_2}F_1} + F_2 = \frac{+10.00\text{D}}{1-\left(\frac{0.35\text{m}}{1.50}\right)(+10.00\text{D})} + 5.00\text{D} = -2.50\text{D}$$

$$F_n = \frac{F_2}{1-\frac{t}{n_2}F_2} + F_1 = \frac{+5.00\text{D}}{1-\left(\frac{0.35\text{m}}{1.50}\right)(+5.00\text{D})} + 10.00\text{D} = -20.00\text{D}$$

The front and back focal lengths are calculated by solving Equations 7-10 and 7-11:

$$\text{ffl} = -\frac{n_1}{F_n} = -\frac{1.00}{-20.00\text{ D}} = +0.05\text{ m} = +5.00\text{ cm}$$

$$\text{bfl} = \frac{n_3}{F_v} = \frac{1.00}{-2.50\text{ D}} = -0.40\text{ m} = -40.00\text{ cm}$$

This means that the secondary focal point is located 40 cm to the left of the second surface, and the primary focal point is located 5 cm to the right of the first surface. Figure 7-8 shows the location of the primary and secondary focal points relative to the front and back surfaces, respectively. The positions of these focal points make sense if you go back to the definitions. Remember that the vergence leaving the back surface (with an object at infinity - parallel incident light) is the back vertex power. In this case, the vergence leaving is negative and therefore divergent. The secondary focal point corresponds to the image position for this divergent light. It should therefore be in front (to the left) of the second surface. You may wish to prove this by using the element-by-element method. The primary focal point is the object position that yields an image at infinity. In this case, there must be convergent light incident on the front surface (aimed at the primary focal point - a virtual object) to have the proper (– 5.00 D) incident light upon the second surface to result in zero vergence (image at infinity) leaving the second surface.

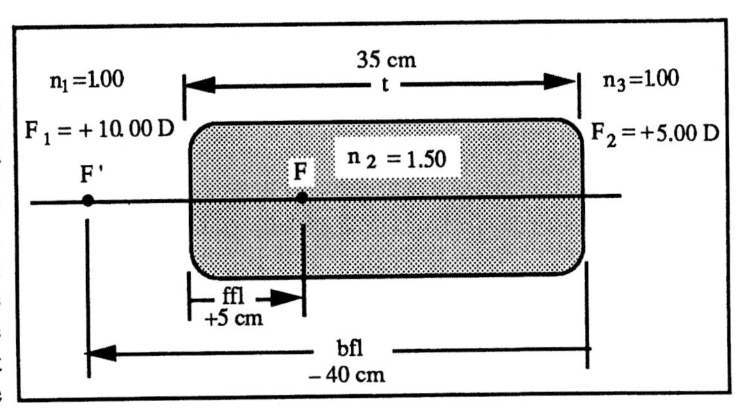

Figure 7-8. *Example 7-f.* Front and back focal lengths.

Principal Planes (H ; H') - conjugate planes in an optical system or lens in which the lateral magnification is + 1. The first principal plane is labeled **H**, and the second principal plane is labeled **H'**. The principal planes are the reference planes in which distances are measured. In establishing our new reference system, the entire optical system is reduced to these two planes. All incident rays strike H, and all emerging rays refract from H'. Rays leaving the second principal plane follow the same path as an emerging ray would refract through the system element-by-element. This is shown in Figure 7-9 (left). The secondary focal length of the system is measured from the second principal plane.

The first principal plane is located using the parallel emergent ray, as shown in Figure 7-9 (right). Incident rays follow the same path as that found using the element-by-element procedure. The primary focal length of the system is measured from the first principal plane. The principal plane location from the front (A_1) or back (A_2) of the

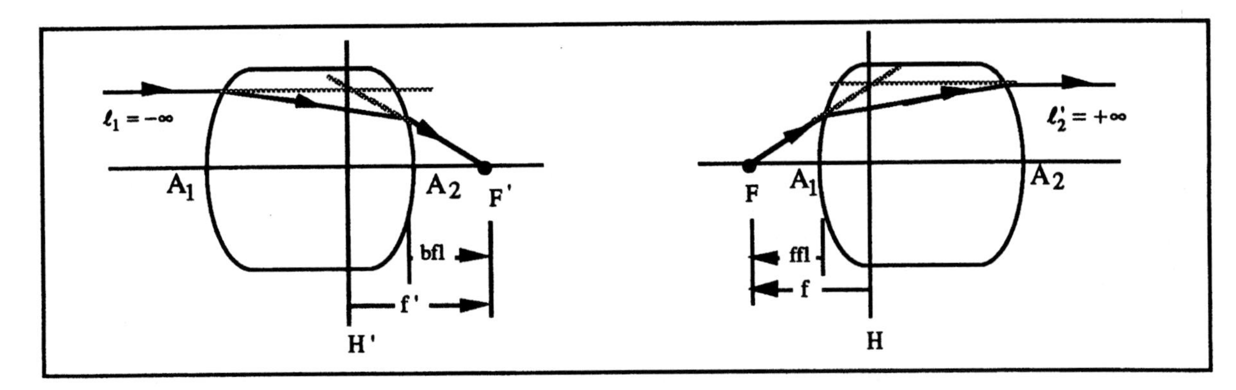

Figure 7-9. Location of the principal planes using *Example 7-d. Left:* The incident parallel ray is extended forward and the emergent ray is extended backward. The intersection of these extensions represents the second principal plane H'. *Right:* The emergent parallel ray is extended backward, and the incident ray (through F) is extended forward. The intersection of these extensions represents the first principal plane. Note that the primary (f) and secondary (f') focal lengths are measured from the first and second principal plane, respectively.

lens is calculated using:

$$A_1H = n_1\left(\frac{t}{n_2}\right)\left(\frac{F_2}{F}\right) \qquad (7\text{-}14) \qquad A_2H' = -n_3\left(\frac{t}{n_2}\right)\left(\frac{F_1}{F}\right) \qquad (7\text{-}15)$$

where F is the true or thick lens power and is given by

$$F = F_1 + F_2 - \left(\frac{t}{n_2}\right)F_1F_2 \qquad (7\text{-}16)$$

The thick lens power is sometimes referred to as the equivalent power. This power is related to the focal lengths of the system by

$$F = -\frac{n_1}{f} = \frac{n_3}{f'} \qquad (7\text{-}17)$$

This is similar to the relationship between the power and focal lengths for single refracting surfaces and thin lenses, however, the focal lengths for a thick lens, are measured from the respective principal planes. (See *Example 7-g*.) If the surrounding indices are the same, the primary and secondary focal lengths have equal magnitude (opposite sign).

Bending a Lens

As stated in Chapter 5, a lens may have many different forms (bi-convex, meniscus, etc.) with the same equivalent power. Changing the form of the lens does not change the separation between the two planes, but it does change the location of the principal planes. (See Figure 7-10.) In the concave or convex form of a lens, the principal plane usually falls within the lens. Changing from this form toward a meniscus form, shifts the principal planes toward the more curved surface.

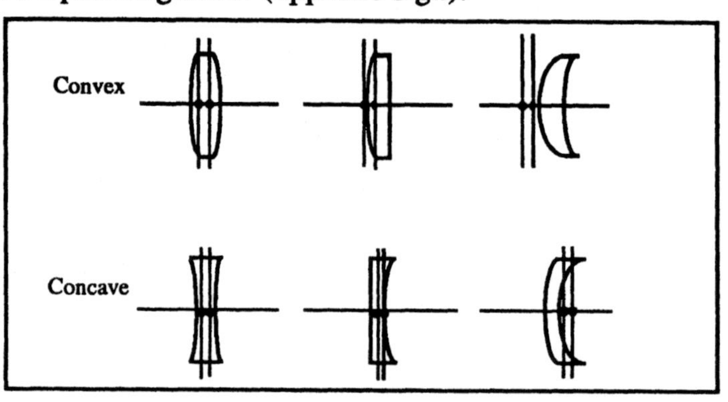

Figure 7-10. The effect of lens form has on principal plane location. (Modified from Fry, G. *Geometrical Optics*, 1969.)

The Effect of Thickness on Power

As seen in Equation 7-16, the thickness of the lens can significantly change the equivalent power. When the lens has zero thickness (as is approximated by thin lenses), the power is simply the sum of the two surface powers. If both surface powers are negative or positive, increasing the thickness will cause the net equivalent power to become more negative (less positive). This is demonstrated in Figure 7-11 for a biconvex lens. If one surface power is negative and one positive, the net effect of adding thickness is to make the lens more negative.

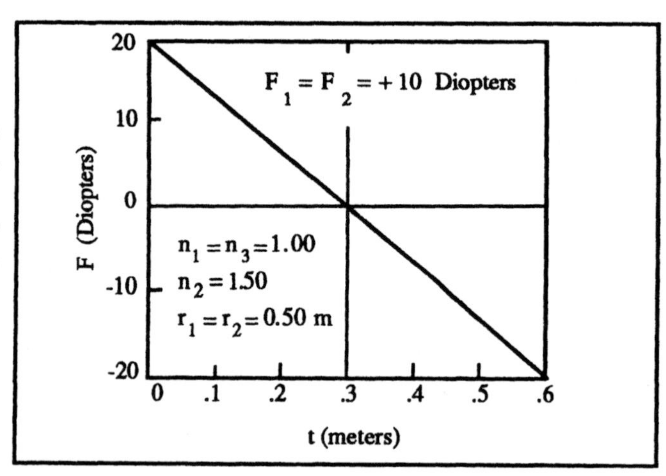

Figure 7-11. The effect of thickness on thick lens power. (Modified from Fry, G. *Geometrical Optics*, 1969.)

The Effect of Thickness on the Principal Planes

Figure 7-12 shows the changes in the principal plane locations as a function of thickness for a biconvex lens. For a thin lens, the principal planes coincide and fall at the center of the lens. As the thickness increases, the principal planes separate with H falling to the left of H', and both falling within the lens. Increasing the thickness further causes the planes to coincide and reverse order (i.e., H' to the left of H). The order of the principal planes does not depend on whether the lens has positive or negative power. When the secondary focal point of the first surface falls at the center of curvature of the second surface, the principal planes fall at the surfaces, and when the secondary focal point of the first surface falls at the primary focal point of the second surface, the principal planes fall at + ∞ and – ∞. In this case, the power is zero. As the thickness further increases, the primary and secondary focal points change places at + ∞ and – ∞ and gradually move toward the lens.

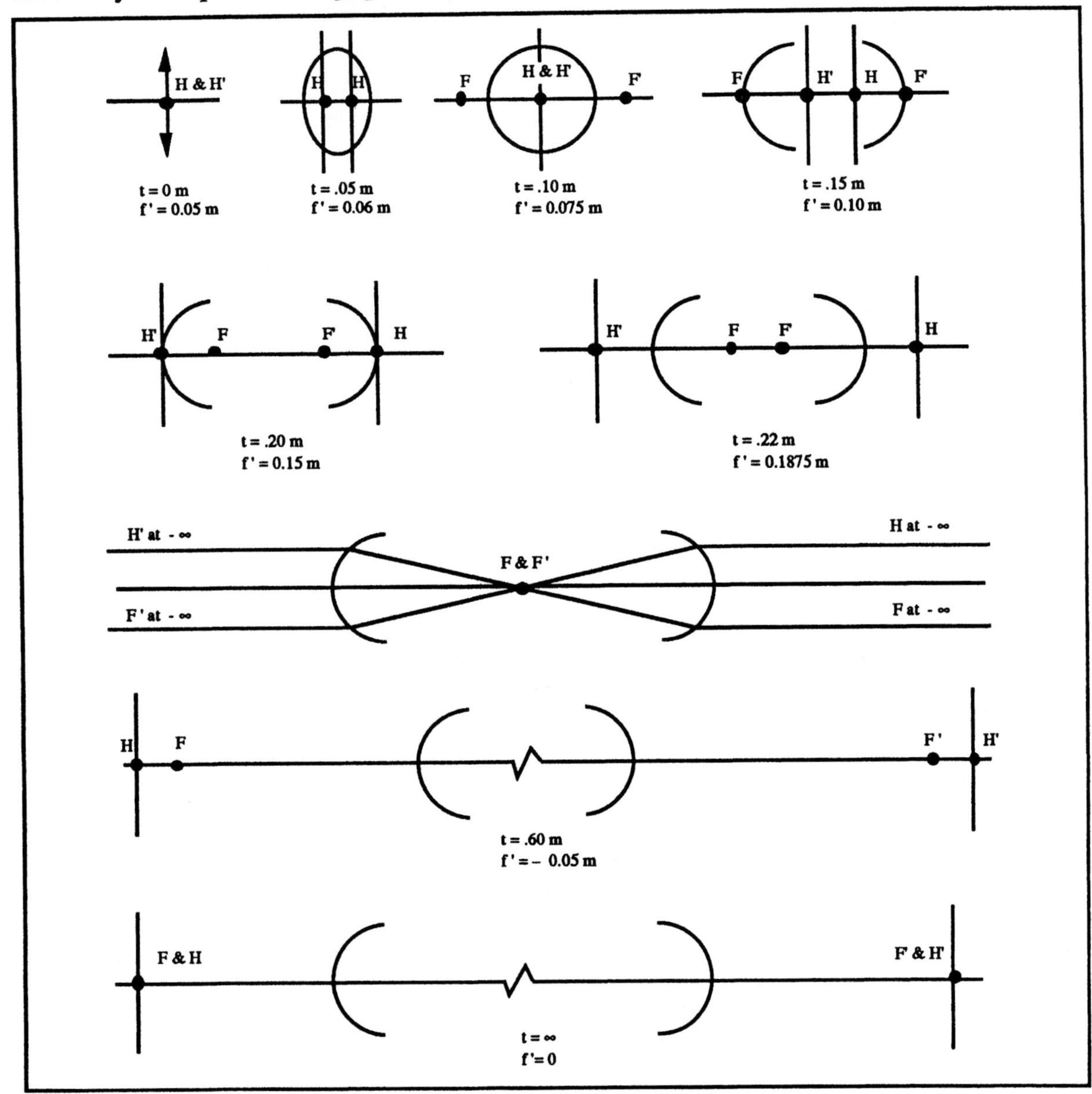

Figure 7-12. The effect of thickness on the location of the principal planes. (Modified from Fry, G. *Geometrical Optics*, 1969.)

Example 7-g

A thick lens has a front surface power of + 7.00 D, a back surface power of + 13.00 D, a center thickness of 10 cm, and an index of refraction of 1.56. Calculate the front, back, and equivalent powers of the lens. Locate the principal planes and focal points. Diagram the system (to scale), and label f, f ', bfl, ffl, H, and H'.

Known	***Unknown***	***Equations/Concepts***
Front surface power: $F = +7.00$ D	Back vertex power: F_v	(7-12), (7-13) Power formula
Back surface power: $F = +13.00$ D	Neutralizing power: F_n	(7-10), (7-11) Focal lengths
Thickness of rod: $t = 10$ cm $= 0.10$ m	Front focal length: ffl	(7-14), (7-15) Principal planes
Index of rod: $n_2 = 1.56$	Back focal length: bfl	(7-16) Equivlent power
Index of surround: n=1.00	Equivalent power: F	(7-17) Focal points
	Primary focal length: f	
	Secondary focal length: f '	

Start by solving the back vertex, neutralizing, and equivalent powers using Equations 7-12,7-13, and 7-16:

$$F_v = \frac{F_1}{1-\left(\frac{t}{n_2}\right)F_1} + F_2 = \frac{+7.00\text{ D}}{1-\left(\frac{0.10}{1.56}\right)(+7.00\text{ D})} + 13.00\text{ D} = +25.70\text{ D}$$

$$F_n = \frac{F_2}{1-\left(\frac{t}{n_2}\right)F_2} + F_1 = \frac{+13.00\text{ D}}{1-\left(\frac{0.10}{1.56}\right)(+13.00\text{ D})} + 7.00\text{ D} = +85.00\text{ D}$$

$$F = F_1 + F_2 - \left(\frac{t}{n_2}\right)F_1 F_2 = +7.00\text{ D} + 13.00\text{ D} - \left(\frac{0.10}{1.56}\right)(+7.00\text{ D})(+13.00\text{ D}) = +14.17\text{ D}$$

Next solve for the location of the principal planes using Equations 7-14 and 7-15:

$$A_1H = n_1\left(\frac{t}{n_2}\right)\frac{F_2}{F} = (1.00)\left(\frac{0.10}{1.56}\right)\left(\frac{+13.00\text{ D}}{+14.17\text{ D}}\right) = +0.059\text{ m} = +5.90\text{ cm}$$

$$A_2H' = -n_3\left(\frac{t}{n_2}\right)\frac{F_1}{F} = -(1.00)\left(\frac{0.10}{1.56}\right)\left(\frac{+7.00\text{ D}}{+14.17\text{ D}}\right) = -0.032\text{ m} = -3.20\text{ cm}$$

Calculate the focal lengths using Equations 7-10, 7-11 and 7-17:

$$\text{ffl} = -\frac{n_1}{F_n} = -\frac{1.00}{+85.00\text{ D}} = -0.0118\text{ m} = -1.18\text{ cm}$$

$$\text{bfl} = \frac{n_3}{F_v} = \frac{1.00}{+25.70\text{ D}} = +0.0389\text{ m} = +3.89\text{ cm}$$

$$f' = \frac{n_3}{F} = \frac{1.00}{+14.17\text{ D}} = +0.0705\text{ m} = +7.05\text{ cm}$$

$$f = -\frac{n_1}{F} = -\frac{1.00}{+14.17\text{ D}} = -0.0705\text{ m} = -7.05\text{ cm}$$

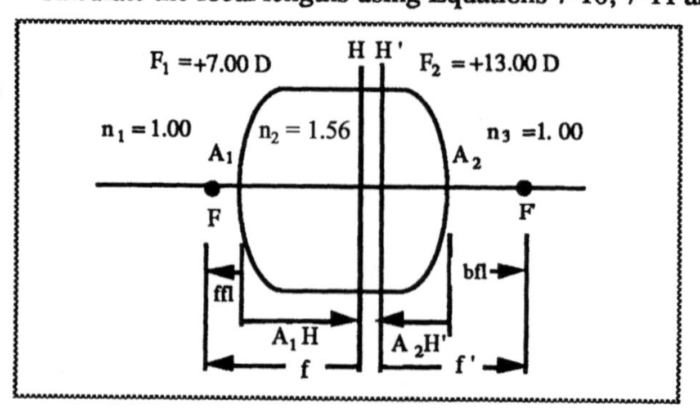

Figure 7-13. *Example 7-g.*

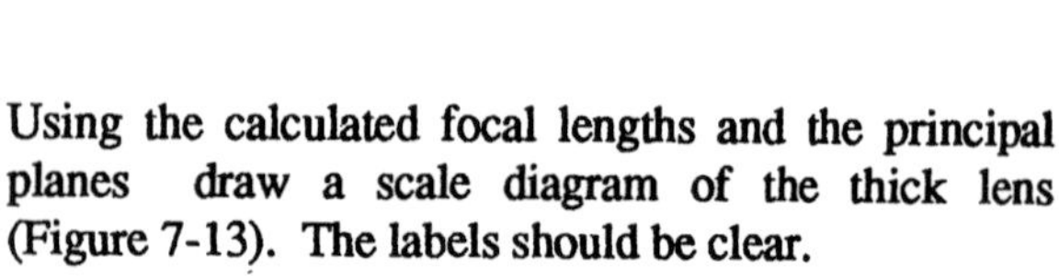

Using the calculated focal lengths and the principal planes draw a scale diagram of the thick lens (Figure 7-13). The labels should be clear.

Optical Center and Nodal Points

There is one more set of points that is used in defining a thick lens system. These points are called the **nodal points** and are labeled **N** and **N'**. An incident ray is aimed toward the first nodal point and, after refraction, leaves the second nodal point undeviated (i.e., with the same angle to the optical axis). The position where this undeviated ray actually crosses the optical axis is called the **optical center** and is labeled **O** in Figure 7-14. In reality, the nodal points actually represent the apparent position of the optical center as viewed from the front and back of the lens.

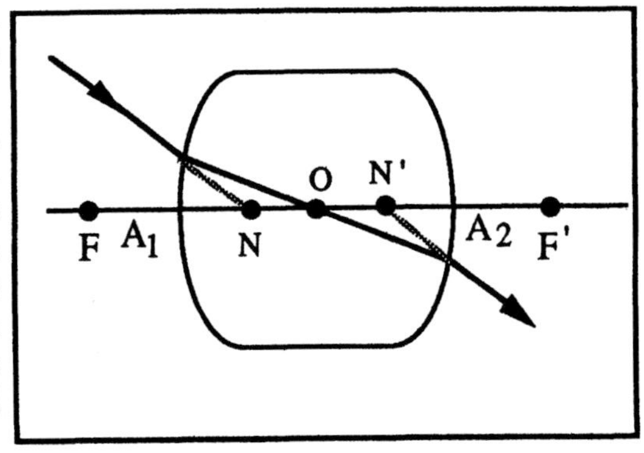

Figure 7-14. Nodal points and optical center.

The optical center may be calculated using the radii and thickness (t) of the thick lens (or the separation between two lenses) with this equation:

$$A_1O = \frac{r_1 t}{r_1 - r_2} \qquad (7\text{-}18)$$

where:

r_1 = the front surface radius

r_2 = the back surface radius

t = the center thickness

A_1O = the distance from the front apex to the optical center

Notice that the optical center is independent of the index of refraction. This is the only point in the optical system that is wavelength independent.

The nodal points coincide with the intersection of the principal planes with the optical axis when the media in front of and behind the lens or system are the same (i.e., the surrounding medium is the same). If the surrounding media are not equal then the nodal points are shifted away from the principal planes. These relationships that are useful in locating the nodal points:

$$HN = H'N' = f + f' \qquad (7\text{-}19)$$

$$FN = f' \qquad F'N' = f \qquad (7\text{-}20)$$

Equation 7-19 shows that the distance from the first principal plane to the first nodal point is equal to the distance from the second principal plane to the second nodal point and equal to the sum of the primary and secondary focal lengths. The sign indicates the direction that the measurement is made. A negative value is measured to the left, and a positive value is measured to the right. Note that the distance from the principal plane to the respective nodal point has the same value. If the focal lengths have equal magnitude (opposite sign), the sum is equal to zero and the nodal points coincide with the principal planes. This is the case when the same medium is on either side of the lens. If the focal lengths are not equal (media are not equal) the points are shifted, as indicated by the equation.

Equation 7-19 also indicates that the distance between the principal planes is equal to the distance between the nodal points:

$$HH' = NN' \qquad (7\text{-}21)$$

Be sure to review ray tracing through thick lenses and lens systems in Chapter 9. It will give you more insight into how the principal planes, nodal points, and focal points are used in imaging through thick lens systems.

Example 7-h

Calculate the location of the optical center for a plano convex lens with a front radius of 10 cm, an equiconvex lens with radii of + 10 cm and – 10 cm, respectively, and a positive meniscus lens with radii of + 10 cm. Assume the thickness of the lens to be 5 cm.

Known	***Unknown***	***Equations/Concepts***
Surface radii - plano convex: $r_1 = +10$ cm; $r_2 = 0$	Location of optical center	(7-18)
Surface radii - equiconvex: $r_1 = +10$ cm; $r_2 = -10$ cm		
Surface radii - meniscus: $r_1 = +10$ cm; $r_2 = +10$ cm		

Substitute the values into Equation 7-18 to calculate the optical center:

plano convex: $$A_1O = \frac{r_1 t}{r_1 - r_2} = \frac{(+10\text{ cm})(5\text{ cm})}{+10\text{ cm} - 0\text{ cm}} = +5\text{ cm} \quad \text{at the plano surface}$$

equiconvex $$A_1O = \frac{r_1 t}{r_1 - r_2} = \frac{(+10\text{ cm})(5\text{ cm})}{+10\text{ cm} - (-10\text{ cm})} = +2.5\text{ cm} \quad \text{middle of lens}$$

meniscus $$A_1O = \frac{r_1 t}{r_1 - r_2} = \frac{(+10\text{ cm})(5\text{ cm})}{+10\text{ cm} - 10\text{ cm}} \to \infty \quad \text{at infinity}$$

From the solutions it can be shown that for a plano lens, the optical center is at the flat surface. This is true no matter what the orientation of the lens. Let's show this by switching the two radii ($r_2 = -10$ cm; $r_1 = 0$) and recalculating:

plano convex: $$A_1O = \frac{r_1 t}{r_1 - r_2} = \frac{(0\text{cm})(5\text{ cm})}{0\text{ cm} - (-10\text{ cm})} = 0\text{ cm} \quad \text{at the plano surface}$$

For the equiconvex lens, the optical center is in the middle of the lens; for other radii, it will be inside of the lens. The meniscus lens has the optical centers outside of the lens; in this case, at infinite.

Example 7-i

Assume that the crystalline lens of the human eye is uniform throughout and have an index of 1.45. It is suspended between aqueous fluid in front (n = 1.33) and vitreous material behind (n = 1.33). If the thickness of the lens is 3.6 mm, and the front and back radii of the lens are 6.0 mm and 10.0 mm, respectively, calculate the following:

a. The primary and secondary focal lengths
b. The power of the lens
c. The location of the principal planes
d. The location of the nodal points
e. Diagram and label the front and back focal lengths.

Known	***Unknown***	***Equations/Concepts***
Front surface radius: $r = +6.0$ mm	f and f '	(7-4) Surface power
Back surface: radius $r = -10.0$ mm	Nodal point location	(7-14), (7-15) Principal planes
Thickness of lens: $t = 3.6$ mm $= 0.0036$ m	Front focal length: ffl	(7-16) Equivalent power
Index of lens: $n = 1.45$	Back focal length: bfl	(7-17) Focal lengths
Index of surround: $n = 1.33$	Equivalent power: F	(7-19) Nodal points

Solve for the front and back surface powers (Equation 7-14), the equivalent power (Equation 7-16), and the focal lengths (Equations 7-17)

$$F_1 = \frac{n_L - n_s}{r_1} = \frac{1.45 - 1.33}{0.006\text{ m}} = +20.00\text{ D} \qquad F_2 = \frac{n_s - n_L}{r_2} = \frac{1.33 - 1.45}{-0.010\text{ m}} = +12.00\text{ D}$$

$$F = F_1 + F_2 - \left(\frac{t}{n_L}\right)F_1F_2 = +20.00\text{ D} + 12.00\text{ D} - \left(\frac{0.0036}{1.45}\right)(+20.00\text{ D})(+12.00\text{ D}) = +31.40\text{ D}$$

$$f = -\frac{n_s}{F} = -\frac{1.33}{+31.40\text{ D}} = -0.0424\text{ m} = -4.24\text{ cm} \qquad f' = \frac{n_s}{F} = \frac{1.33}{+31.40\text{ D}} = +0.0424\text{ m} = +4.24\text{ cm}$$

The principal planes are located using the Equations 7-14 and 7-15:

$$A_1H = n_s\left(\frac{t}{n_L}\right)\left(\frac{F_2}{F}\right) = (1.33)\left(\frac{0.0036\text{ m}}{1.45}\right)\left(\frac{+12.00\text{ D}}{+31.40\text{ D}}\right) = +0.00126\text{ m} = +1.26\text{ mm}$$

$$A_2H' = -n_s\left(\frac{t}{n_L}\right)\left(\frac{F_1}{F}\right) = -(1.33)\left(\frac{0.00336\text{ m}}{1.45}\right)\left(\frac{+20.00\text{ D}}{+31.40\text{ D}}\right) = -0.00210\text{ m} = +2.10\text{ mm}$$

The nodal points coincide with the principal planes because the surround index is the same (n = 1.33). As a check, add the focal lengths and the result is zero.

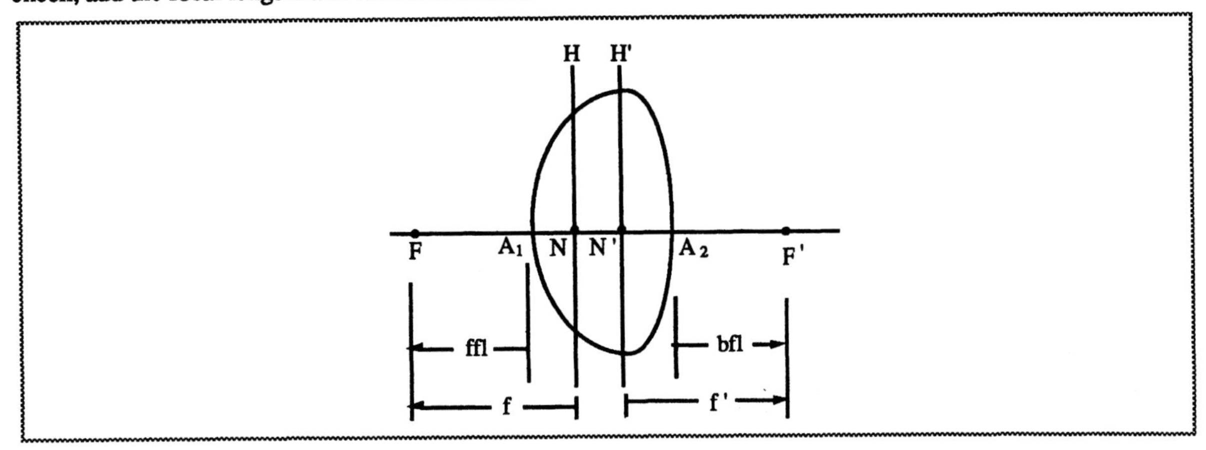

Figure 7-14. *Example 7-h.*

Cardinal Points Method: Image-Object Location, Size, and Orientation

Once the principal planes have been located, standard imaging formulas can be used to determine the image-object location, size, and orientation relationships if measurements are made relative to the principal planes. Object distances are measured from the first principal plane, and image distances are measured from the second principal plane. The system is reduced to the principal planes and the equivalent power of the lens. All incident rays strike the first principal plane; all emergent rays leave as though they were coming from the second principal plane.

The lateral magnification formulas also hold for thick lenses. The resulting sign yields similar information about the orientation of the final image. In summary, once the principal planes are located, treat the lens as a thin lens with the same relationships.

Example 7-j

A lens system consists of two thin lenses with powers of + 5.00 D and a + 8.00 D. If a real object 10 cm high is located 4 m from the first principal plane of the system, a real image is formed 10 cm from the second principal plane of the system. What is the refracting power of the system? What is the separation between the lenses? What is the size of the image?

This problem can be solved directly. First, use Equations 7-2, 7-3, and the Gaussian Imaging Equation to determine the power of the lens. Remember that the object and image distances are measured from the principal planes and the sign convention still holds.

$$L = \frac{n}{\ell} = \frac{1.00}{-4\text{ m}} = -0.25\text{ D} \qquad L' = \frac{n'}{\ell'} = \frac{1.00}{+0.10\text{ m}} = +10.00\text{ D}$$

$$F = L' - L = +10.00\text{ D} - (-0.25\text{ D}) = +10.25\text{ D}$$

The separation between the lenses may be calculated with the equivalent lens power formula (Equation 7-16):

$$F = F_1 + F_2 - \left(\frac{t}{n_2}\right)F_1F_2$$

$$t = \frac{n_2(F_1 + F_2 - F)}{F_1F_2} = \frac{1.00(+5.00\text{ D} + 8.00\text{ D} - 10.25\text{ D})}{(+5.00\text{ D})(+8.00\text{ D})} = 0.0688\text{ m} = 6.88\text{ cm}$$

The lateral magnification is calculated using the object and image distances relative to the principal planes (Equation 7-17):

$$LM = \frac{h'}{h} = \frac{\ell'}{\ell} \qquad LM = \frac{\ell'}{\ell} = \frac{0.10\text{ m}}{-4.0\text{ m}} = -0.025$$

$$h' = h(LM) = (10\text{ cm})(-0.025) = -0.25\text{ cm}$$

The image is inverted and minified.

Example 7-k

A real object is located 19.1 cm from the front of the lens in *Example 7-g*. Where is the image located relative to the back of the lens?

See *Example 7-g* (page 142) for the cardinal points. The first principal plane is +5.90 cm from the front of the lens and therefore the object distance from the principal plane is – 25 cm (19.1 cm + 5.90 cm = 25 cm). See Figure 7-13. Distances are measured from the principal plane to the object (or image) and our sign convention holds. The object distance is negative since the object is to the left of the principal plane and in object space. Use the Gaussian Imaging Equation to calculate the image position (relative to the second principal plane). The equivalent power (+ 14.17 D) from *Example 7-g* is used in the equation.

$$L = \frac{n}{\ell} = \frac{1.00}{-0.25\text{ m}} = -4.00\text{ D} \qquad L' = F + L = +14.17\text{ D} + (-4.00\text{ D}) = +10.17\text{ D}$$

$$\ell' = \frac{n'}{L'} = \frac{1.00}{+10.17\text{ m}} = +0.0983\text{ m} = +9.83\text{ cm from the second principal plane}$$

The second principal plane is – 3.20 cm from the back surface of the lens so the image relative to the back of the lens is + 6.63 cm (+ 9.83 cm – 3.20 cm = + 6.63 cm). The image is real and inverted.

Note that only the principal planes and equivalent power are required to solve for the image-object positions formed by a thick lens or lens system. See Chapter 9 for more details on thick lens problems and ray tracing.

Supplemental Problems

Neutralizing, Back Vertex, and Equivalent Powers; Focal Length, Thickness

7-1. A thick plastic lens (n = 1.49) has a front surface power of + 2.00 D and a back surface power of – 7.00 D. If the equivalent power is – 4.50 D, what is the thickness of the lens?
ANS. 5.32 cm

7-2. What is the back vertex power (Fv) of a lens system consisting of two thin lenses of powers + 8.00 D and – 1.00 D, separated by 4 cm?
ANS. + 10.76 D

7-3. What is the equivalent power of a glass lens (n = 1.5) that has a front surface power of + 10.00 D and a back surface power of – 5.00 D and is 6 mm thick?
ANS. + 5.20 D

7-4. The back focal length of a lens under water is – 40.00 cm. What is the back vertex power of the lens under water?
ANS. – 3.33 D

Principal Planes, Nodal Points, Focal Points

7-5. A thick biconvex lens with front radius of 8 cm and a back radius of 4.31 cm has a back surface power of + 13.00 D and a back vertex power of + 25.70 D. Given that HH' = + 0.90 cm, calculate the following (Answers in parentheses):
a. The index of refraction of the lens (n = 1.56)
b. The power of the front surface (F_1 = + 7.00 D)
c. The thickness of the lens (t = 10 cm)
d. The neutralizing power (F_n) (F_n = + 85.00 D)
e. The total power of the lens (F = + 14.17 D)
f. The primary and secondary focal lengths (f = – 7.06 cm, f' = + 7.06 cm)
g. The front and back focal lengths (ffl = – 1.18 cm, bfl = + 3.89 cm)
h. Locate the first principal plane relative to the front surface (A_1H = + 5.88 cm)
i. Locate the second principal plane relative to the front surface (A_1H' = + 6.83 cm)
j. Locate the nodal points (the nodal points coincide with the principal planes)

7-6. A thick lens system in air has a back vertex power of + 10.91 D, a neutralizing power of + 4.47 D and a secondary focal length of + 19.72 cm. The lens is 5 cm thick and has an index of 1.50. Calculate the following (answers in parentheses):
a. The front and back focal lengths (ffl = – 22.37 cm; bfl = + 9.17 cm)
b. The equivalent power (F = + 5.07 D)
c. Locate the first principal plane relative to the front surface (A_1H = – 2.65 cm)
d. Locate the second principal plane relative to the back surface (A_2H' = – 10.56 cm)
e. Locate the nodal points (N = – 2.65 cm from A_1; N' = – 10.56 cm from A_2)

7-7. A lens system has a back vertex power of + 5.00 D, a neutralizing power of + 4.00 D, and an equivalent power of + 10.00 D. The back surface is in water, and the front surface is in air. Find the following (answers in parentheses):
a. The first principal plane relative to the front surface (A_1H = – 15 cm)
b. The second nodal point located relative to the back surface (H'N' = + 3.3 cm)
c. The secondary focal length of the system (f' = + 13.3 cm)

Image-Object Relationships, Magnification

7-8. If a real object were placed 94.1 cm in front of the lens (to the left) in Problem 7-5, where would the image be formed relative to the back surface?
ANS. +4.42 cm from the back surface

7-9. Locate the object position if an image is formed 4.44 cm behind the lens in Problem 7-3.
ANS. The object is 62.6 cm behind the principal plane

7-10. Use the element-by-element method to determine the image location for the following thick lens system: the front surface power is + 10.00 D, the back surface power is + 5.00 D, the index of refraction and the thickness of the lens are 1.50 and 5 cm, respectively, and the object is placed 50 cm from the front surface.
ANS. The image is formed 6.28 cm behind the back surface

7-11. A real object is placed 40 cm from the first principal plane of a lens system. If the back vertex power is + 2.00 D, the neutralizing power is + 3.00 D, and the equivalent power is + 6.00 D, where is the image located relative to the second principal plane?
ANS. + 29 cm

Chapter 8

Reflection at Plane and Curved Surfaces

Reflecting surfaces follow the rules that have been developed for refracting surfaces with minor modifications. One major difference is that rays, after interacting at the reflecting surface, travel in the opposite direction. Therefore, all rays, both incident and reflected, travel in the same medium in front (to the left) of the interface. The space to the left of the interface becomes **real space** for both objects and images. Rays will never actually enter the space to the right of the interface; they either converge toward or diverge as though they came from this space. This region is labeled **virtual space** for both objects and images. (See Figure 8-1.) Using these definitions, the type, orientation, and position of an image may be determined using refracting surface procedures.

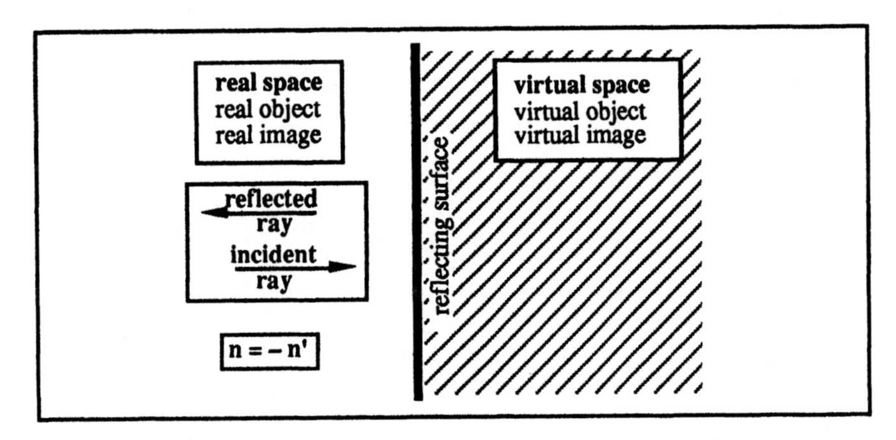

Figure 8-1. Real space (left of) and virtual space (right of) a reflecting surface. The incident and reflected rays travel in the same medium and index in real space.

Plane Mirrors: Image-Object Relationship

The Law of Reflection states that the incident and reflected angles are equal (Chapter 2). Using this law, the image-object relationship may be formulated. In Figure 8-2, two diverging rays from an object are incident on a plane mirror. The incident angles at the mirror (measured from the normal to the surface) and the reflected angles are labeled. The rays leaving the surface diverge as though they came from a position to the right of the mirror. Therefore the image of a real object reflected from a plane mirror appears to come from a position behind the mirror. The image formed is virtual, erect, and equal to the object in both size and shape. The image distance is equal in magnitude to the object distance but opposite in sign ($\ell' = -\ell$). This means that if you stand 1 meter in front of a plane mirror, you will see your image 1 meter behind the mirror.

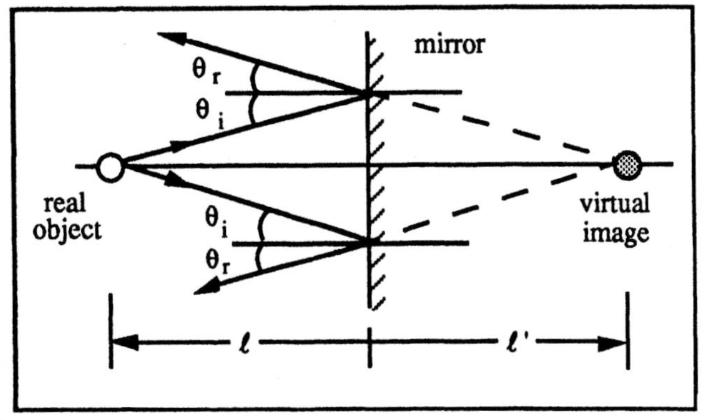

Figure 8-2. Object-image relationship with a plane mirror.

Your image (although reversed left to right) will have the same size and orientation as the object (yourself). Here is a classic example of this relationship.

Example 8-a

A man 162.55 cm (≈ 5'4") tall can just see a total image of himself in a vertical plane mirror 274.32 cm (≈ 9') away. His eyes are 152.40 cm (≈ 5') from the floor. Determine the height and position of the mirror.

Known	***Unknown***	***Equations/Concepts***
Height of man: h = 162.55 cm	Height of mirror: a = ?	Image distance equals object distance
Object distance: $\ell = -274.32$ cm	Position of mirror from floor: b = ?	Image virtual and behind mirror
Distance from floor to eyes: e = 152.40 cm		

Draw a diagram of the object and the image and label the distances (Figure 8-3). First draw the mirror with the object to the left, then draw the image the same distance to the right of the mirror. The eyes must see the head and feet of the image. Draw a line from the top of the head of the image toward the eyes. This is not the actual

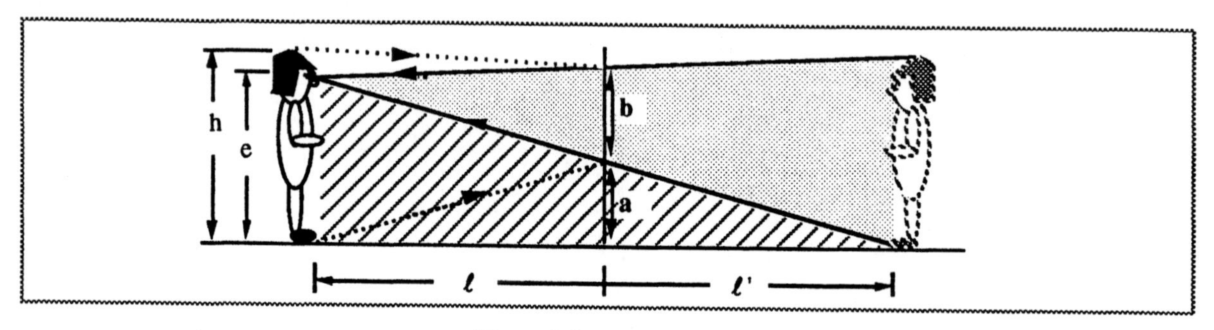

Figure 8-3. *Example 8-a.*

path of the rays. Draw a ray from the intersection of this line with the mirror to the top of the head of the object. This is the path of the ray, as indicated by the arrows. Draw a line from the feet of the image to the eyes of the object, and then construct the ray in the same manner as described for the other ray.

In the diagram, h = the height of the man or 162.55 cm, ℓ and ℓ' are the object and image distances (± 274.32 cm), e is the distance from the floor to the eyes (152.40 cm), a is the distance from the floor to the bottom of the mirror, and b is the size of the mirror. There are two large triangles in the diagram. The large shaded triangle is made up of two similar triangles with the following relationship:

$$\frac{h}{(\ell'-\ell)}=\frac{b}{\ell} \quad \text{because } \ell'=-\ell, \text{ this becomes} \quad \frac{h}{(-2\ell)}=\frac{b}{\ell} \quad \text{or} \quad b=\frac{h}{-2}$$

This equation shows that the size of the mirror required to see the entire image of the man is half of his height, independent of the distance from the mirror. The sign in front of the 2 is not needed. For our problem, the size of the mirror becomes:

$$b=\frac{h}{2}=\frac{162.55}{2}=81.275 \text{ cm}$$

The striped similar triangles are used to determine the distance of the mirror is from the floor:

$$\frac{e}{(\ell'-\ell)}=\frac{a}{\ell'} \quad \text{because } \ell=-\ell', \text{ this becomes} \quad \frac{e}{(2\ell')}=\frac{a}{\ell'} \quad \text{or} \quad a=\frac{e}{2}$$

The distance that the mirror must be from the floor is equal to half of the distance from the floor to the eyes, or

$$a=\frac{e}{2}=\frac{152.40}{2}=76.20 \text{ cm}$$

Example 8-b
Due to the cost, you must build a 12' long examination room instead of the conventional 20' long room. The patient will sit 10' from the front wall, with a floor-to eye-distance of 4.5'. The bottom of a 2' x 2' eye chart will be placed 6' above the floor on the wall behind the patient. The patient will view the eye chart reflecting off a front surface mirror positioned on the front wall. What is the actual testing distance? How large must the mirror be so that the patient can see the entire chart? At what distance from the floor must the mirror be placed?

Known	***Unknown***	***Equations/Concepts***
Distance chart to mirror: $\ell = -12'$	Size of mirror	Image virtual and behind mirror
Distance from floor to eyes: e = 4.5'	Position of mirror from floor	Image distance equal object distance
Patient to mirror: d = 10'	Actual testing distance	
Distance floor to bottom of chart: d = 6'		
Size of chart: s = 2'		

Draw a diagram, and label with all the information (Figure 8-4).

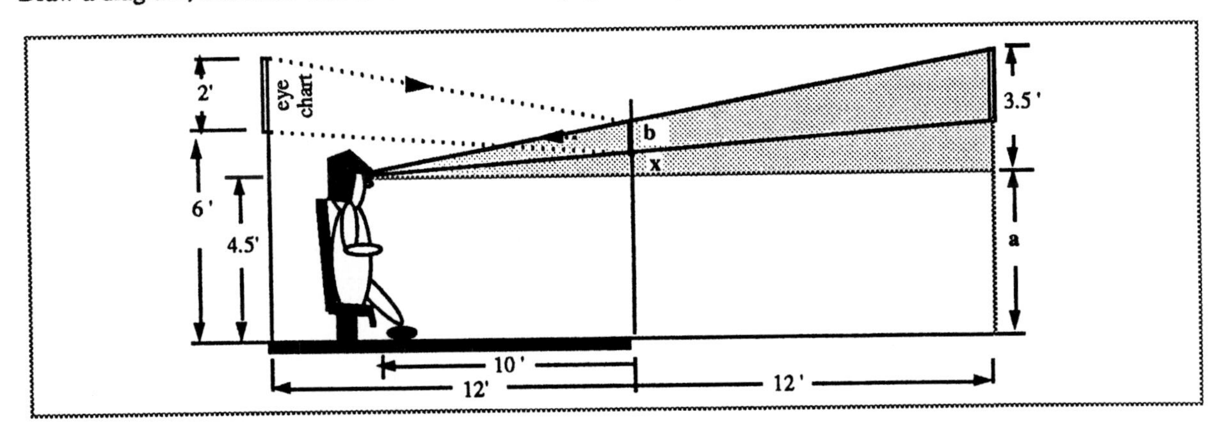

Figure 8-4. *Example 8-b.* Diagram of room and eye chart with distances labeled.

From the diagram, it should be clear that the patient is sitting 10' from the mirror and the image of the eye chart is 12' behind the mirror. The total testing distance is 22' (10' + 12' = 22'). Using the distances labeled in the diagram, determine the similar triangles that are required to solve the problem. The darkly outlined triangle, which has as its base the image of the eye chart (2') and the eyes at the apex, contains the size of the mirror b. The ratios used to find the size of the mirror needed to see the entire chart are:

$$\frac{2'}{22'} = \frac{b}{10'} \qquad b = \frac{20'}{22'} = 0.91' = 10.9''$$

The distance from the mirror to the floor is found by using shaded triangle. The horizontal side of the triangle is the floor-to-eye distance. The component x is the distance from this horizontal line to the mirror. Using the similar triangles, the ratios are

$$\frac{3.5'}{22'} = \frac{b+x}{10'} = \frac{0.91'+x}{10} \qquad 22'(0.91'+x) = 35' \qquad x = 0.68' = 8.17''$$

The total distance from the floor to the bottom of the mirror is 5.18' = 5' 2" (4.5' + 0.68' = 5.18'). It is unlikely that you would buy a mirror this small, but this problem could help in determining the minimum mirror size required.

Properties of Plane Reflecting Surfaces: Mirrors

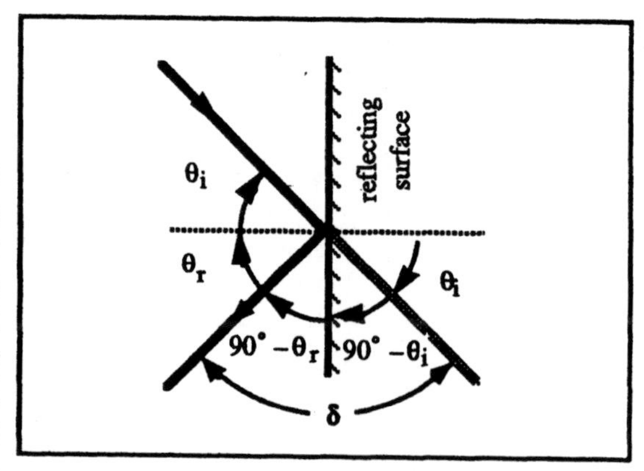

Figure 8-5. Deviation of a reflected ray.

Deviation of Reflected Rays at a Plane Mirror

An incident ray reflected from a plane surface undergoes a deviation δ, as shown in Figure 8-5. The deviation may be calculated by using the labeled angles in the figure. The angle between the normal and the incident ray extended to the right of the reflecting surface is equal to the incident angle. The angle from the extended incident ray to the surface is $(90° - \theta_i)$, and the angle from the reflected ray to the surface is $(90° - \theta_r)$. The deviation of the reflected ray is the sum of these two angles:

$$\delta = (90 - \theta_i) + (90 - \theta_r) = 180 - \theta_i - \theta_r$$

Because $\theta_i = \theta_r$, this can be written as

$$\delta = 180 - 2\theta_i = 180 - 2\theta_r \qquad (8\text{-}1)$$

Rotation of a Reflecting Surface

If a reflecting surface is rotated, the incident and reflected angles are changed. In Figure 8-6, a reflecting surface is rotated by an angle a, with the surface moving from position P1 to position P2 and the normal moving from N1 to N2. The incident and reflected angle increase by (a) to $(\theta_i + a)$ and $(\theta_r + a)$, respectively. The total rotation of the reflected ray is therefore increased by 2a. Thus a small rotation of a mirror will be doubled in the reflected ray rotation. This idea has been incorportated in a method for measuring of eye movements. A mirror is mounted on a contact lens off of a small post. The contact lens is fitted so that it moves with the eye (not the best conditions for the cornea, but measurements are only made for short times). Light reflected off the mirror will be rotated twice the amount of the eye movement. If the light were reflected on a wall a large distance from the eye, a small eye movement would make a large linear change. This eye movement device has not been the method of choice since infrared instruments became available.

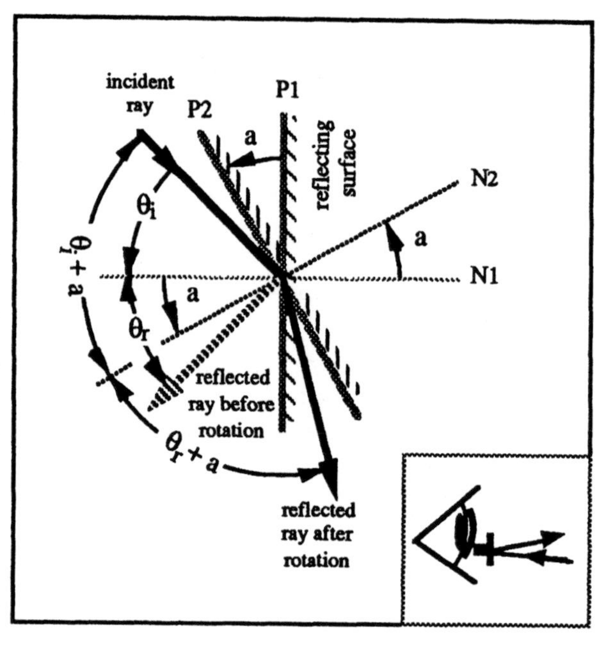

Figure 8-6. Rotation of a reflecting surface. The insert shows a mirror mounted on a contact lens used to measure eye movements.

Two-Mirror Deviation

The deviation of a ray reflected from two inclined mirrors depends only on the the angle between the mirrors and is independent of the angle of incidence. Rotation of both mirrors, keeping the angle between them constant, will not change the direction of the rays. In Figure 8-7, two mirrors are inclined at an angle α. The incident ray makes an angle β_1 with the first mirror surface (this is not the incident angle). The reflected ray makes an angle β_1 with the surface also. The angle β_1 is also labeled behind the first mirror for the extended incident ray (opposite internal angles are equal). At the second mirror, the angles made with the surface of the mirror and the incident and reflected rays are β_2 and the extended incident ray is also labeled. The deviation of the incident ray at the first mirror is $2(\beta_1)$, and the deviation of the ray at the second mirror is $2(\beta_2)$. The total deviation of the ray is the sum of these two deviations:

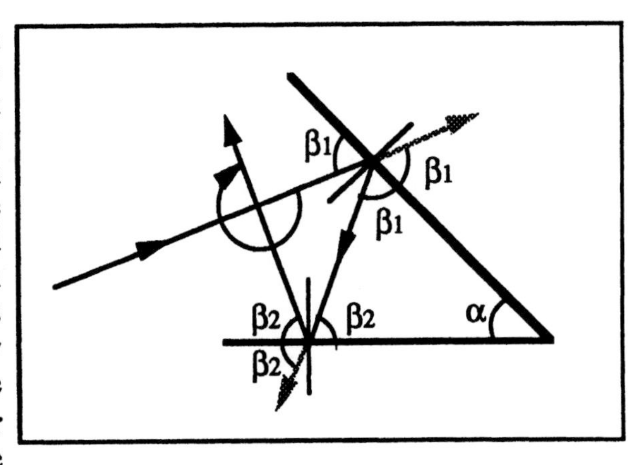

Figure 8-7. Two-mirror deviation.

$$\delta = 2(\beta_1) + 2(\beta_2) = 2(\beta_1 + \beta_2)$$

Since the sum of the three angles in a triangle are equal to 180°:

$$\beta_1 + \beta_2 + \alpha = 180^\circ \quad \text{or} \quad \beta_1 + \beta_2 = 180^\circ - \alpha$$

Substituting this becomes:

$$\delta = 2(\beta_1 + \beta_2) = 2(180^\circ - \alpha) \qquad (8\text{-}2)$$

Multiple Images

Two inclined or facing mirrors will form multiple images of an object placed between them. Each image acts as an object for the successive reflection. In the case of the inclined mirror, a finite number of images are formed. In Figure 8-8, the images formed by two inclined mirrors are shown. The box object is between the two mirrors. Because an image formed by a plane mirror is the same distance from the mirror as the object, the images all fall on a circle centered at the apex of the mirrors. By drawing lines perpendicular to the mirrors, each image may be found. For example, the line drawn up to the top mirror is extended until it intersects the circle where the first image is found. This image acts as an object for the bottom mirror, and the process is repeated. The same type of reflection occurs at the bottom mirror.

For mirrors that face each other, a similar process is used. Theoretically, facing mirrors could form an infinite number of images. This *infinity mirror* is sold as a novelty.

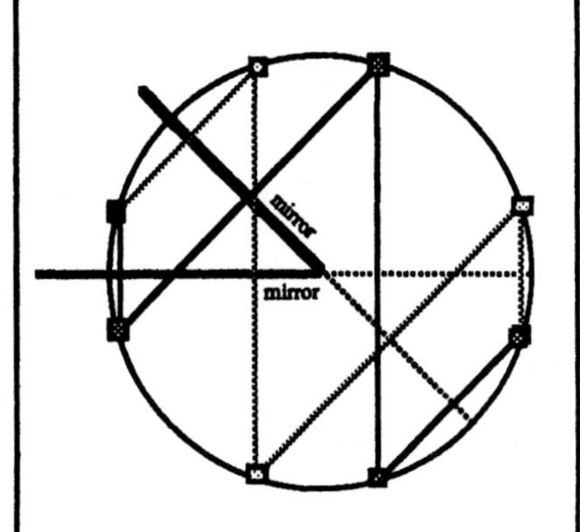

Figure 8-8. Images formed by two inclined mirrors.

Example 8-c

A ray undergoes a deviation of 120° after reflection off of a plane mirror. What is the incident angle?

Known	***Unknown***	***Equations/Concepts***
Deviation of the ray: $\delta = 120°$	Incident angle	(8-1) Deviation of a mirror
		Law of Reflection

Refer to Figure 8-5 for the deviation of the ray. Equation 8-1 is used to solve for the incident angle:

$$\delta = 180 - 2\theta_i \qquad \theta_i = \frac{180-\delta}{2} = \frac{180-120°}{2} = 30° = \text{incident angle}$$

Example 8-d

If the mirror in *Example 8-c* is rotated by 45°, how much angular change will the incident ray undergo ? What is the new deviation of the incident ray?

Known	***Unknown***	***Equations/Concepts***
Angle of incident ray: $\theta = 30°$	Change in incident ray with rotation	(8-1) Deviation of a mirror
Rotation of mirror: 45°	Deviation of incident ray	Rotation of reflecting surface

Because the ray is incident at an angle of 30°, with a rotation of 45°, the incident angle will increase to 75° (sum of incident angle and rotation). The reflected ray will also increase by 45° to 75°. The total change the ray will undergo is two times the rotation, or 90°. The deviation of the ray is calculated using Equation 8-1 with the new incident angle:

$$\delta = 180 - 2\theta_i = 180 - 2(75°) = 30°$$

Thus the larger the incident angle, the larger the reflected angle and the lower the deviation.

Example 8-e

Two mirrors are inclined at an angle of 30° to each other. What is the deviation of the reflected ray? At what angle of incidence at the first mirror will cause the ray to reflect off the second mirror and retrace its path?

Known	***Unknown***	***Equations/Concepts***
Mirror inclinded at: $\alpha = 30°$	Deviation of incident ray	(8-2) deviation inclined mirrors
	Incident angle to retrace itself	Ray normal to surface reflects on itself

Start by calculating the deviation of the reflected ray using Equation 8-2:

$$\delta = 2(180-\alpha) = 2(180°-30°) = 300° = \text{deviation by inclined mirrors}$$

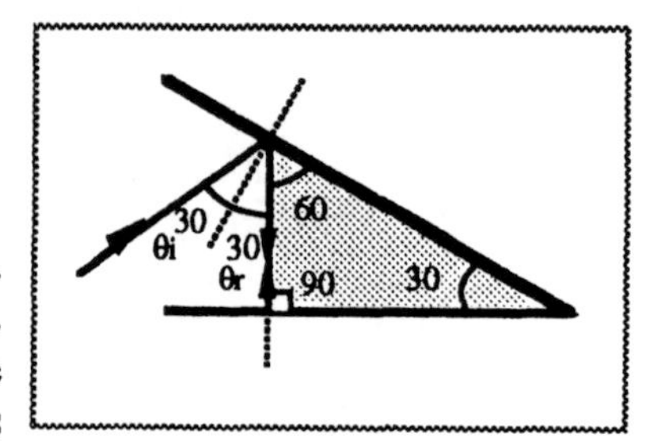

Figure 8-9. *Example 8-e.*

The next part of the problem is best solved with a diagram (Figure 8-9). For a ray to reflect back on itself, it must hit the second surface at 90°. Label the inclined angle and reflected angle at the second mirror. Because 180° is the sum of the angles in a triangle (shaded), the unknown angle is calculated to be 60°. From the normal to the first mirror surface is 90°, 60° of which is in the triangle. The reflected angle measured from the normal must then be 30°, which is equal to the incident angle of 30°.

Curved Mirrors

Curved mirrors have many of the imaging properties of single refracting surfaces and thin lenses. Recall the properties of reflecting surfaces shown in Figure 8-1. The reflected ray travels in the same medium as and in the opposite direction of the incident ray. This may be compensated by simply substituting:

$$n' = -n \tag{8-3}$$

into all the formulas developed for single refracting surfaces. With this substitution (keeping in mind real and virtual space location), our sign conventions holds.

Curved mirrors may have a variety of surfaces, including spherical, elliptical, and parabolic. To stress imaging properties, the spherical mirror will be used in this text. Other types of mirrors, however, yield better image quality. Spherical mirrors usually have an excessive amount of **spherical aberration** (paraxial rays focus at a different location than peripheral rays). (See Figure 8-10.) Because we are confining our imaging to a paraxial region, and using a flat surface approximation, the effect of the aberration is minimized. The terms used to define the spherical surface in Chapter 4 (**radius, power, center of curvature, optical axis, sag,** etc.) apply to a spherical mirror. In fact, a spherometer may be used to determine the radius of the mirror just as with a spherical refracting surface.

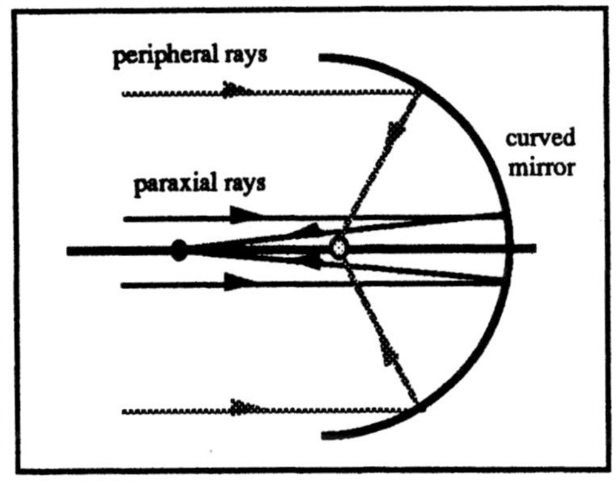

Figure 8-10. Spherical aberration of a curved mirror where peripheral and paraxial rays focus at different locations.

Radius-Power Relationship

The radius and power of a single refracting surface were defined by Equation 4-4. By substituting n' = –n in the formula, this relationship may be developed for a spherical reflecting surface:

$$F = \frac{n'-n}{r} = \frac{-n-n}{r} = \frac{-2n}{r} \tag{8-4}$$

The radius of the mirror must maintain the sign convention. Looking at the formula, it can be seen that a negative radius will yield a positive power, and a positive radius, a negative power. This means that a concave mirror surface has positive power and a convex mirror surface has negative power. (See Figure 8-11.)

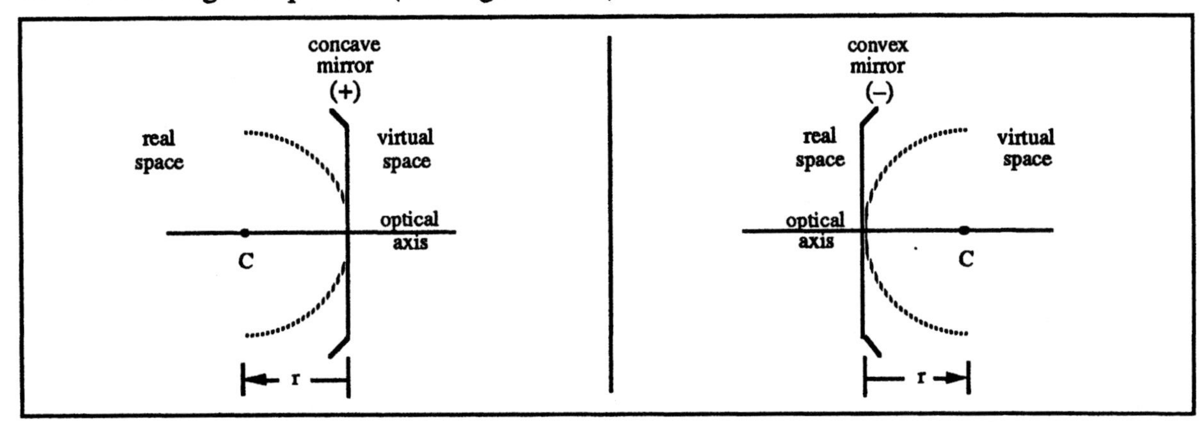

Figure 8-11. Radius and power relationship for curved mirrors. Concave mirrors have positive power, and convex mirrors have negative power. This differs from refracting surfaces.

Example 8-f
The surface power of a curved mirror is measured with a lens clock as + 5.00 D. What is the type and actual power of the mirror? Assume the lens clock is calibrated for an index of 1.53.

Known	***Unknown***	***Equations/Concepts***
Power reading of lens clock: F = + 5.00 D	Power of mirror	Curved surfaces and lens clock
Index of lens clock calibration : n = 1.53	Type of mirror	(8-4) Refracting and reflecting surfaces

The lens clock measures power for a material with an index 1.53. By using Equation 8-4 for refracting surfaces, the radius of the surface may be determined:

$$F = \frac{n_L - n_s}{r} \qquad r = \frac{n_L - n_s}{F} = \frac{1.53 - 1.00}{+5.00\ D} = +0.1060\ m = +10.60\ cm$$

Use this radius in Equation 8-4 to calculate the power for a curved mirror:

$$F = \frac{-2n_s}{r} = \frac{-2.00}{0.106\ m} = -18.87\ D$$

The mirror has negative power and is convex.

Image-Object Relationships: Position

Just as with refracting surfaces, the image-object relationship for a curved mirror is a function of the power. This relationship is expressed in terms of the incident and emergent reduced vergence in the Gaussian Imaging Formula:

$$L' = L + F \qquad (8\text{-}5)$$

The definition of the incident object vergence L is the same as for refracting surfaces. However, the emergent image vergence L' must be modified by substituting the index of refraction relationship (Equation 8-3):

Incident reduced object vergence: $$L = \frac{n}{\ell} \qquad (8\text{-}6)$$

Emergent reduced image vergence: $$L' = \frac{n'}{\ell'} = \frac{-n}{\ell'} \qquad (8\text{-}7)$$

The position, type, and orientation of the object and image may be determined using the sign convention and definitions already established for refracting surfaces. Be sure to review the section on ray tracing through curved mirrors in Chapter 9.

Example 8-g
A concave mirror with a radius of 20 cm, forms a real image 50 cm from the mirror. Where is the object located?

Known	***Unknown***	***Equations/Concepts***
Concave mirror - positive power	Power of mirror	(8-4) Power of mirror
Radius mirror: r = 20 cm = 0.20 m	Object location: ℓ = ?	(8-5), (8-6), (8-7)
Real image distance: ℓ' = – 50 cm = – 0.50 m		Real image in front of mirror

As usual, the first step is to draw a diagram and label the mirror. (Figure 8-12.) A concave mirror has positive power. The index to the left of the mirror is assumed to be air. Measurements from the surface into real space are negative, and therefore the radius and the real image distance have negative values. First calculate the power of the mirror using Equation 8-4:

$$F = \frac{-2n}{r} = \frac{-2(1.00)}{-0.20\ \text{m}} = \frac{-2}{-0.20\ \text{m}} = +10.00\ \text{D}$$

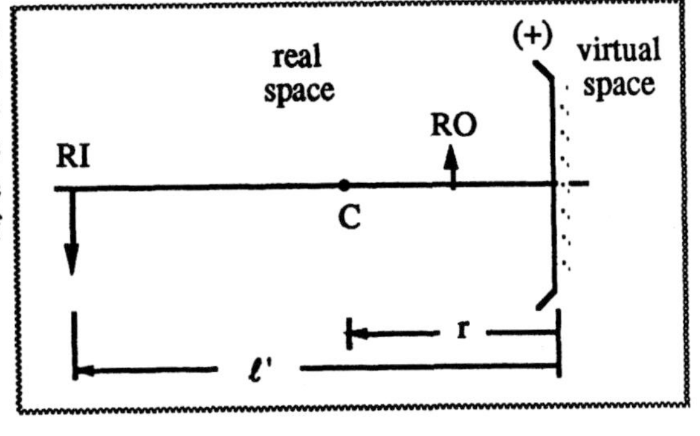

Figure 8-12. *Example 8-g.*

Image vergence is calculated using Equation 8-7:

$$L' = \frac{-n}{\ell'} = \frac{-1.00}{-0.50\ \text{m}} = +2.00\ \text{D}$$

The object location is calculated using the Imaging Formula and Equation 8-6:

$$L = L' - F = +2.00\ \text{D} - (+10.00\ \text{D}) = -8.00\ \text{D}$$

$$\ell = \frac{n}{L} = \frac{1.00}{-8.00\ \text{D}} = -0.125\ \text{m} = -12.50\ \text{cm}$$

The object is real and located 12.50 cm in front of the mirror.

Example 8-h

The outside mirror on my truck is inscribed, "Objects Are Closer Than Then They Appear To Be." Show what type of mirror this is by example.

This inscription states that the objects shown in the mirror (i.e., the image) appear to farther away than they actually are. If unaware of this perception, changing lanes could result in disaster. The objects in front of the mirror are real and a large distance away. The images are erect and smaller than the object. As with refracting elements, if the image is erect, it must be virtual. Concave mirrors, like positive refracting elements, form virtual images when the object is between the element and the primary focal point. We know that for a positive lens, the result is a magnified image, and so the concave mirror is eliminated as a choice. The convex mirror acts as a negative refracting element. A real object anywhere in front of the element will result in a minified virtual image closer to the element than the object.

Let's test this theory with a – 1.00 D convex mirror and an object distance of 4 m (ℓ = – 4.00 m). Using the Equations 8-6, 8-5, and 8-7, find the image location:

$$L = \frac{n}{\ell} = \frac{1.00}{-4.00\ \text{m}} = -0.25\ \text{D}$$

$$L' = F + L = -1.00\ \text{D} + (-0.25\ \text{D}) = -1.25\ \text{D}$$

$$L' = \frac{n'}{\ell'} = \frac{-n}{\ell'} \qquad \ell' = \frac{-n}{L'} = \frac{-1.00}{-1.25\ \text{D}} = +0.80\ \text{m} = +80.00\ \text{cm}$$

This calculation shows that a real object located 4 m in front of the mirror appears to be 80 cm behind the mirror. This would mean the image is closer than the object - not what is printed on the mirror. Perhaps there is more to the statement than just the apparent object position. We'll come back to this question after the discussion on lateral magnification.

Image-Object Relationships: Focal Points

With spherical mirrors, as with refracting surfaces, an infinite object forms an image at the secondary focal point. (See Figure 8-13.) Similarly, an object located at the primary focal point yields an emerging vergence of zero resulting in an infinite image position. (See Figure 8-14.) The power of the surface and the primary and secondary focal points are modified for a spherical mirror with the substitution (Equation 8-3) for the index:

$$F = \frac{-n}{f} = \frac{n'}{f'} \qquad \text{Substituting } n' = -n \qquad F = \frac{-n}{f} = \frac{-n}{f'} \qquad (8\text{-}8)$$

This equation indicates that the primary and secondary focal lengths have the same magnitude and sign; the primary and secondary focal points coincide with each other. By substituting Equation 8-4 for the power, the focal lengths may be related to the mirror radius:

$$\frac{-2n}{r} = \frac{-n}{f} = \frac{-n}{f'}$$

Dividing by $-n$ and taking the reciprocal this becomes

$$\frac{r}{2} = f = f' \qquad (8\text{-}9)$$

The focal lengths of a spherical mirror are equal to half the radius of the mirror, independent of the index of the medium. A mirror in water has the same focal length as a mirror in air. The power, however, is dependent on the surrounding medium (Equation 8-4).

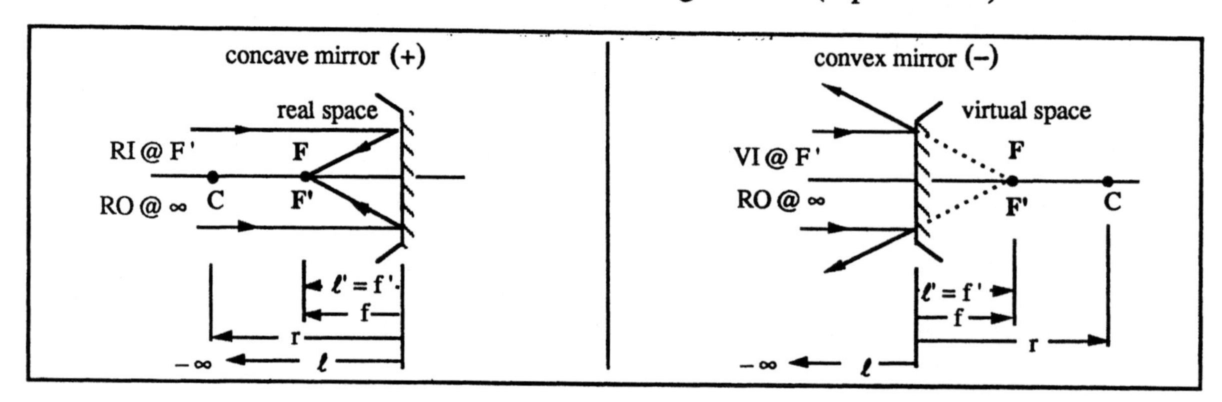

Figure 8-13. An infinite object forms an image at the secondary focal point of concave and convex mirror. For a concave mirror the image is real; for a convex mirror, the image is virtual.

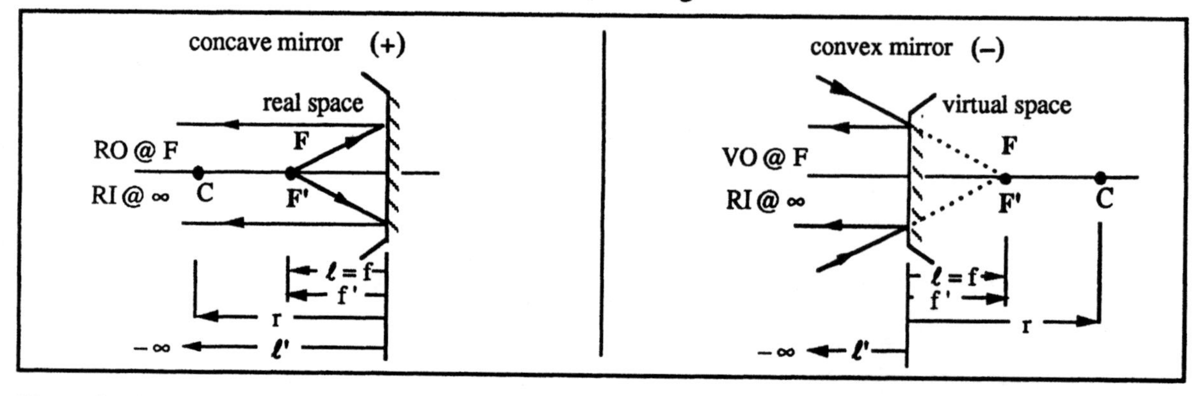

Figure 8-14. An object located at the primary focal point of convex and concave mirrors results in an infinite image. For a concave mirror, the object is real; and for a convex mirror, the object is virtual.

Example 8-i
A curved mirror has a power of –6.00D under water. What is the focal length, radius and power of the mirror in air.

Known	***Unknown***	***Equations/Concepts***
Power of mirror in water: $F = -6.00$ D	Focal lengths of mirror	(8-4) Power-radius
Convex mirror (negative power)	Radius of mirror	(8-8), (8-9) Focal length
Index of water: $n=1.33$	Power of mirror in air	

The power of the mirror under water may be related to the radius of the mirror by Equation 8-4:

$$F = \frac{-2n}{r} \qquad r = \frac{-2n}{F} = \frac{-2(1.33)}{-6.00\text{ D}} = +0.4433\text{ m} = +44.33\text{ cm}$$

The primary and secondary focal lengths are equal to half of the radius:

$$f = f' = \frac{r}{2} = \frac{0.4433\text{ m}}{2} = +0.2217\text{ m} = +22.17\text{ cm}$$

The power in air is calculated with the radius of the mirror and using an index of 1.00:

$$F = \frac{-2n}{r} = \frac{-2(1.00)}{+0.4433\text{ m}} = -4.51\text{ D}$$

For curved mirrors, the higher the index in front of the mirror the greater the power.

For relationships between the image and object positions and the focal points see Chapter 9 for ray tracing with spherical mirrors.

Image-Object Relationships: Size and Orientation

Lateral magnification using image and object distances (Equation 4-15) may be used for spherical mirrors with the substitution of Equation 8-3 for the image space index:

$$LM = \frac{h'}{h} = \frac{n\ell'}{n'\ell} = \frac{n\ell'}{-n\ell} = -\frac{\ell'}{\ell} \qquad (8\text{-}10)$$

The extrafocal length formula may also be used to calculate lateral magnification.

$$LM = \frac{h'}{h} = -\frac{f}{x} = -\frac{x'}{f'} \qquad (8\text{-}11)$$

The sign convention indicates an inverted (–) or erect (+) image-object relationship.

Example 8-h revisited
In the outside truck mirror, what is the lateral magnification? How big would a 10 ' high van appear to be?

Known	***Unknown***	***Equations/Concepts***
Object distance: $\ell = -4\text{ m} = -400$ cm	Lateral magnification	(8-10) Lateral magnification
Virtual image distance: $\ell' = +80$ cm	Image height of truck: $h' = ?$	
Object size of van: $h = 10'$		

Calculate the lateral magnification using Equation 8-10:

$$LM = \frac{h'}{h} = -\frac{\ell'}{\ell} = -\frac{+80\text{ cm}}{-400\text{ cm}} = +0.20\text{ x}$$

The image size can be calculated by substituting h in Equation 8-10:

$$LM = \frac{h'}{h} \qquad h' = (h)(LM) = (10')(+0.20) = 2'$$

A 10'-high van appears to be 2' high. If we use size as a judgment of distance, the object appears to be much smaller than the actual truck. Because size and distance are related in vision, perhaps the manufacturer is worried that someone will mistake the distance because the objects look so small.

Example 8-j

A 3.2-cm-tall object is reflected at a curved surface and a 2.4-cm-tall, inverted image is formed. If the object extra-focal distance is –12 cm, what is the power of the mirror? Where are the object and image located?

Known	***Unknown***	***Equations/Concepts***
Object height:: $h = 3.2$ cm	Power of mirror	(8-10), (8-11) Lateral magnification
Image height: $h' = -2.4$ cm	Object location: $\ell = ?$	(8-8) Focal length-power
Extra-focal distance: $x = -12$ cm	Image location: $\ell' = ?$	(4-16) Extrafocal distance

Using Equation 8-10 and 8-11, solve for the lateral magnification and the focal length. Note that the lateral magnification is negative because the image is inverted:

$$LM = \frac{h'}{h} = -\frac{f}{x} \qquad f = -x\left(\frac{h'}{h}\right) = -(-12\text{ cm})\left(-\frac{2.4\text{ cm}}{3.2\text{ cm}}\right) = -9.00\text{ cm} = -0.09\text{ m}$$

Use Equation 8- 8 to determine the power of the mirror:

$$F = \frac{-n}{f} = \frac{-n_s}{f} = \frac{-1.00}{-0.09\text{ m}} = +11.11\text{ D}$$

To determine the location of the object, use the relationship between focal length and extrafocal distance (Equation 4-16):

$$\ell = f + x = -9.00\text{ cm} + (-12.00\text{ cm}) = -21.00\text{ cm}$$

Use the lateral magnification (Equation 8-10) to solve for the image distance:

$$LM = \frac{h'}{h} = -\frac{\ell'}{\ell} \qquad \ell' = -\ell\frac{h'}{h} = -(-21.00\text{ cm})\left(\frac{-2.4\text{ cm}}{3.2\text{ cm}}\right) = -15.75\text{ cm}$$

Example 8-k

A convex mirror with a power of – 4.75 D forms a virtual image 50 cm behind the mirror. If the object is 2.0 cm tall, find the lateral magnification, and the image size and orientation.

Known	***Unknown***	***Equations/Concepts***
Power of mirror: $F = -4.75$ D	Lateral magnification: LM = ?	(8-10), (8-11) Lateral magnification
Image location: $\ell' = +50.00$ cm	Image size: $h' = ?$	(8-5), (8-6), (8-7) Imaging formula
Object size: $h = 2.00$ cm	Image orientation	Erect: + LM; inverted: – LM

First, we need to find the object location using Equations 8-7, 8-5, and 8-6:

$$L' = \frac{n'}{\ell'} = \frac{-n_s}{\ell'} = \frac{-1.00}{+0.50 \text{ m}} = -2.00 \text{ D}$$

$$L' = F + L \qquad L = L' - F = -2.00 \text{ D} - (-4.75 \text{ D}) = +2.75 \text{ D}$$

$$L = \frac{n}{\ell} = \frac{n_s}{\ell} \qquad \ell = \frac{n_s}{L} = \frac{1.00}{+2.75 \text{ D}} = +0.3636 \text{ m} = +36.36 \text{ cm}$$

The object is virtual because the incident vergence is positive and the object position is to the right of the mirror in virtual space. Use Equation 8- 10 to determine the lateral magnification and image size.

$$LM = \frac{h'}{h} = -\frac{\ell'}{\ell} = -\frac{+50.00 \text{ cm}}{+36.36 \text{ cm}} = -1.38 \text{ x}$$

$$LM = \frac{h'}{h} \qquad h' = h(LM) = 2.00 \text{ cm}(-1.38) = -2.76 \text{ cm}$$

From the negative sign of the lateral magnification, we know that the image is inverted.

Lens Mirror

At times, a system uses both refraction and reflection at curved surfaces. One example of this is the lens mirror. If a thin lens is silvered on the back surface, the incident light will refract at the front surface, reflect off the back surface, and then refract through the front surface again. The power of the lens mirror takes into account this path of the ray. The lens mirror power may be calculated by summing the front lens power (through which the ray is refracted twice) and the mirror power:

$$F_{lm} = 2F_1 + F_m \qquad (8\text{-}12)$$

where:

F_{lm} = the power of the lens mirror

F_1 = the power of the front surface of the lens

F_m = the power of the mirror

The power of the mirror is calculated using the index of the lens materal. This is shown in the next example.

Example 8-1

An equiconvex glass lens (n =1.523) has a power of +10.00 D. The back surface of the lens is coated so that it will reflect light. What is the power of the lens mirror?

Known	***Unknown***	***Equations/Concepts***
Power of lens: F = + 10.00 D	Power of lens mirror	(8-4) Power-radius
Equiconvex lens surface powers: + 5.00 D		(8-12) Lens mirror power
Index of lens: n = 1.523		

Solve for the radius of the back surface of the lens using Equation 8-4 for refracting surfaces:

$$F_2 = \frac{n_s - n_L}{r} \qquad r = \frac{n_s - n_L}{F_2} = \frac{1.00 - 1.523}{+5.00\ \text{D}} = -0.1046\ \text{m} = -10.46\ \text{cm}$$

Use this radius to calculate the mirror power:

$$F_m = \frac{-2n}{r} = \frac{-2(1.523)}{-0.1046\ \text{m}} = +29.12\ \text{D}$$

Substitute the mirror and front surface powers into Equation 8-12:

$$F_{lm} = 2F_1 + F_m = 2(+5.00\ \text{D}) + 29.12\ \text{D} = +39.12\ \text{D}$$

The lens mirror has considerable more power than the lens itself.

Example 8-m

Using the lens mirror from *Example 8-l*, where would the image of a virtual object 1 m from the lens be formed? What is the magnification and the orientation and type of the image formed?

Known	***Unknown***	***Equations/Concepts***
Power of lens miror: $F = +39.12$ D	Power of mirror	(8-5), (8-6), (8-7) Lens Maker's Formula
Index of lens: $n = 1.523$	Image location: $\ell' = ?$	(8-10) Lateral magnification
Object location: $\ell = +1.0$ m	Image type and orientation	

For the thin lens mirror, once the power is found, treat the system as though it were a mirror with the total power: in this case, + 39.12 D. The surround medium is the index used to calculate vergence, just as with thin lenses. Solve the problem by using the mirror imaging equations (8-6, 8-5, 8-7).

$$L = \frac{n}{\ell} = \frac{n_s}{\ell} = \frac{1.00}{+1.00\ \text{m}} = +1.00\ \text{D}$$

$$L' = F + L = +1.00\ \text{D} + 39.12\ \text{D} = +40.12\ \text{D}$$

$$L' = \frac{-n'}{\ell'} = \frac{-n_s}{\ell'} \qquad \ell' = -\frac{n_s}{L'} = \frac{-1.00}{+40.12\ \text{D}} = -0.0249\ \text{m} = -2.49\ \text{cm}$$

The image is real because positive vergence leaves the mirror, and the image is in front of the lens in real space.

Solve for the lateral magnification using Equation 8-10, again using the index of the surrounding medium or air, in this case:

$$LM = -\frac{\ell'}{\ell} = -\frac{-2.49\ \text{cm}}{+100\ \text{cm}} = +0.025\ \text{x}$$

The image is erect and smaller than the object.

Supplemental Problems

Plane Mirrors

8-1. With a virtual object 20 cm from a mirror, a real image is formed 20 cm from the mirror. What is the power of the mirror?
ANS. Power of mirror is plano

8-2. Two plane mirrors are inclined at an angle θ so that the deviation of any ray reflected once at each mirror is 316°. What is the incline, θ?
ANS. $\theta = 22°$

8-3. If a ray reflected at a plane mirror has a deviation of 100°, what is the angle of reflection?
ANS. Angle of reflection = 40°

8-4. If a 7'4"-tall man wishes to purchase a mirror in which he can see himself from the top of his head to the bottom of his shoes, what is the smallest mirror he should buy?
ANS. 3' 8"

8-5. Convergent light with a vergence of + 2.00 D is incident on a plane mirror. Where is the image formed after reflection?
ANS. – 50 cm

8-6. You wish to hang a mirror in your office so that you can see the doorway behind you. If you sit 12' from the door in a 15' long office, and your eyes are 4' above the floor when you are seated, where should you place the mirror on the wall so that you can see the entire 7' doorway?
ANS. 3.2' above the floor

8-7. What is the total rotation of a reflected ray if the incident ray before rotation is 60° and the reflected ray after rotation makes an angle of 18° with the surface?
ANS. The total rotation = 24°

8-8. Find the emergent vergence for a real object positioned 75 cm in front of a plane mirror.
ANS. + 1.33 D (convergent light)

Curved Mirrors

8-9. An object is located 125 cm from a convex mirror, forming a virtual image 20 cm behind the mirror. What is the power of the mirror in diopters?
ANS. – 4.20 D

8-10. After reflection, a curved reflector forms a parallel pencil of a point source. If the mirror has a radius of 40 cm, what is the power of the mirror?
ANS. + 5.00 D

8-11. The index of refraction of carbon bisulfide is 1.629. What is the focal power of a concave mirror of radius 15 cm in contact with this liquid?
ANS. + 21.70 D

8-12. If you hold a spoon with a radius of 3 cm at a distance 40 cm in front of you, where will your image be located? Solve for both the front and back surfaces.
ANS. The concave surface will form a real image 1.56 cm in front of the spoon; the convex surface will form a virtual image 1.45 cm behind the spoon.

8-13. A friend wants to know the index of refraction of an unknown liquid. You determine that a point source placed 10 cm in front of a concave mirror submerged in the liquid forms a real image 25 cm in front of the mirror. If the mirror has a power of + 10.00 D in air, what is the unknown index of refraction?
ANS. n = 1.40

8-14. Convergent light of + 1.00 D is incident on a spherical mirror and a real image is formed. If the incident light is increased to + 3.00 D the new image is 0.75 times closer to the mirror than was the first image. What is the power of the mirror? Is it concave or convex?
ANS. + 5.06 D; the mirror is concave

8-15. As the radius is increased two-fold, and the object position held constant, an image moves from a position 10 cm behind a curved mirror to a position 20 cm behind the mirror. If the original power was – 10.00 D, find the object location.
ANS. The object is at infinity

Focal Points of Spherical Mirrors

8-16. Find the index of refraction of the surround, as well as the primary and secondary focal lengths of a convex mirror with a radius of 97.1 cm and power of – 3.50 D.
ANS. n = 1.70; primary focal length = secondary focal length = + 48.6 cm

8-17. If the primary focal point of a concave mirror is – 12 cm in air (n = 1.00) and the secondary focal length is – 12 cm under water (n = 1.33), what is the radius and the power of the mirror?
ANS. Radius = 24 cm; power in air = + 8.33 D; power under water = + 11.08 D

Lateral Magnification

8-18. An object is placed at the center of curvature of a – 2.00 D mirror. What is the resulting lateral magnification and orientation ?
ANS. Lateral magnification = – 1.00; inverted image

8-19. An object 2 cm high is placed 50 cm in front of a concave mirror with a radius of 15 cm. What is the size of the image formed and what is its orientation?
ANS. – 0.35 cm; inverted

8-20. Find the lateral magnification for an object placed 50 cm in front of a + 5.00 D concave mirror.
ANS. Lateral magnification = – 0.67x

8-21. What size image would result if a 5-cm-high object were positioned 1 m in front of a convex mirror with a primary focal length of 5 cm?
ANS. 23.8 cm

8-22. You are designing a mirror that will magnify (12 times) an object placed 10 cm in front of it and form an erect image. Will the mirror be concave or convex, and how large should you make the radius?
ANS. The mirror should be concave with a radius of – 21.8 cm

Chapter 9

Ray Tracing

Ray tracing is a graphical technique for determining the image-object relationship of imaging elements and systems. The procedure is a useful check for mathematical calculations. In this chapter, one technique for ray tracing, the *parallel-ray method,* is presented. The first section of the chapter defines general rules for ray tracing. The subsequent sections outline and illustrate ray tracing for positive and negative power single refracting surfaces, lenses, mirrors, multiple lens systems, and thick lenses. Refer to each when studying the material in the other chapters.

To be efficient with the ray tracing technique, you must practice several different imaging situations. Review the regions of the imaging element and the type of image that would be expected for each region. Work problems backward (i.e., with a known image position, locate the object). Ray tracing may also be used to determine lateral magnification and orientation by measuring values directly off the graph and substituting into the appropriate magnification formula.

General Procedures for Ray Tracing

The first step in ray tracing is to draw a scale diagram of the optical element or system. The vertical and horizontal scales may differ; in fact, the vertical scale is not important in many instances. For curved single refracting surfaces, thin lenses, and curved mirrors, each element may be represented by a vertical line. Indicate the power of the element and label it with a (+) or (–). If known, indicate the location of the primary and secondary focal points (F and F'), the optical center or nodal points (N and N'), the principal planes (H and H'), the object, and the image. Label the type of object or image, i.e., a virtual object (VO), a real image (RI), etc. You may also want to indicate object and image space. Remember that for the sign convention, *light travels from left to right.* Real objects must be placed to the left of the imaging element.

In general, three rays are required for the location of an image or object in ray tracing. These rays will refract or reflect at the imaging element. All three rays must pass through, come from, or travel toward the same off-axis object position, and they must cross (or appear to cross) at the corresponding off-axis image position. Arrows should be drawn on the incident and emergent rays.

The three rays are outlined below for positive elements *(a +),* negative elements *(b –),* and both positive and negative elements *(c ±).*

Specific considerations are discussed in each section.

1. Incident Ray Parallel to the Optical Axis

Figure 9-1. *(a +)* An incident ray leaving from an extra-axial object point parallel to the optical axis will leave converging toward the secondary focal point.

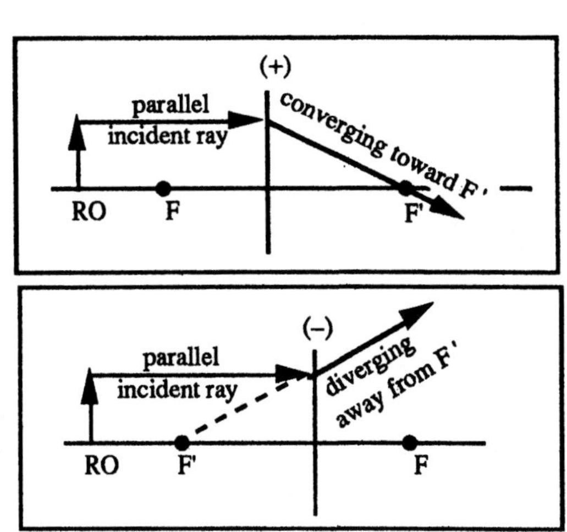

Figure 9-2. *b (–)* An incident ray leaving from an extra-axial object position parallel to the optic axis will leave diverging as though it came from the secondary focal point. For refracting elements, this ray should be dotted back toward object space.

Figure 9-3. *(±)* If the object is virtual, the incident parallel ray is aimed toward the off-axis object position (extended into image space as a dotted line). The emergent ray will diverge (–) or converge (+) from F', depending on the surface power. This is shown with a positive element in the figure.

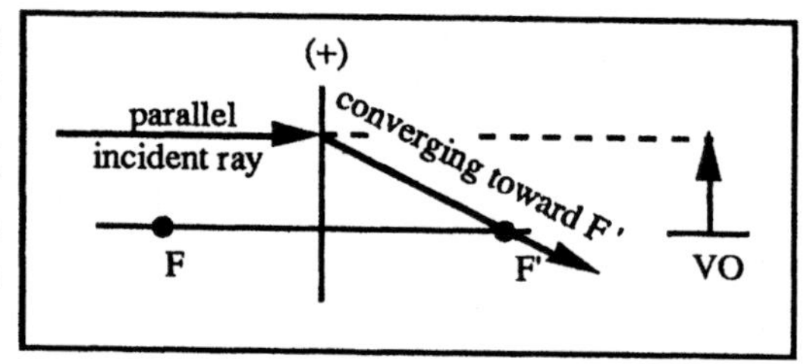

2. *Emergent Ray Parallel to Optical Axis*

Figure 9-4. *(+)* An incident ray leaving an extra-axial object point passing through the primary focal point will emerge parallel to the optical axis.

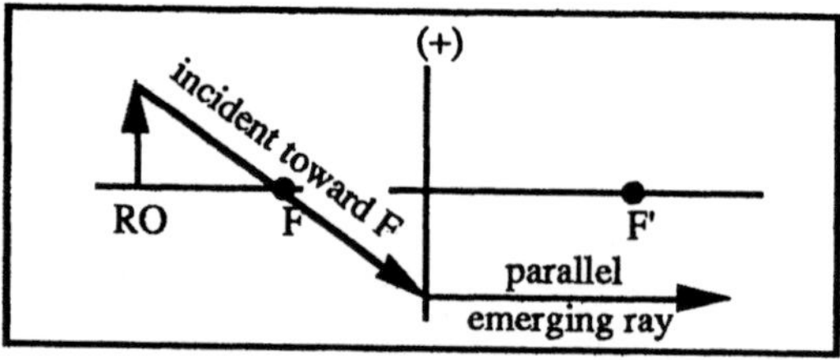

Figure 9-5. *(–)* An incident ray leaving an extra-axial object point directed toward the primary focal point will emerge parallel to the optical axis. The incident ray should be dotted in image space.

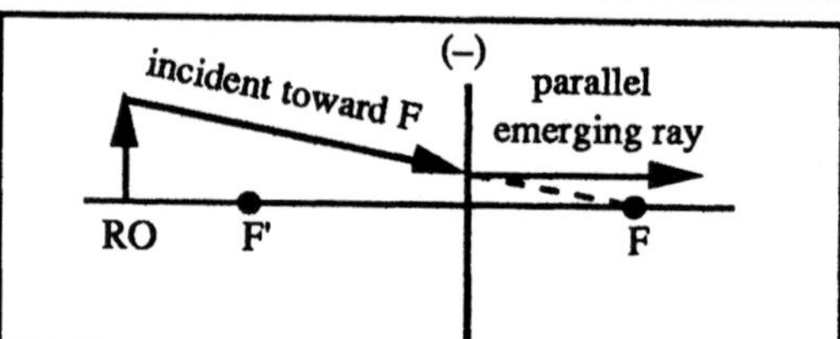

Figure 9-6. *(±)* For a virtual object, the incident ray passing through the primary focal point is aimed toward the off-axis object position (extended into image space with a dotted line. The emergent ray will leave parallel to the axis. This is shown with a positive element in the figure.

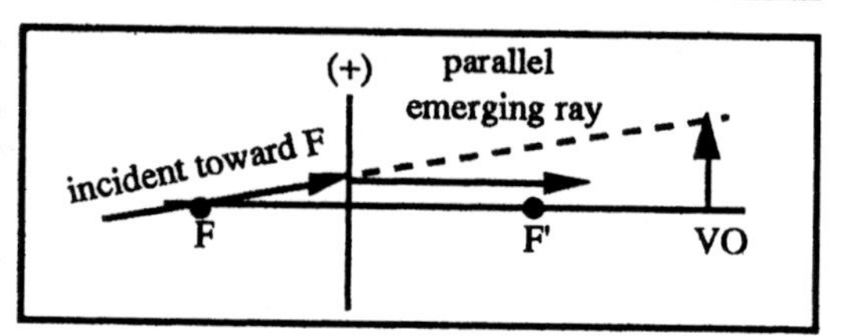

3. *The Undeviated Ray*

Figure 9-7a, b. *(±)* A ray that passes through the image element without changing direction is called the **chief, principal, or undeviated ray.** This ray leaves an off-axis object position aimed toward the nodal points of the element. For single refracting surfaces, thin lenses, and mirrors, the two nodal points coincide with each other and with the optical center. For the single refracting surface and mirror, this point is the center of curvature of the surface. For the thin lens, this point is located at the intersection of the optical axis with the element. Thick lenses will be discussed in later sections. The undeviated ray is shown for a negative single refracting surface (Figure 9-7a) and positive thin lens (Figure 9-7b). This ray is never dotted. It is the same for positive and negative elements. You may want to draw this ray first because it is the easiest.

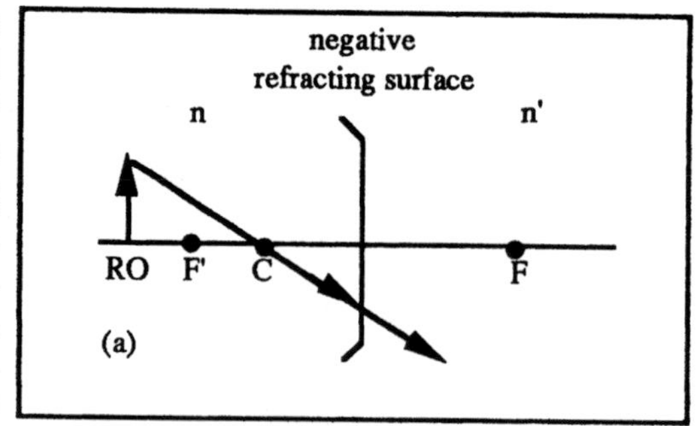

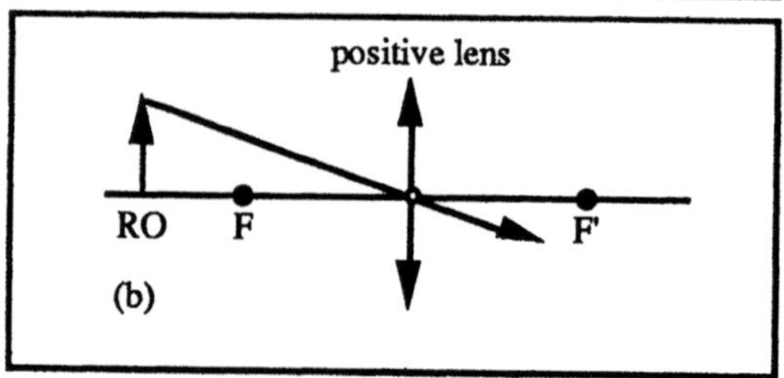

Single Refracting Surfaces

The general rules apply directly for ray tracing through single refracting surfaces. The region to the left of the interface is (in my definition) object space and the region to the right of the interface is image space. Object space and image space have different indices. Thus the primary and secondary focal lengths have different magnitudes and should be diagrammed accordingly. The center of curvature must be located so that the undeviated ray may be drawn. In the diagram, the sum of the focal lengths (use the appropriate signs) should be equal to the radius of the surface (f + f ' = r). Label distances on ray tracing diagrams with an arrow that indicates the sign by the direction of the measurement (to left - negative; to right - positive).

Positive Single Refracting Surfaces

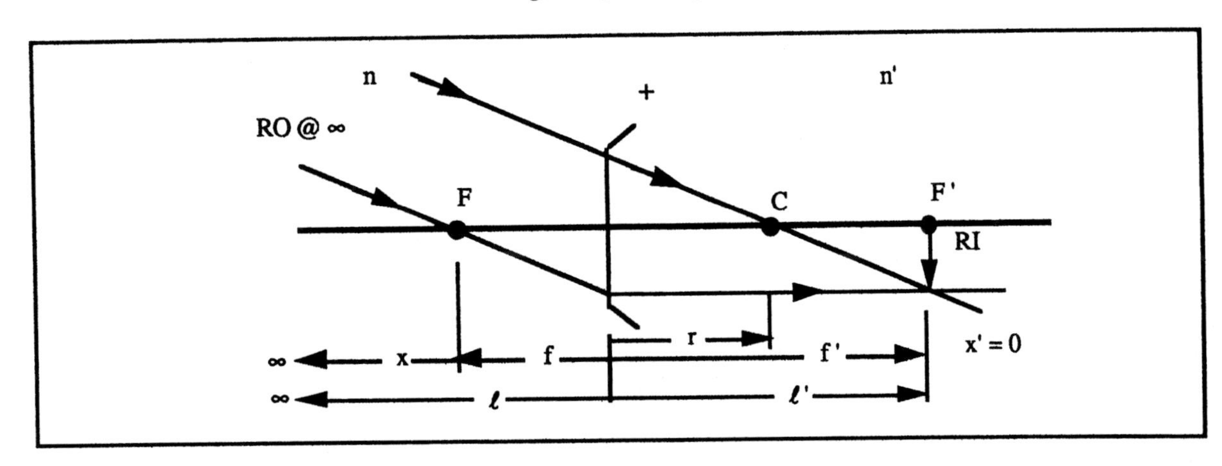

Figure 9-8. A real object is located at an infinite object position. Two rays are drawn from the same off-axis object position (i.e., the rays are parallel to each other and oblique to the optical axis). One ray is directed through F and leaves parallel to the optical axis, the other is directed toward the center of curvature and leaves undeviated. A real, inverted image is formed in the secondary focal plane.

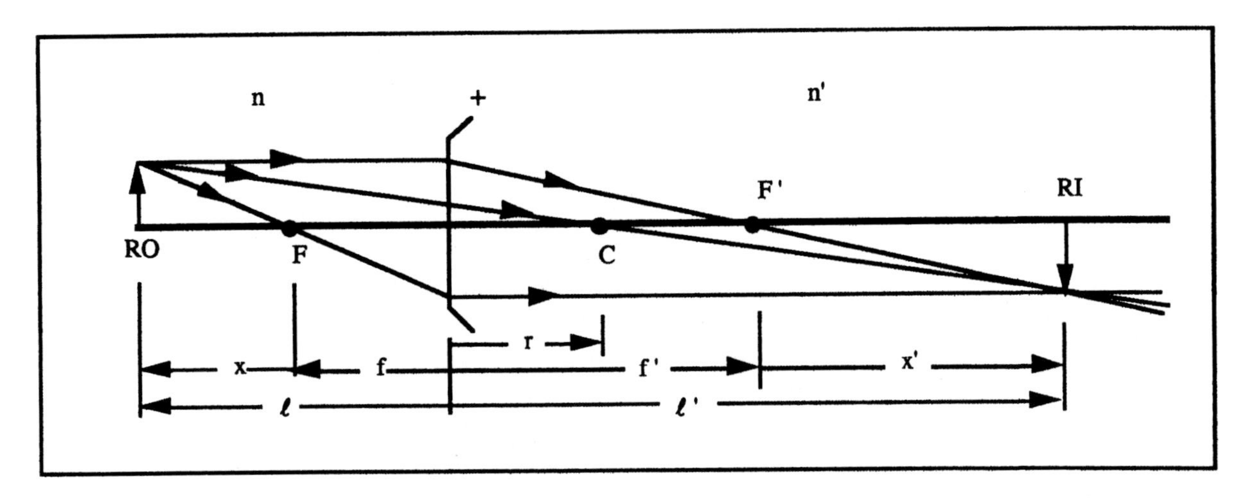

Figure 9-9. A real object is located at a finite position to the left of the primary focal point. Three rays are drawn from the same off-axis object position. A real, inverted image is formed to the right of the secondary focal point.

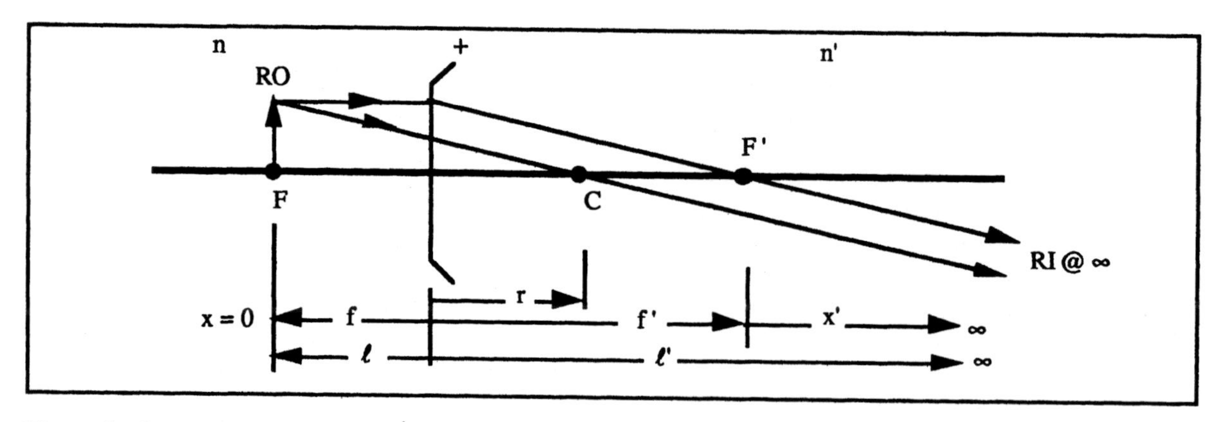

Figure 9-10. A real object is located at the primary focal point. Two rays are drawn from the same off-axis object position. One ray is directed parallel to the optical axis and is refracted through F'. The other is directed toward the center of curvature and leaves undeviated. A real, inverted image is formed at infinity.

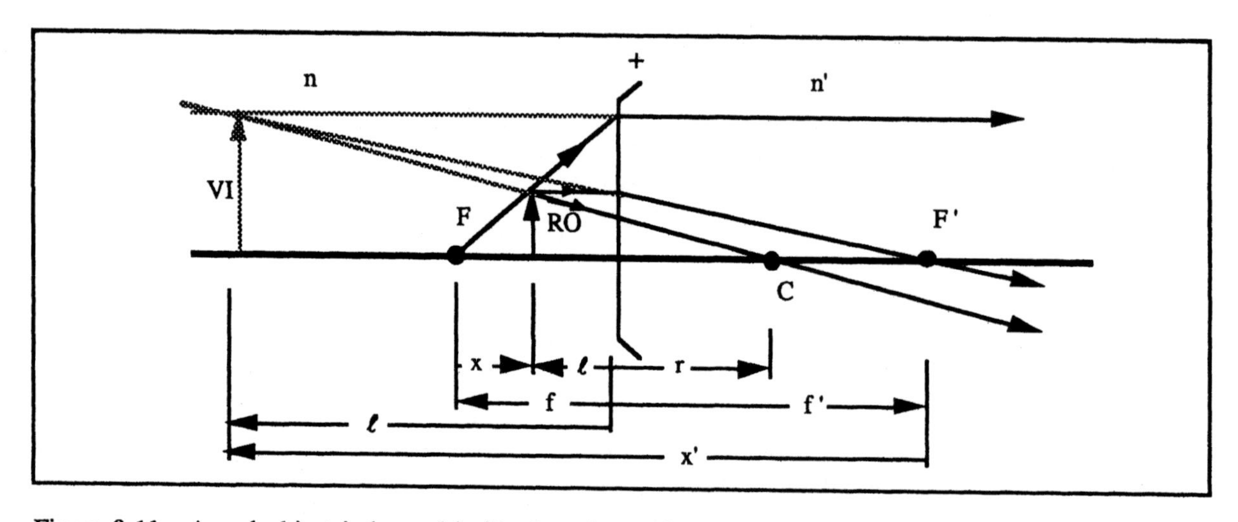

Figure 9-11. A real object is located inside the primary focal point. An erect, magnified, virtual image is formed to the left of the primary focal point. To locate the image, dot the divergent emerging rays back into object space.

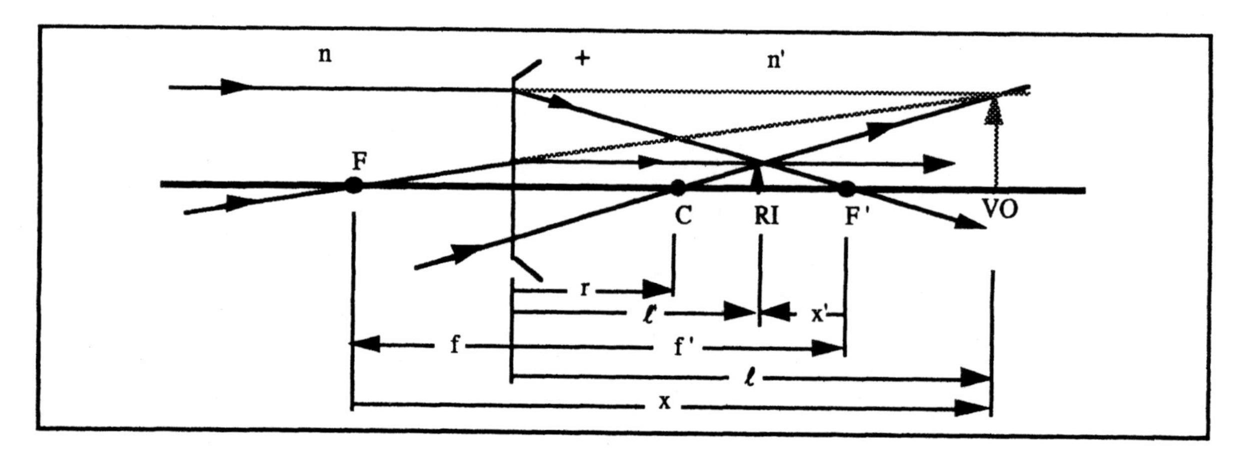

Figure 9-12. A virtual object formed with convergent incident rays. Aim incident rays toward the top of the virtual object. The refracted rays follow the same tracing rules. A real, erect image is formed.

Negative Single Refracting Surfaces

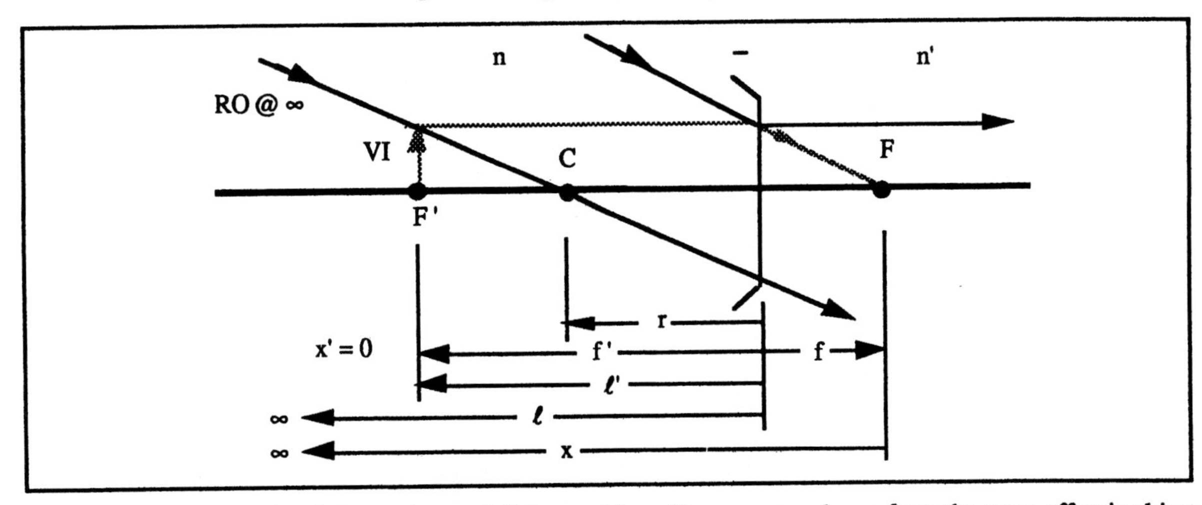

Figure 9-13. A real object is located at an infinite position. Two rays are drawn from the same off-axis object position (i.e., the rays are parallel to each other and oblique to the optical axis). The ray aimed toward F leaves parallel to the axis. An erect, virtual image is formed in the secondary focal plane.

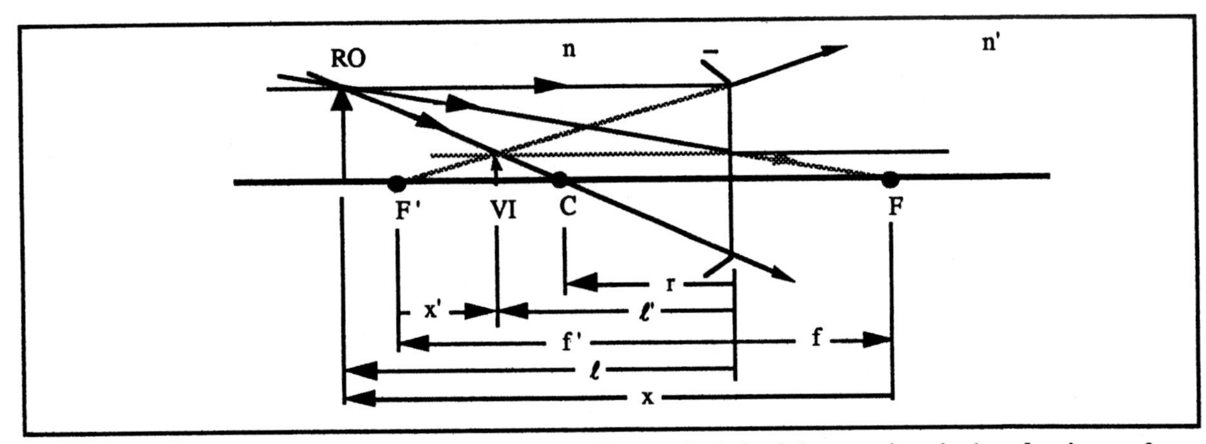

Figure 9-14. A real object is located at a finite position to the left of the negative single refracting surface. Three rays are drawn from the same off-axis object position using the ray tracing rules. A virtual, erect, minimized image is formed.

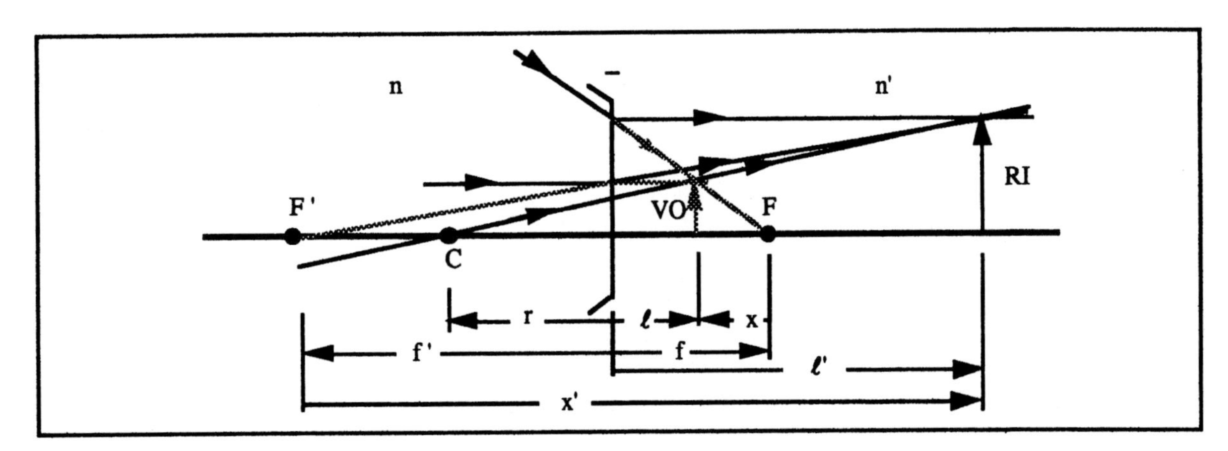

Figure 9-15. A virtual object is located inside the primary focal point. Aim the incident rays to the top of the virtual object. The shaded lines indicate extensions of lines in object or image space. An erect, magnified, real image is formed.

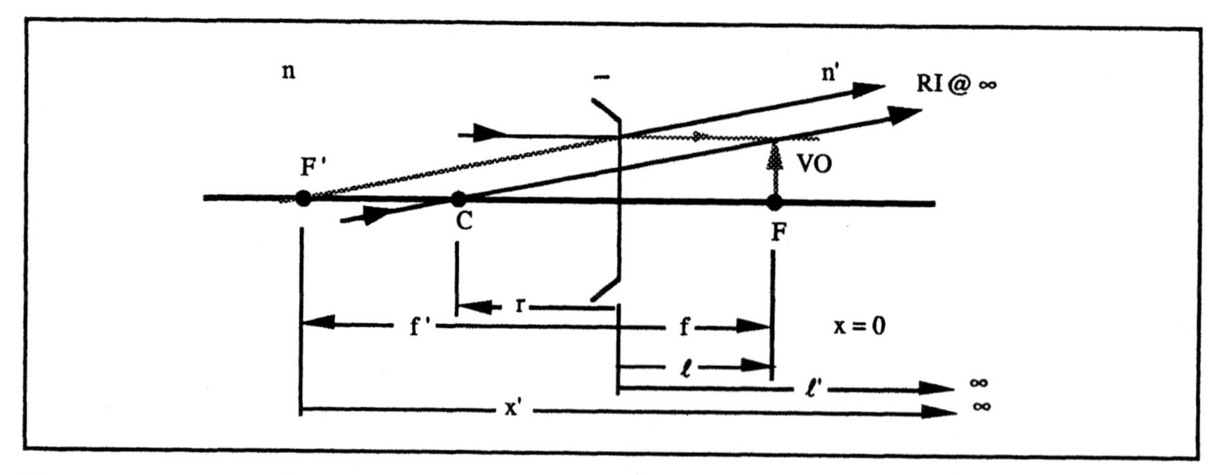

Figure 9-16. A virtual object is located at the primary focal point. Two rays are drawn toward the same off-axis object position. One ray is incident parallel to the optical axis and aimed toward the top of the object and is refracted appearing to leave from F'. The other is directed toward the center of curvature and leaves undeviated. An erect, real image is formed at infinity.

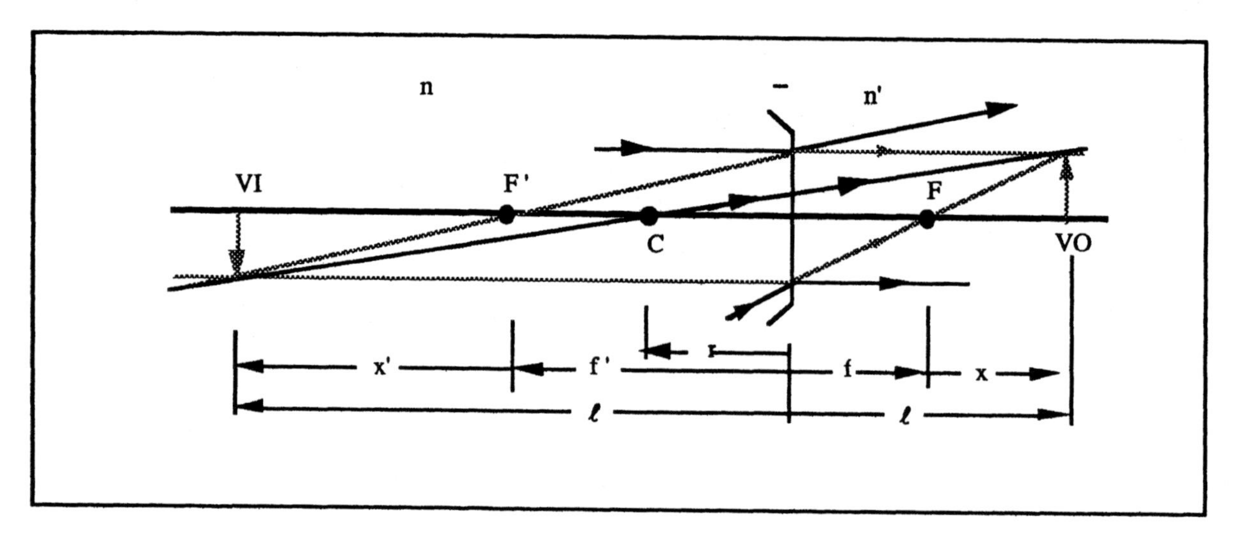

Figure 9-17. A virtual object is located outside (to the right of) the primary focal point. A virtual, inverted image is formed to the left of the surface. Dot the incident and emerging rays if they are extensions of the actual rays.

Thin Lenses

Ray tracing through thin lenses is simplier than through single refracting surfaces. Because object and image space have the same index (surrounding index), the primary and secondary focal lengths have the same magnitude (opposite sign) and are therefore drawn the same distance from the lens. (If the indices of object space and image space are different, treat the element as a single refracting surface). The lens is represented as a vertical line with end arrows pointed away from the axis to indicate positive power and end arrows pointed toward the axis to indicate negative power. The undeviated ray passes through the optical center at the intersection of the lens with the optical axis. The general ray tracing rules apply.

Positive Thin Lenses

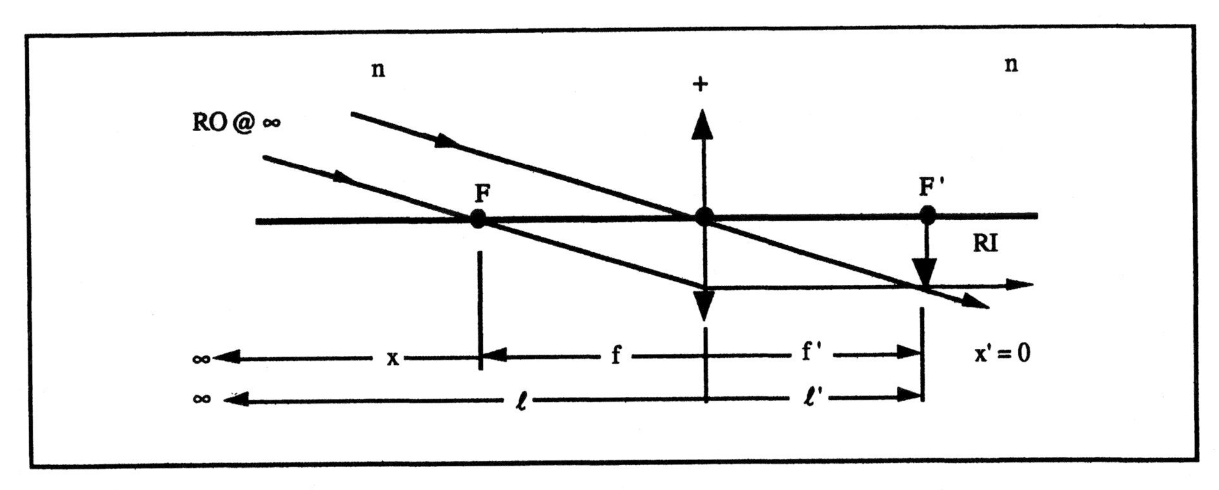

Figure 9-18. A real object is located at an infinite object position. Two rays are traced from the same off-axis object position. A real, inverted image is formed in the secondary focal plane.

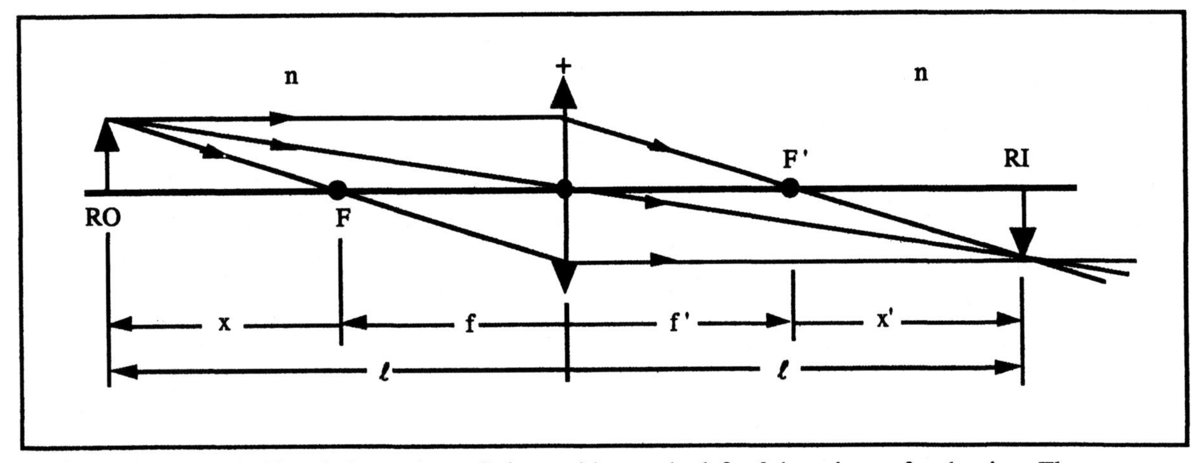

Figure 9-19. A real object is located at a finite position to the left of the primary focal point. Three rays are drawn from the same off-axis object position. A ray parallel to the axis is refracted through F'. A ray directed through F is refracted parallel to the axis. The undeviated ray passes through the center of the lens. A real, inverted image is formed to the right of the secondary focal point.

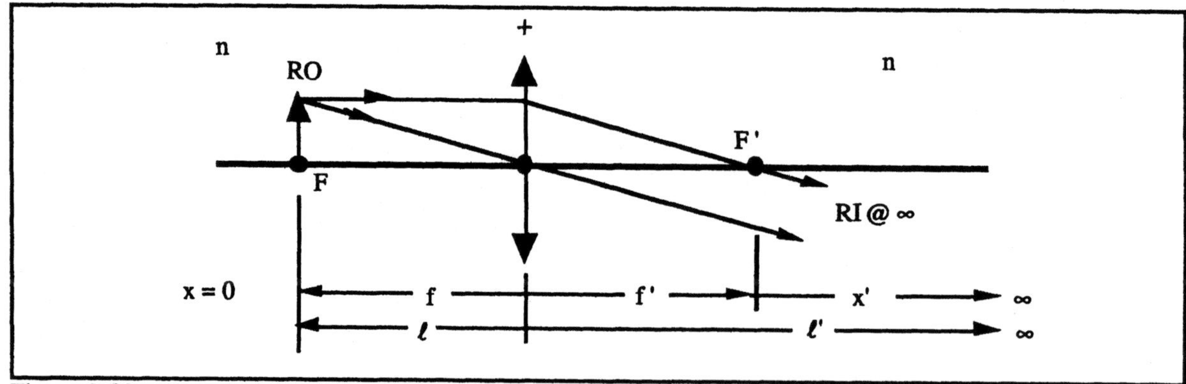

Figure 9-20. A real object is located at the primary focal point. Two rays are drawn from the same off-axis object position. One ray is directed parallel to the optical axis and is refracted through F'. The other is directed toward the center of curvature and leaves undeviated. A real, inverted image is formed at infinity.

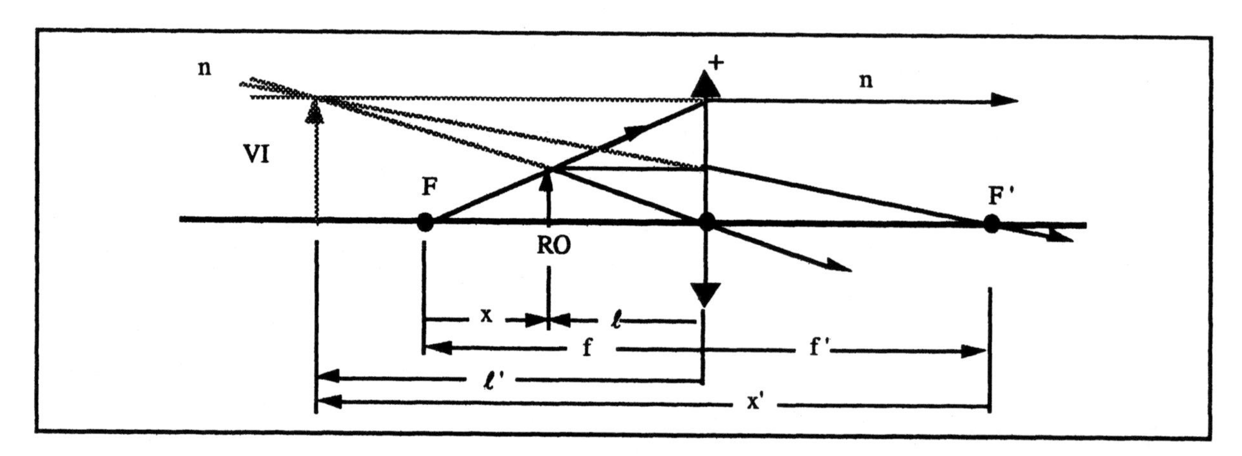

Figure 9-21. A real object is located inside the primary focal point. The three rays are traced as shown. An erect, magnified, virtual image is formed to the left of the primary focal point. This is an example of a hand held magnifier.

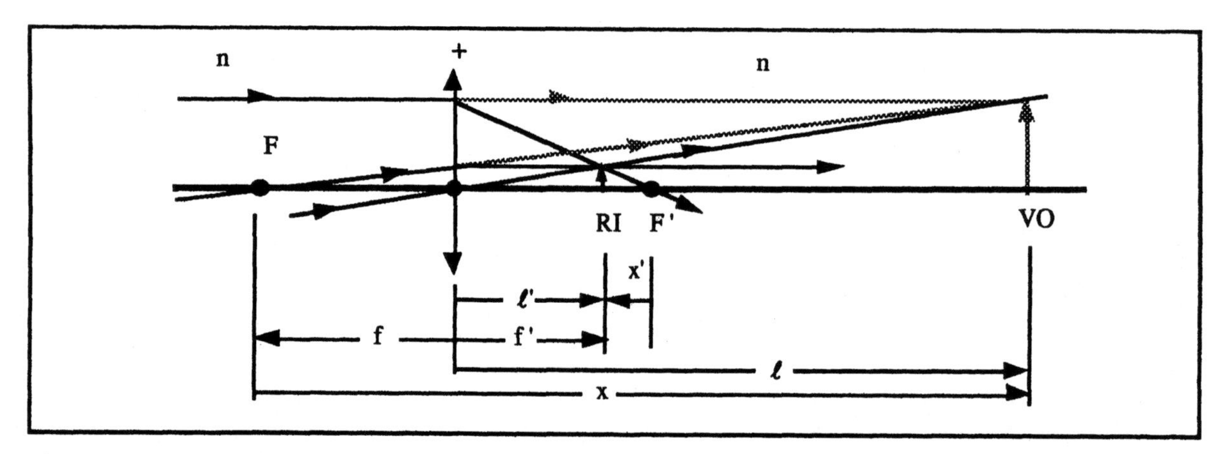

Figure 9-22. A virtual object formed with convergent incident rays. Aim the three incident rays toward the top of the virtual object using the ray tracing rules. A real, erect image is formed.

Negative Thin Lenses

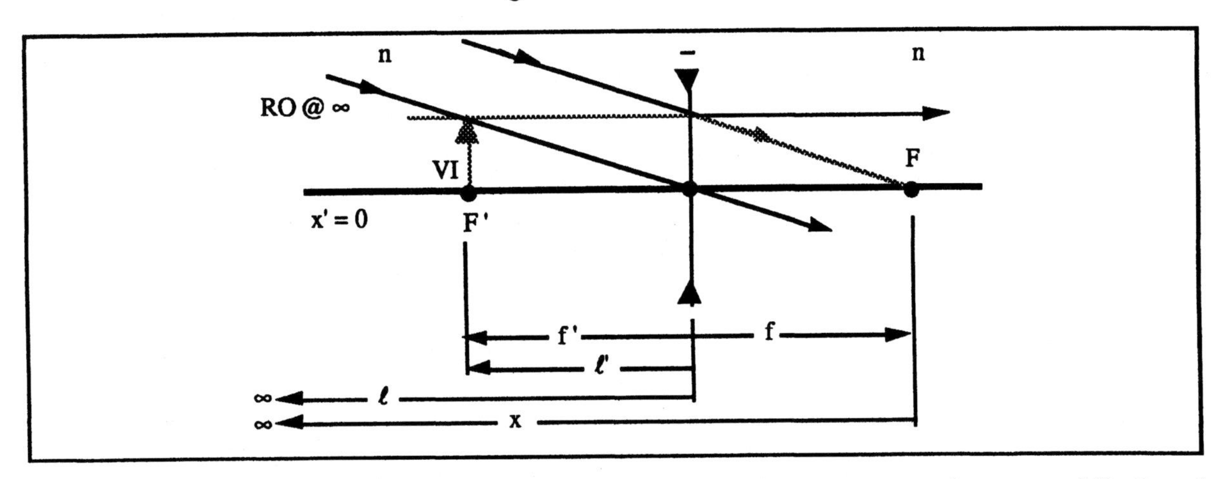

Figure 9-23. A real object is located at an infinite object position. Two rays are traced; one toward F refracted parallel to the axis; the other is the undeviated ray. A virtual, erect image is formed in the secondary focal plane.

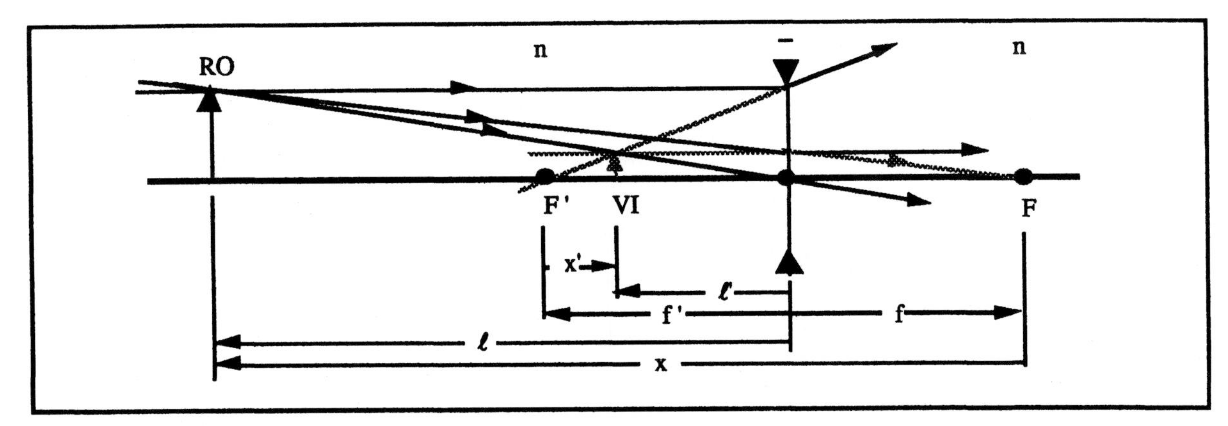

Figure 9-24. A real object is located at a finite position to the left of the negative lens. Three rays are drawn from the same off-axis object position using the ray tracing rules. A virtual, erect, minimized image is formed.

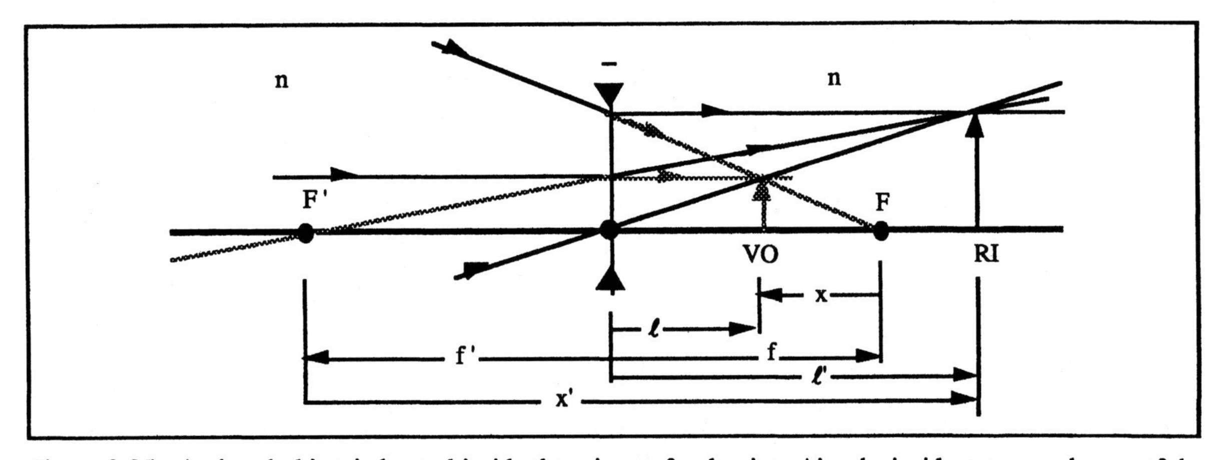

Figure 9-25. A virtual object is located inside the primary focal point. Aim the incident rays to the top of the virtual object. An erect, magnified, real image is formed.

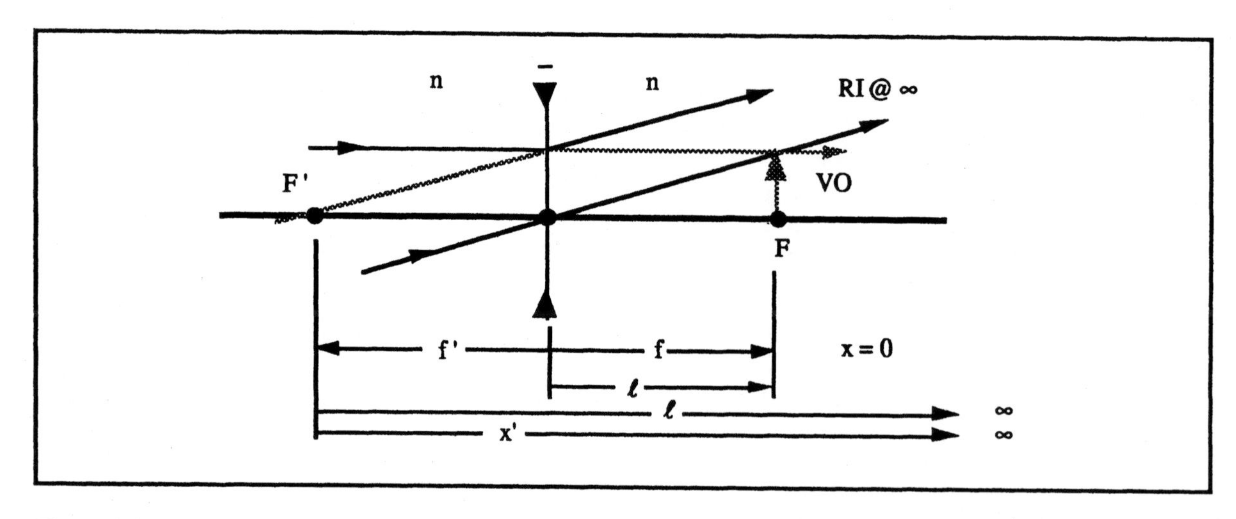

Figure 9-26. A virtual object is located at the primary focal point. An erect, real image is formed at infinity.

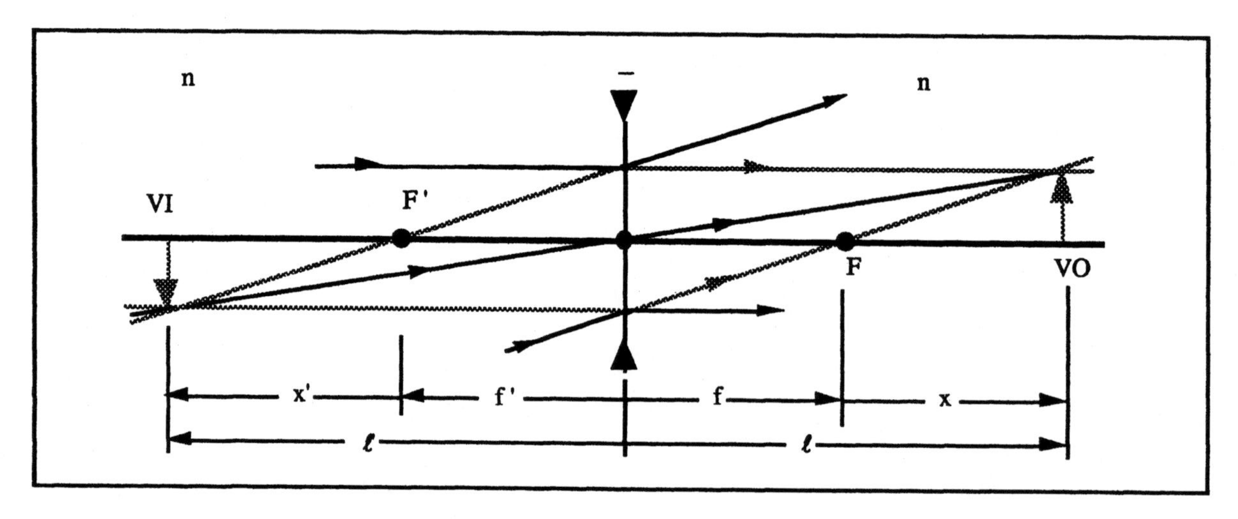

Figure 9-27. A virtual object is located outside (to the right of) the primary focal point. A virtual, inverted image is formed to the left of the lens. Dot the incident and emerging rays if they are extensions of the actual rays.

Curved Mirrors

For the curved mirror, the primary and secondary focal points coincide and are on the same side of the mirror as the center of curvature. A convex mirror has negative power, and a concave mirror has positive power. These surfaces are drawn as a vertical line, with ends representing the type of mirror. The power is also indicated by the sign. The region to the left of the mirror (in front) is *real space* for both objects and images; the region to the right of the mirror (behind) is *virtual space* for both objects and images. Rays in real space are solid, and rays in virtual space are dotted. All incident rays must represent the same extra-axial object position. The ray tracing rules can be summarized as follows:

1. An incident ray (solid line) parallel to the axis reflects through (or as though it came from) the secondary focal point.
2. A ray aimed toward (or passes through) the primary focal point reflects parallel to the optical axis.
3. The undeviated ray is aimed toward (or passes through) the center of curvature. This ray reflects on itself and leaves the surface after reflection along the same path as the incident ray.
4. A fourth ray may be drawn with curved mirrors using the law of reflection. The incident ray from the extra-axial object position is aimed toward the vertex of the mirror (intersection of the center of the mirror and the optical axis) and is reflected making the same angular subtense with the optical axis as the incident ray.

The position where the reflected rays cross represents the extra-axial image position. This may require extending divergent reflected rays (dotted) back into virtual space.

The following are some representative examples of ray tracing through curved mirrors.

Positive (Concave) Curved Mirrors

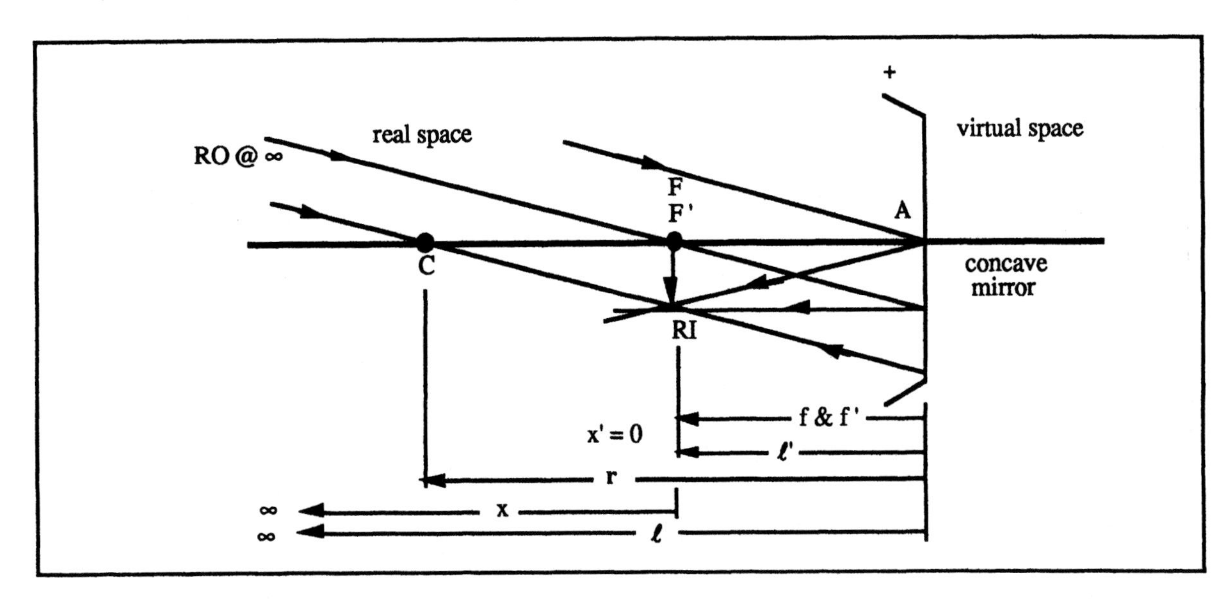

Figure 9-28. A real object is located at infinity. Three rays are drawn from the same off-axis object position (i.e., the rays are parallel to each other and oblique to the optical axis). One ray is directed through F and reflects parallel to the optical axis. Another is directed toward the center of curvature and reflects back on the same path. The third ray stikes the vertex of the mirror above the axis and leaves with the same angular subtense below the axis. A real, inverted image is formed in the secondary focal plane.

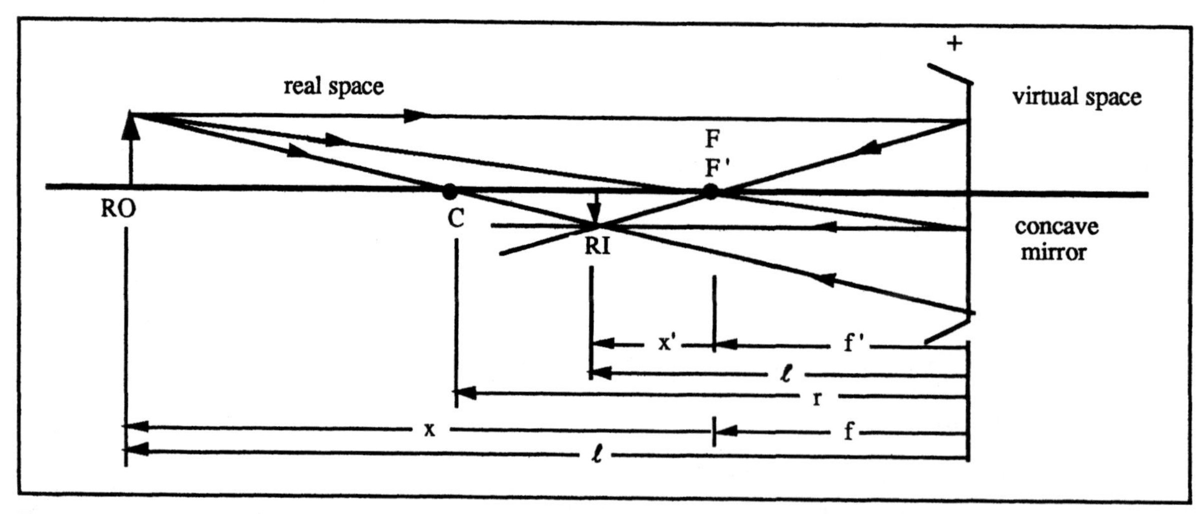

Figure 9-29. A real object is located at a finite position to the left of the primary focal point. Three rays are drawn from the same off-axis object position using the ray tracing rules. A real, inverted image is formed to the left of the secondary focal point.

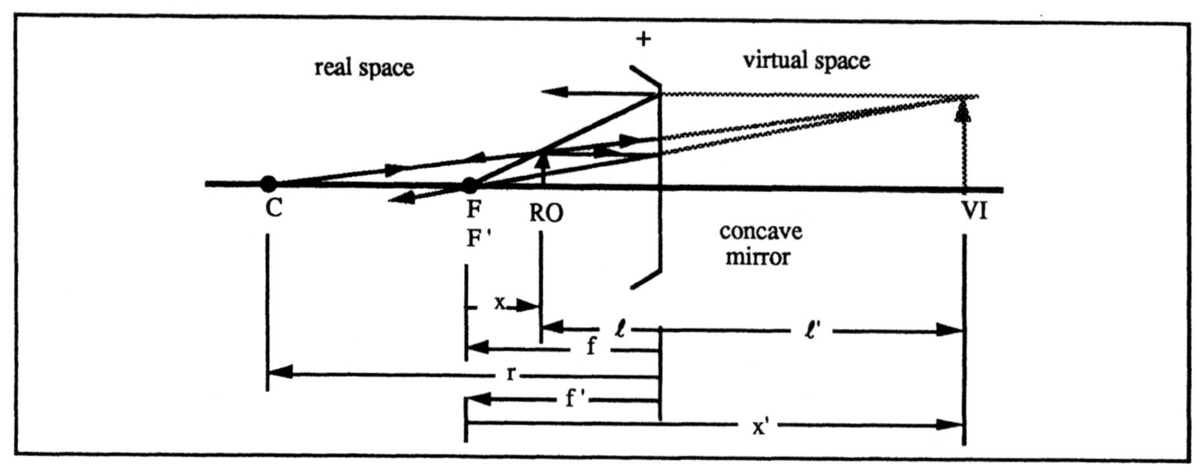

Figure 9-30. A real object is located inside the primary focal point. An erect, magnified, virtual image is formed to the right of the mirror. This is an example of a make-up mirror magnifier.

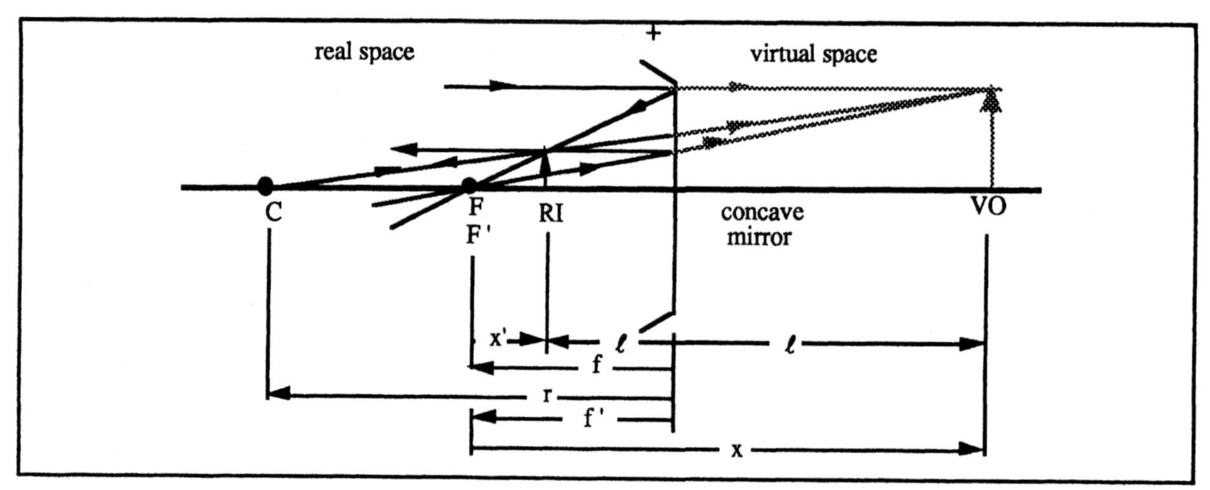

Figure 9-31. A virtual object is located to the right of the mirror. An erect, minified, virtual image is formed in real space (to the left of the mirror).

Negative (Convex) Curved Mirrors

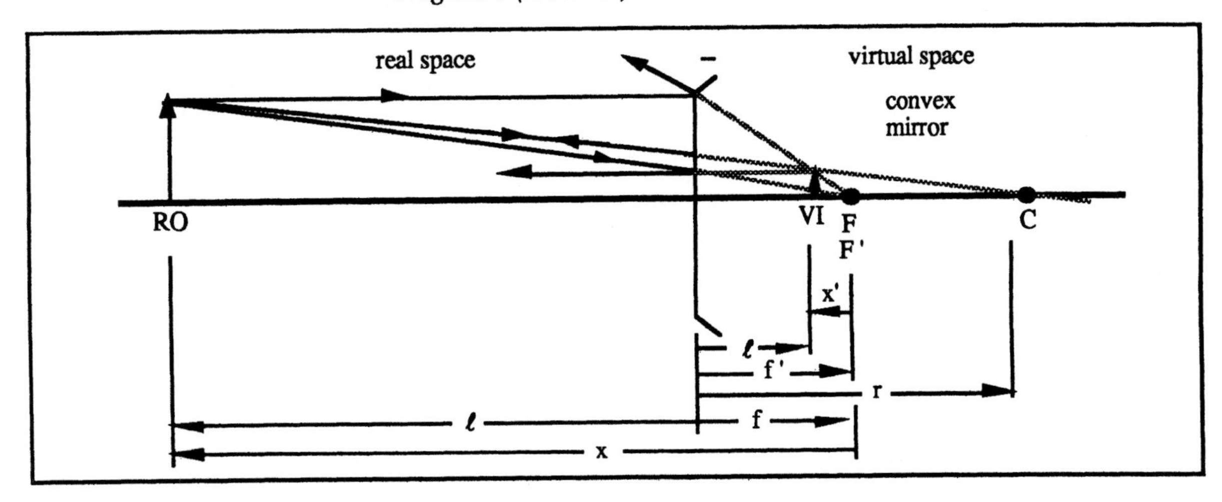

Figure 9-32. A real object is located at a finite position to the left of the concave mirror. A virtual, erect, minimized image is formed to the right of the mirror. This is the example of a outside car mirror.

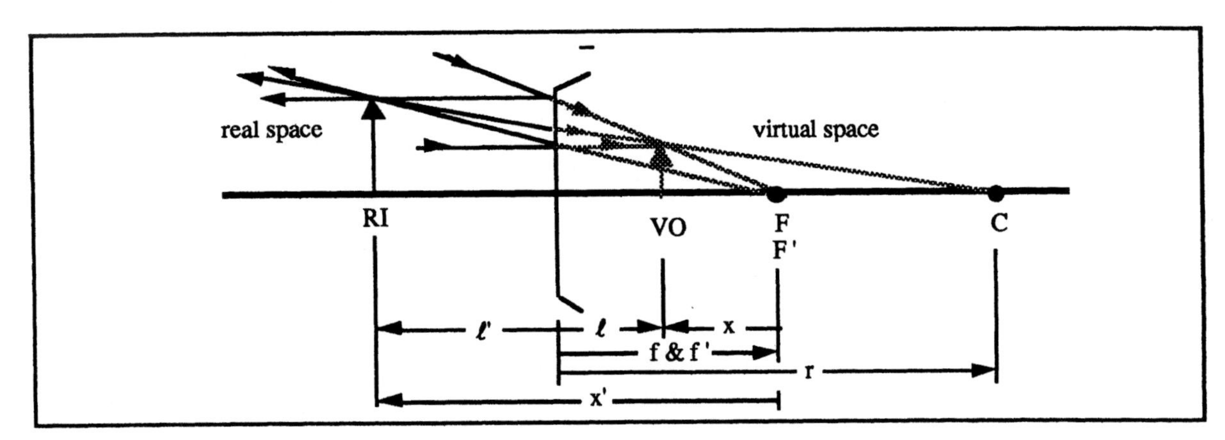

Figure 9-33. A virtual object is located inside the primary focal point. An erect, magnified, real image is formed.

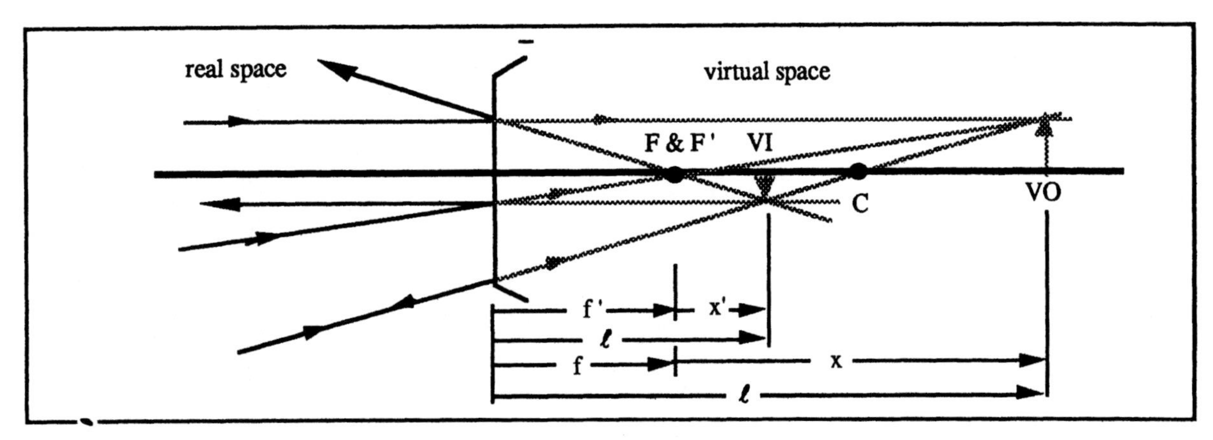

Figure 9-34. A virtual object is located outside (to the right of) the primary focal point. A virtual, inverted image is formed.

Multiple Lens Systems (Element-by-Element Method)

Ray tracing may be used to determine the final image location formed by a multiple lens system by using each image as the object for the next element. An image formed in front of the next element (divergent light incident) is treated as a real object; an image formed behind the next element (convergent light incident) is treated as a virtual object. Draw several diagrams and transfer each image to the next diagram. The lateral magnification of the system is the ratio of the final image size to the original object size measured off the diagram.

System 1: Two Positive Thin Lenses

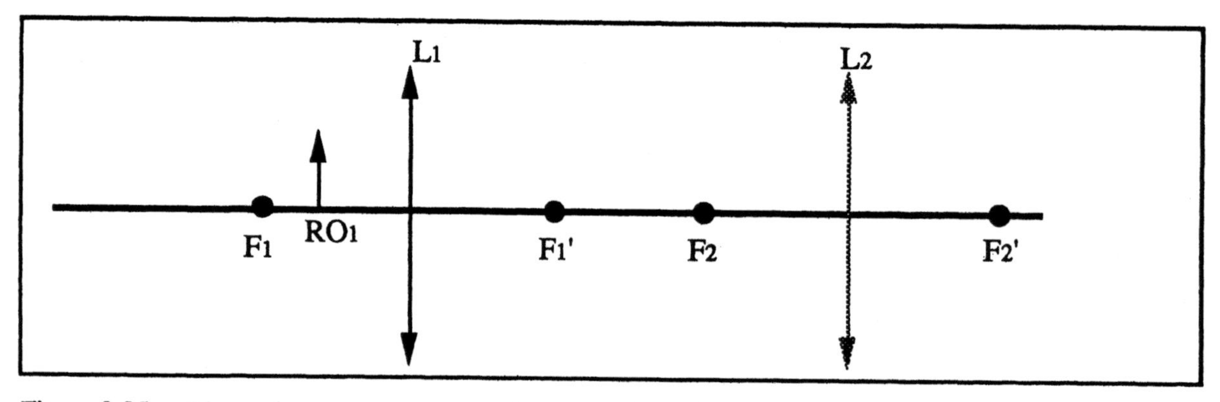

Figure 9-35a. The optical system consists of two positive thin lenses. A real object is located inside the primary focal point of the first lens

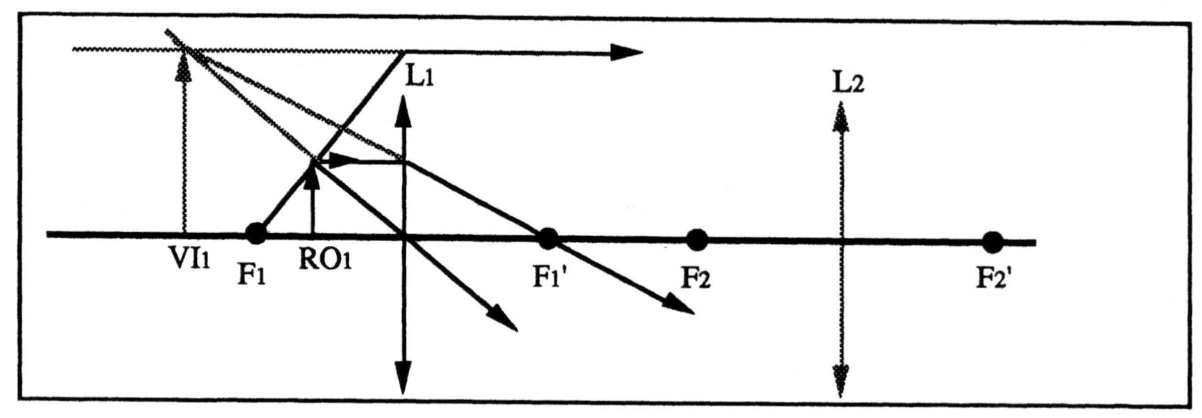

Figure 9-35b. Trace the rays to find the virtual, erect image formed by the first lens. Ignore the second lens.

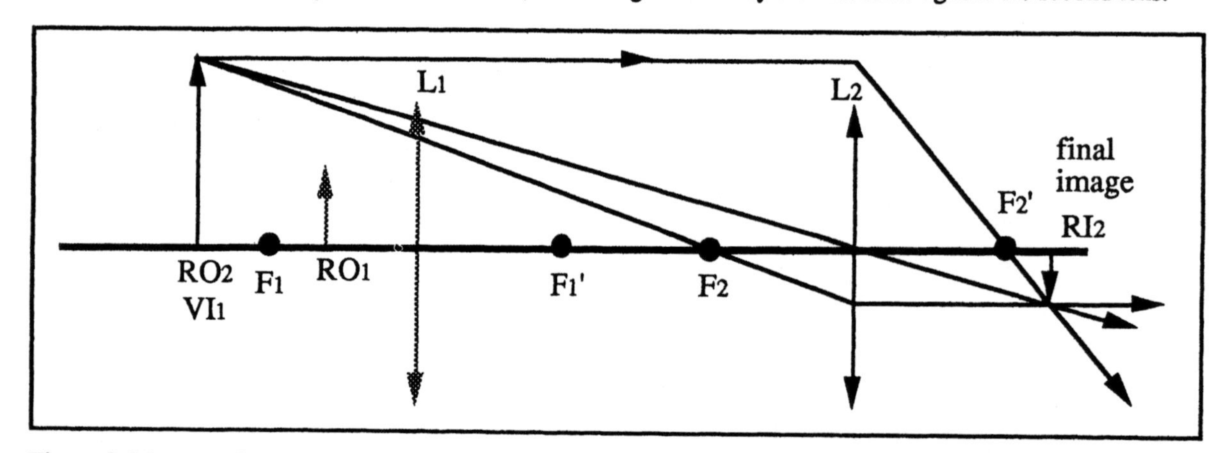

Figure 9-35c. The first image acts as a real object for the second lens because divergent light is incident. Trace the rays through the second lens ignoring the first lens. The final image is real and inverted.

System 2: Thin Lens and Single Refracting Surface (Eye-Lens Model)

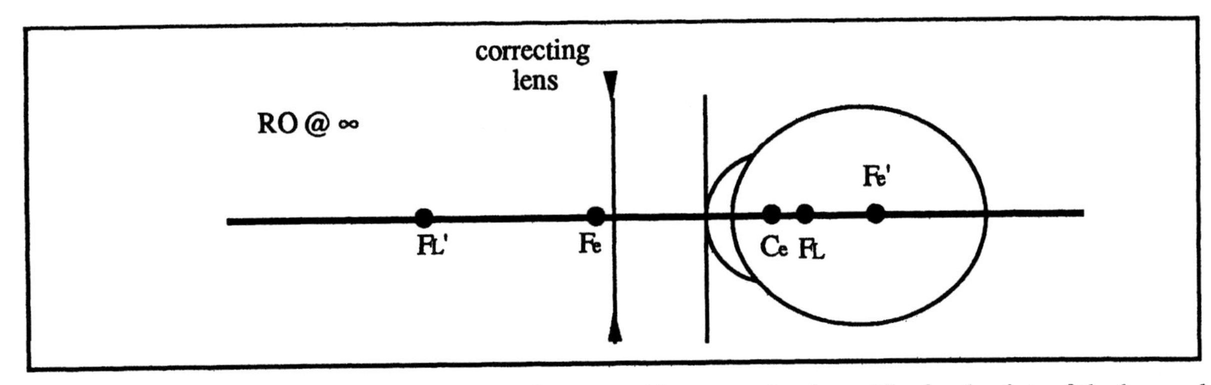

Figure 9-36a. An eye-lens system consisting of an eye with a correcting lens. The focal points of the lens and eye and the center of curvature of the cornea are labeled. The object is real and located at an infinite distance in front of the system.

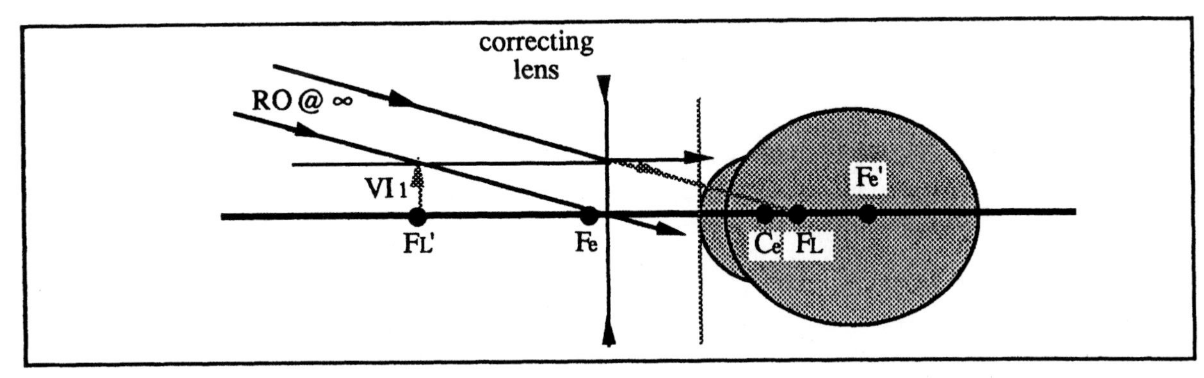

Figure 9-36b. Two rays are traced through the lens from an off-axis object position: the undeviated ray through the center of the lens and the ray that is aimed toward the primary focal point of the lens. A virtual image is found by dotting the diverging rays leaving the lens back to the left. This image acts as a real object for the eye.

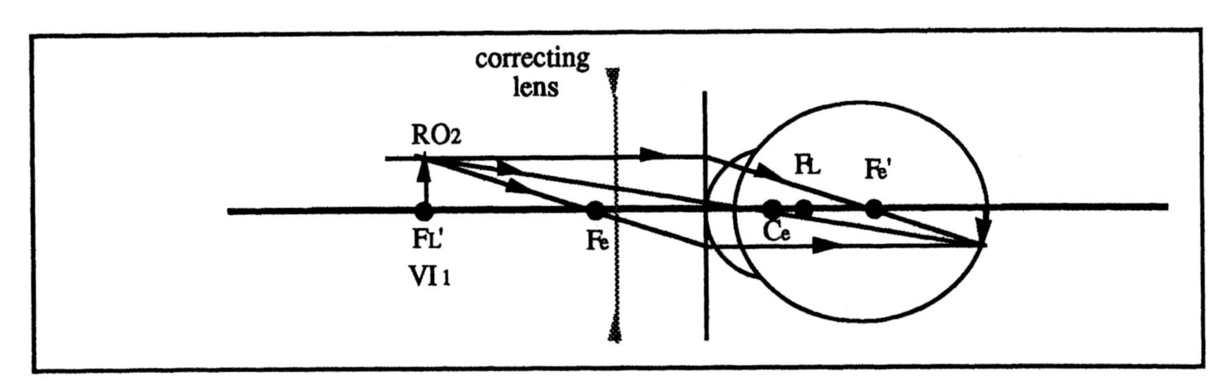

Figure 9-36c. The real object is traced through the single refracting surface (eye), and a real, inverted image is formed on the retina. This is how refractive correction of an eye works.

System 3: Thin Lens and Curved Mirror

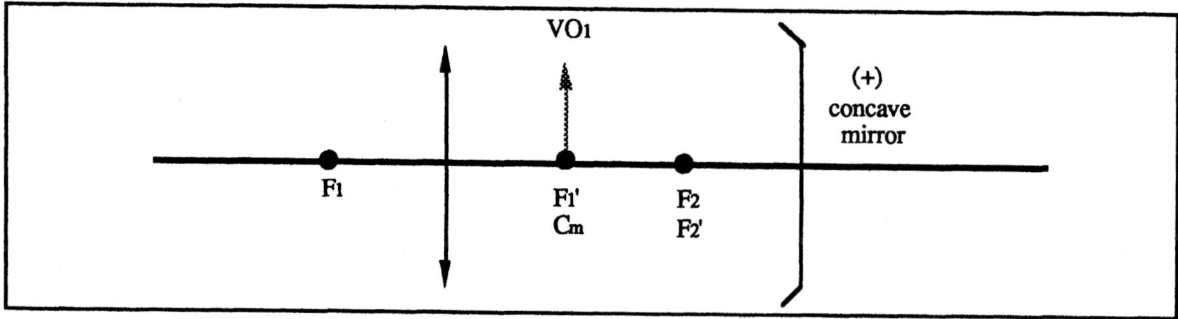

Figure 9-37a. The optical system consists of a lens and a concave mirror. The object is virtual.

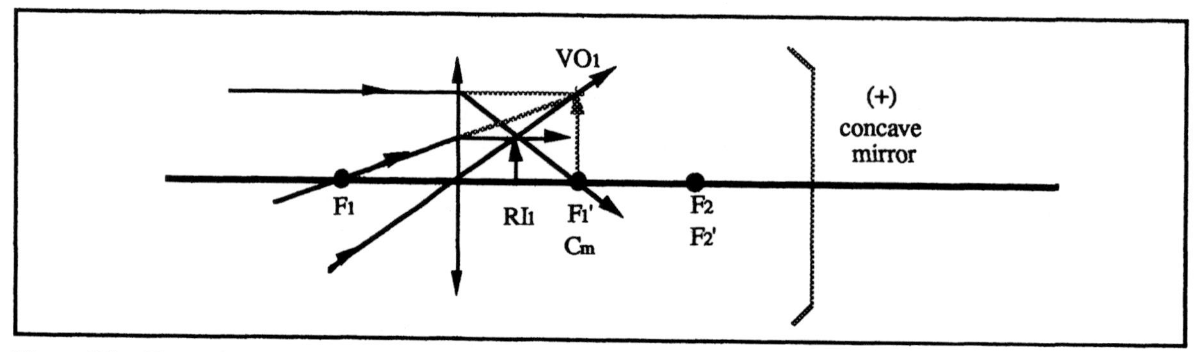

Figure 37b. Trace the rays through the lens to locate the real, erect image in front of the mirror.

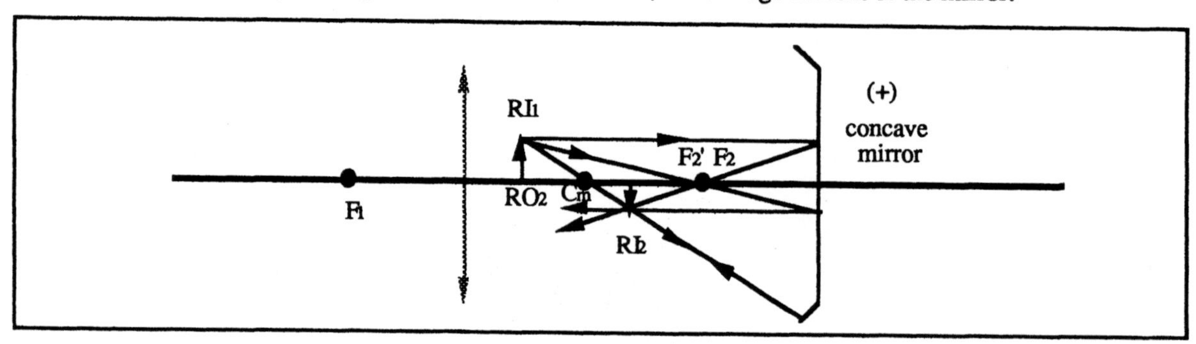

Figure 9-37c. The real image formed by the lens acts as a real object for the mirror. The reflected rays found by ray tracing form a real, inverted image. The rays continue through the system.

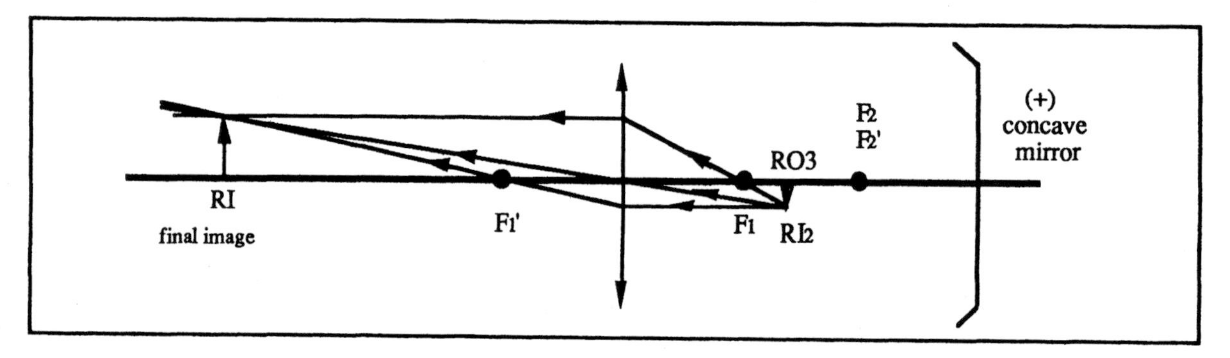

Figure 9-37d. The divergent rays from the mirror act as if they came from a real object. The rays enter the lens backward. Switch F and F' (turn the paper upside down), and trace the rays using the same rules for lenses. A real, inverted image is found in front of the lens. When compared to the original object, the system formed an erect image.

Thick Lenses (Principal Planes and Nodal Points)

In this section, ray tracing using the cardinal points is demonstrated. The location of the principal planes (H and H'), the nodal points (N and N'), the primary and secondary focal points (F and F'), and the object location relative to the first principal plane must be known and labeled on the diagram. It is assumed that the location of these points will be found with the techniques and calculations described in Chapter 7.

Although the rules for ray tracing are followed, the rays will be drawn differently. The location of the lens surface need not be known because it will not be used in the procedure. All rays represent the same extra-axial object position. The rules are outlined below:

1. The ray leaving the object parallel to the optical axis is extended to the first principal plane (H). Because the principal planes are of unit magnification, this parallel ray is dotted to the second principal plane (H'). The ray leaves the second principal plane aimed toward (or diverging from) the secondary focal point.

2. The ray leaving the object aimed toward the primary focal point is extended to the first principal plane. A dotted line parallel to the axis is drawn from the first principal plane to the second principal plane. The ray leaves the second principal plane parallel to the axis.

3. The undeviated ray is aimed toward the first nodal point (N) and stops at the first principal plane. A line is dotted to the second principal plane at the height of the incident ray. The ray leaves the second principal plane aimed toward (or coming from) the second nodal point (N'). The emergent ray is made parallel to the incident ray (both rays form the same angle with the optic axis). If the surrounding media are the same, the nodal points coincide with the principal plane at the optical axis. If the media differ, the location of the nodal points may be found graphically by measuring the focal lengths and summing:

$$HN = H'N' = f + f'$$

This indicates that the distance from the first principal plane to the first nodal point is equal to the distance from the second principal plane to the second nodal point and to the sum of the focal lengths. Remember, the sign of the focal length is important. If HN or H'N' is negative, the nodal points are located to the left of the principal planes; if positive, they are located to the right; if zero, the nodal points coincide with the principal planes.

Thick Lens System 1: Positive Power System

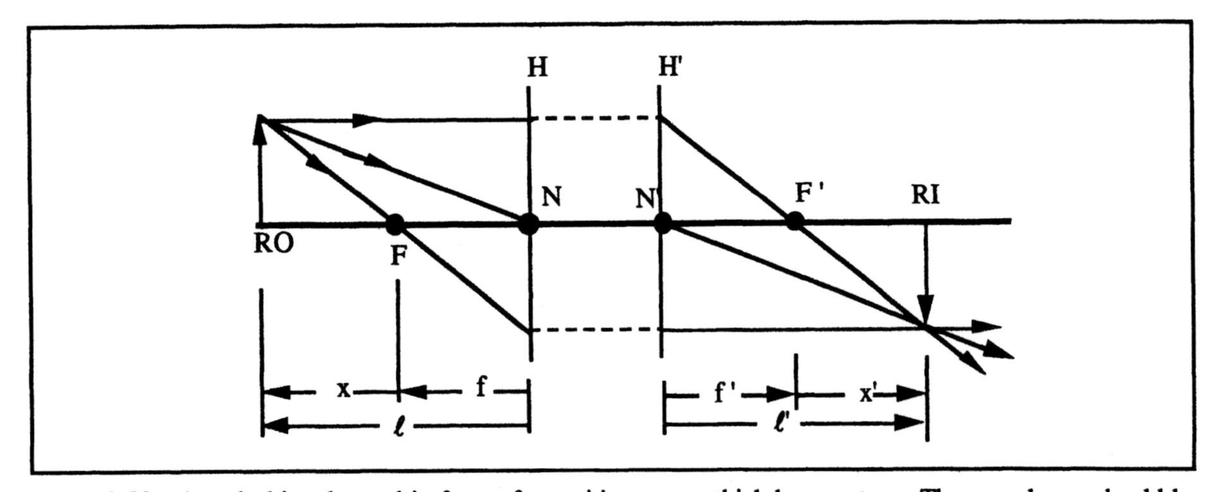

Figure 9-38. A real object located in front of a positive power thick lens system. The rays drawn should be self-explanatory. A real, inverted image is formed. If the location relative to the back surface of the last element is required, the surfaces of the elements that make up the system should be drawn.

Thick Lens System 2: Negative Power System

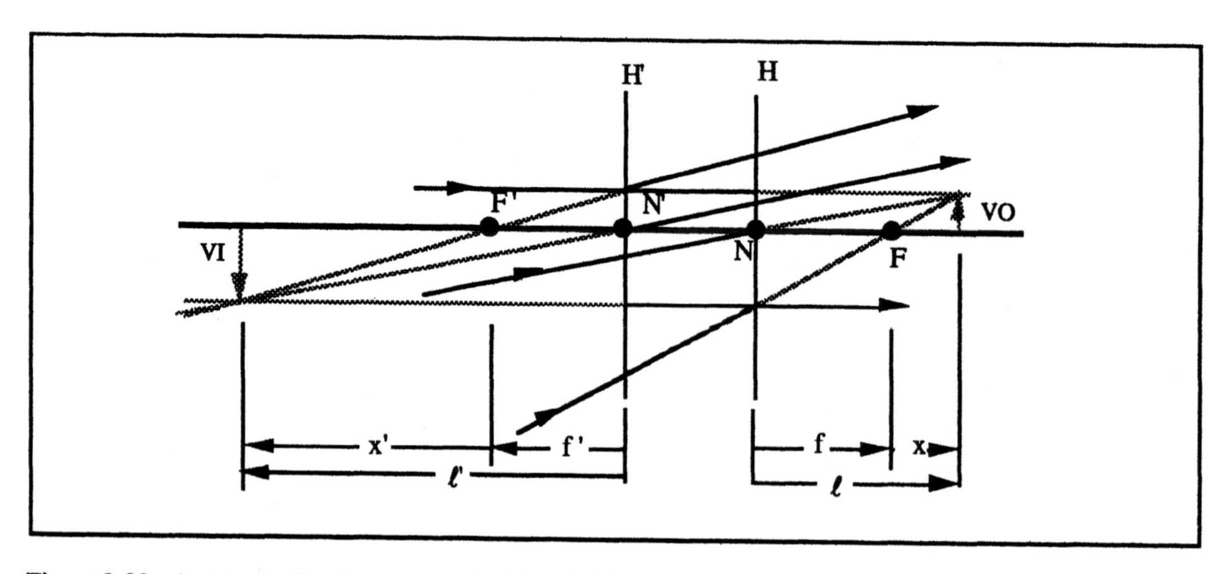

Figure 9-39. A virtual object (convergent incident light) is imaged through a negative power thick lens system. The principal planes shown here are reversed, but this is not the rule with negative power lenses. An inverted virtual image is formed.

Thick Lens System 3: Nodal Points Do Not Coincide with Principal Planes

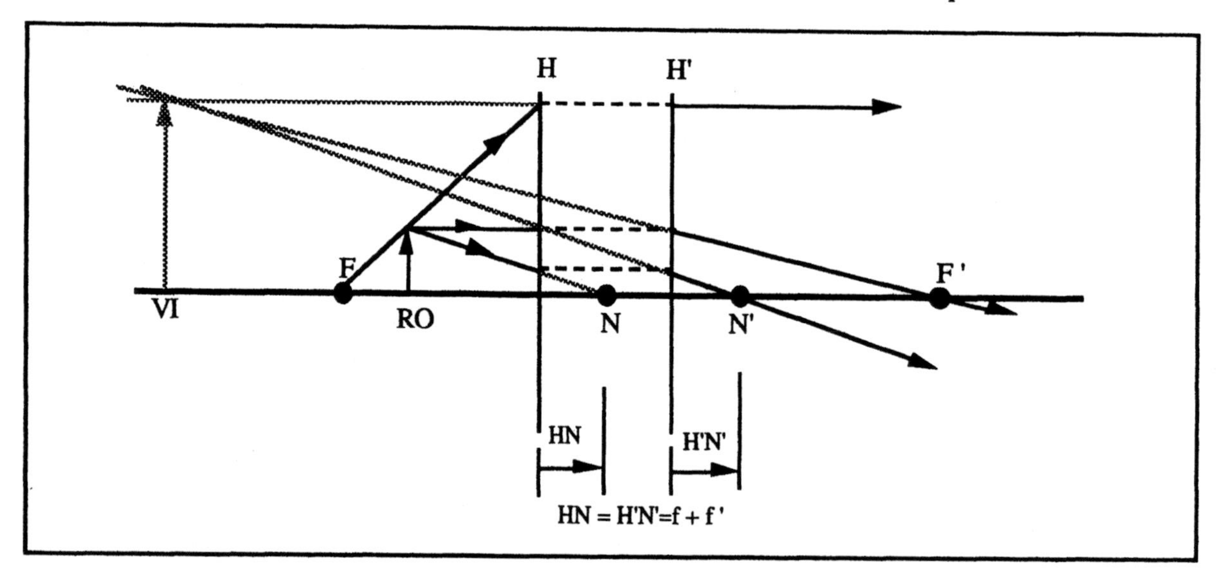

Figure 9-40. A system in which the medium in object space differs from that in image space. The focal lengths do not have the same magnitude (f = – 27 mm and f' = 37 mm), and the nodal points are shifted from the principal planes to the right. The sum of the focal lengths is equal to the distance from the principal plane to the respective nodal point (HN = H'N' = f + f' = – 27 + 37 = + 10 mm). A real object inside the primary focal point yields a virtual, erect image.

Supplemental Problems

Answers for the ray tracing problems are not given, however, you can check your answers by comparing the rays to those traced in the examples.

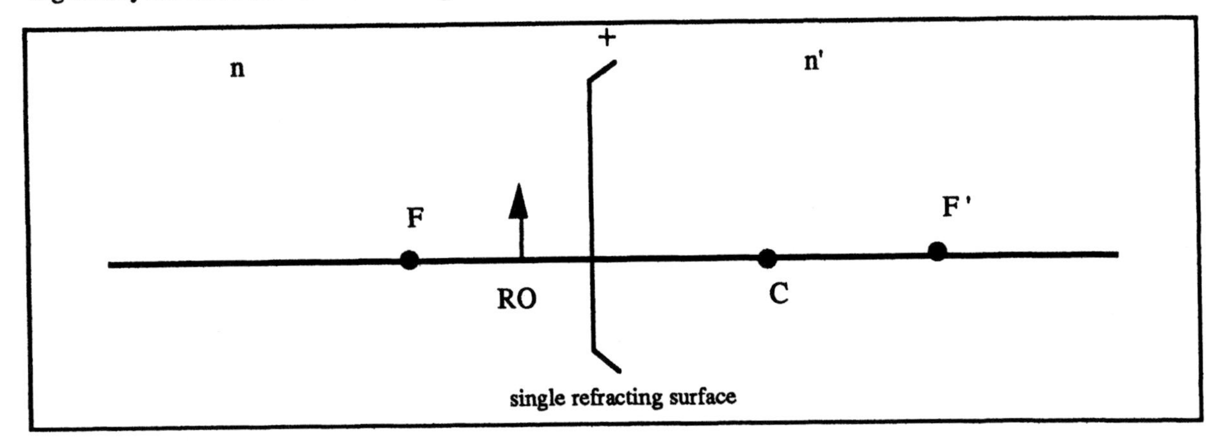

9-1. Locate the image by tracing three rays. Label the diagram.

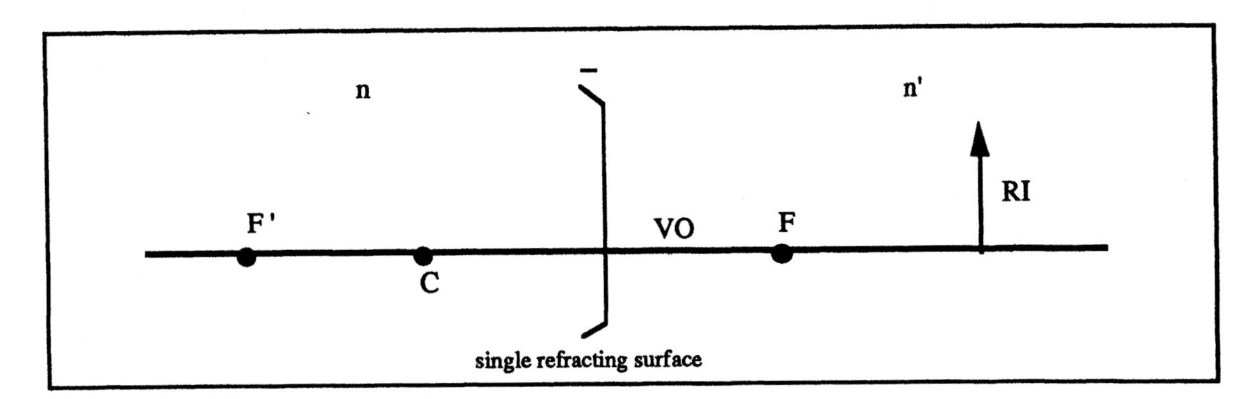

9-2. Locate the object by tracing three rays. Think about how the rays leave. Label the diagram.

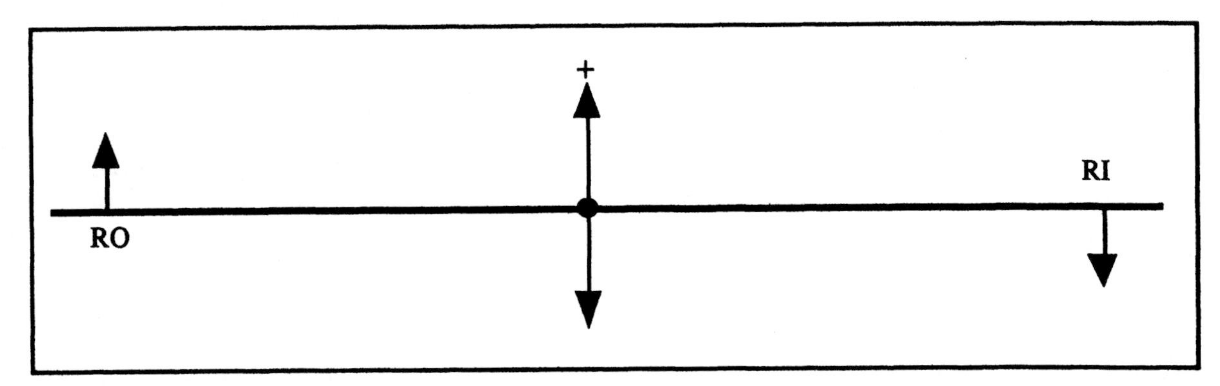

9-3. Locate the focal points by tracing three rays. What is the lateral magnification from the diagram?

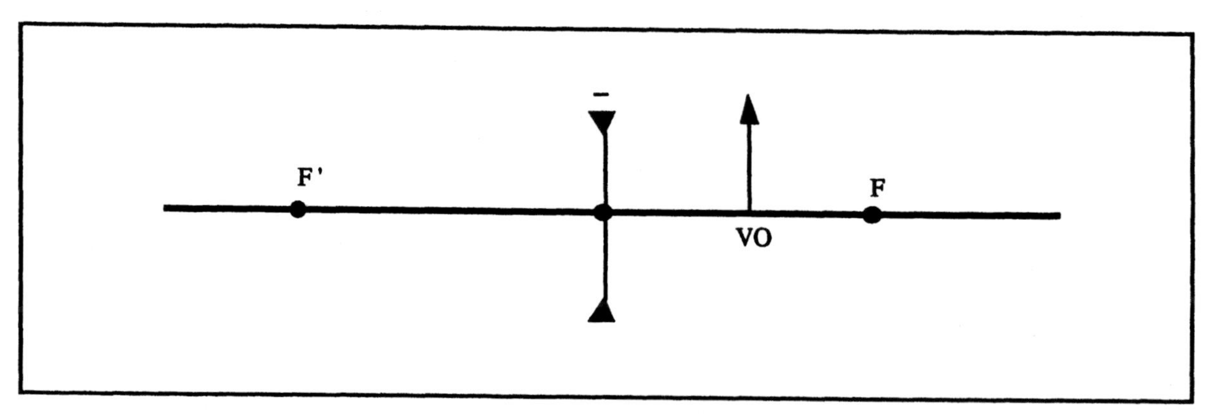

9-4. Find the image location by tracing three rays. Label the diagram.

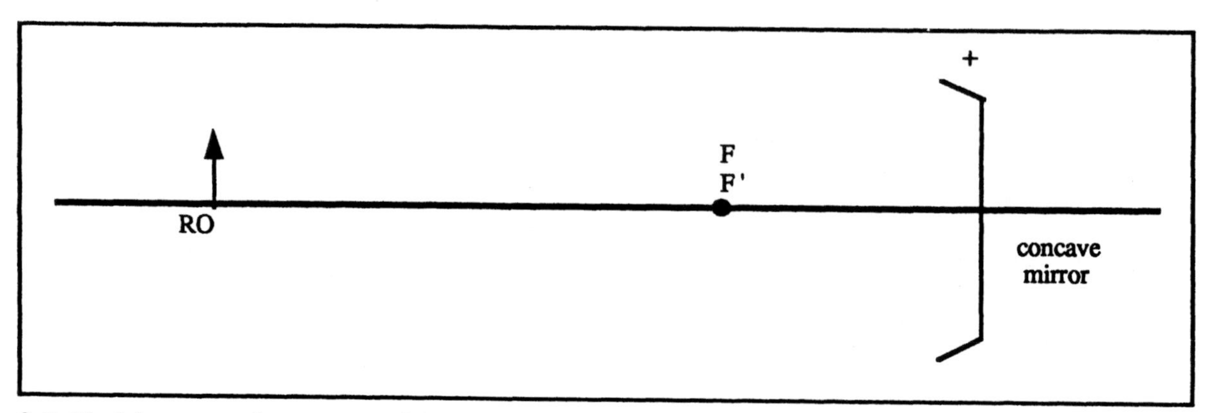

9-5. Find the center of curvature and the image location by tracing four rays. Label the diagram.

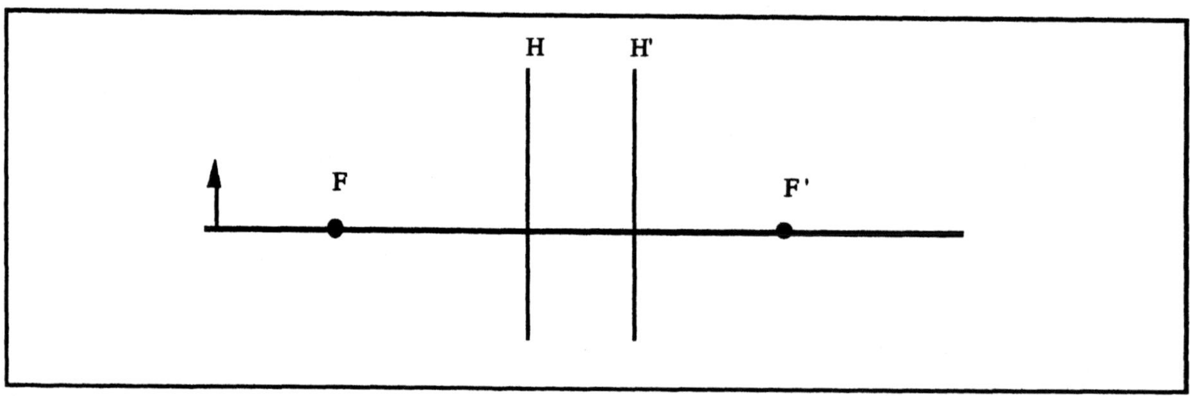

9-6. Find the nodal points and the image location by tracing three rays. Note that the focal lengths do not have equal magnitude.

9-7. An object is located 40 cm from a + 5.00 D lens. Set up a diagram and trace the rays to locate the image. What is the lateral magnification ?
ANS. Image location = + 40 cm; lateral magnification = – 1.00 (Do this with ray tracing)

Chapter 10

Summary for Quick Reference

This chapter summarizes the important equations used in geometrical optics. For more details, go to the chapter that covers the material.

Basic Relationships

v	= velocity (meters per second)
c	= velocity of light in vacuum (3×10^8 m/s)
f	= frequency (cycles per second)
λ	= wavelength
V_D	= vergence (diopters)
R_D	= curvature (diopters)
r_m	= radius (meters)
n	= index of medium in front of refracting surface
n'	= index of medium behind refracting surface
θ	= incident angle at refracting surface
θ'	= refracted angle at refracting surface
θ_i	= incident angle at plane surface
θ_r	= reflected angle at plane surface
θ_c	= critical angle
t	= time (seconds)
t	= thickness
d	= dis tan ce
d	= displacement
ℓ	= object distance or thickness of medium
ℓ'	= image distance

Velocity, Frequency, and Wavelength of Electromagnetic Waves

$$v = f\lambda \qquad (1\text{-}1)$$

Sign Convention

Light travels left to right; measurements in the direction of light travel are positive; measurements in the direction opposite light travel are negative. Measurements are from the wavefront to the source or from the wavefront to the image.

Vergence

$$V_D = R_D = \frac{1}{r_m} \quad (1\text{-}2)$$

Index of Refraction

$$n = \frac{\text{velocity in vacuum}}{\text{velocity in medium}} = \frac{c}{v_m} = \frac{3\text{x}10^8 \text{ m/s}}{v_m} \quad (2\text{-}2)$$

Law of Reflection

$$\theta_i = \theta_r \quad (2\text{-}3)$$

Fresnel's Law of Reflection

$$\%\text{Reflected} = \left(\frac{n'-n}{n'+n}\right)^2 \text{x}\,100 \quad (2\text{-}4)$$

Snell's Law (Law of Refraction)

$$n \sin\theta = n' \sin\theta' \quad (2\text{-}5)$$

Critical Angle - Total Internal Reflection

$$n\sin\theta = n'\sin\theta' \quad \text{or} \quad n\sin\theta_c = n'\sin 90°$$

$$\sin\theta_c = \frac{n'}{n} \quad (2\text{-}6)$$

$$\theta_c = \sin^{-1}\left(\frac{n'}{n}\right)$$

Fermat's Principle (Principle of Least Time)

$$t = \frac{nd}{c} = \frac{(c/v)d}{c} = \frac{cd}{cv} = \frac{d}{v} \quad (2\text{-}7)$$

Time to Travel through Several Media

$$t = \frac{1}{c}\sum_{i=1}^{m} n_i d_i = \frac{1}{c}\{n_1 d_1 + n_2 d_2 + n_3 d_3 + \ldots n_m d_m\} \quad (2\text{-}8)$$

Displacement through a Parallel-Sided Element

$$d = (t\cos\theta'_1)\tan\{\theta_1 - \theta'_1\} \quad (2\text{-}11)$$

Apparent Position

$$\frac{n}{\ell} = \frac{n'}{\ell'} \tag{2-12}$$

Apparent Longitudinal Displacement

$$d = \frac{t(n'-n)}{n'} \tag{2-14}$$

Reduced Distance

$$\text{reduced distance} = \frac{t}{n} = \frac{\ell}{n} \tag{2-15}$$

Apparent Position Viewed Through Several Media

$$\frac{\ell'}{n'} = \sum_{i=1}^{m} \frac{\ell_i}{n_i} = \frac{\ell_1}{n_1} + \frac{\ell_2}{n_2} + \frac{\ell_3}{n_3} \ldots + \frac{\ell_m}{n_m} \tag{2-16}$$

NOTES AND YOUR OWN EQUATIONS

Prisms

n_1	= index of medium in front of (left of) first refracting surface
n_2	= index of medium behind (right of) second refracting surface
n_p	= index of prism
n_s	= index of surrounding medium
θ_1	= incident angle at first refracting surface
θ'_1	= refracted angle at first refracting surface
θ_2	= incident angle at second refracting surface
θ'_2	= refracted angle at second refracting surface
γ	= apical or refracting angle of prism
θ_c	= critical angle of material
δ	= deviation of prism
δ_m	= minimum deviation of prism
δ^Δ	= deviation of prism in prism diopters

Refraction through a Prism

$$n_1 \sin\theta_1 = n_p \sin\theta'_1 \quad (3\text{-}1a)$$

$$\gamma = \theta_1 + \theta'_2 \quad (3\text{-}2)$$

$$n_p \sin\theta_2 = n_3 \sin\theta'_2 \quad (3\text{-}3a)$$

Deviation of a Prism

$$\delta = \theta_1 + \theta'_2 - \gamma \quad (3\text{-}7)$$

Limitations on Refraction through a Prism

$$\gamma > 2\theta_c \quad (3\text{-}9)$$

Minimum Deviation of a Prism

$$\theta'_1 = \theta_2 = \frac{\gamma}{2} \quad (3\text{-}12)$$

$$\theta_1 = \theta'_2 = \frac{\delta_m + \gamma}{2} \quad (3\text{-}13)$$

Deviation of an Ophthalmic Prism

$$\delta = \left(\frac{n_p}{n_s} - 1\right)\gamma \qquad (3\text{-}15)$$

Deviation of an Ophthalmic Prism Surrounded by Air

$$\delta = (n_p - 1)\gamma \qquad (3\text{-}16)$$

Deviation of an Ophthalmic Prism in Prism Diopters

$$\delta^{\Delta} = 100(n_p - 1)\tan\gamma^{\circ} = 100(n_p - 1)\gamma^{rad} \qquad (3\text{-}19)$$

NOTES AND YOUR OWN EQUATIONS

Curved Surfaces and Single Refracting Surfaces

n	= index of medium to the left of refracting surface
n'	= index of medium to the right of refracting surface
r	= radius of surface
s	= sag of curved surface
y	= half of the chord of the surface
F	= power of the surface in diopters
ℓ	= object distance
ℓ'	= image distance
L	= incident (object) reduced vergence
L'	= emergent (image) reduced vergence
f	= primary focal length
f'	= secondary focal length
x	= object extrafocal distance
x'	= image extrafocal distance
h	= height of object
h'	= height of image
LM	= lateral magnification

Sag-Radius Relationship

$$r = \frac{y^2}{2s} + \frac{s}{2} \qquad (4\text{-}1)$$

if r << s then

$$r = \frac{y^2}{2s} \qquad (4\text{-}2)$$

Power of a Curved Surface

$$F = \frac{n' - n}{r} \qquad (4\text{-}4)$$

Reduced Object Vergence or Incident Vergence

$$L = \frac{n}{\ell} \qquad (4\text{-}5)$$

Reduced Image Vergence or Emergent Vergence

$$L' = \frac{n'}{\ell'} \qquad (4\text{-}6)$$

Gaussian Imaging Equation

$$L' = L + F \qquad (4\text{-}7)$$

Primary and Secondary Focal Length and Power

$$F = -\frac{n}{f} = \frac{n'}{f'} \tag{4-10}$$

Lateral Magnification for Single Refracting Surface

$$LM = \frac{h'}{h} = \frac{\ell' - r}{\ell - r} = \frac{n\ell'}{n'\ell} = -\frac{f}{x} = -\frac{x'}{f'}$$

(4-11)
(4-12)
(4-15)
(4-18)
(4-19)

Newton's Relationship for Single Refracting Surfaces

$$xx' = ff' \tag{4-20}$$

NOTES AND YOUR OWN EQUATIONS

Thin Lenses

See Single Refracting Surfaces for symbols

n_L = index of lens

n_s = index of surrounding medium

r_1 = radius of first surface

r_2 = radius of second surface

F_1 = power of first surface

F_2 = power of second surface

F = power of the lens in diopters

F_x = effective power

F_c = correcting power

ω = apparent size of angle that distant object makes with axis

δ^{Δ} = prismatic effect of lens

y = distance from axis

Thin Lens Power

$$F = F_1 + F_2 \qquad (5\text{-}3)$$

Lens Makers Formula

$$F = (n_L - n_s)\left(\frac{1}{r_1} - \frac{1}{r_2}\right) \qquad (5\text{-}4a)$$

Lateral Magnification for Lenses

$$LM = \frac{h'}{h} = \frac{\ell'}{\ell} = -\frac{f}{x} = -\frac{x'}{f'} \qquad (5\text{-}10)\ (5\text{-}11)$$

Newton's Relationship for Thin Lenses

$$xx' = -(f')^2 \qquad (5\text{-}14)$$

Size of Distant Object for Single Refracting Surfaces or Lenses

$$h' = -f' \tan\omega \qquad (5\text{-}18)$$

Prismatic Effect of Lenses

$$\delta^{\Delta} = \frac{y_{cm}}{f'_m} = y_{cm}F_D \qquad (5\text{-}20)$$

Effective Power

$$F_x = \frac{F}{1 - dF} \qquad (5\text{-}21)$$

Correcting Lens Power with Displacement

$$F_c = \frac{F}{1 + dF} \qquad (5\text{-}22)$$

NOTES AND YOUR OWN EQUATIONS

Cylinders and Spherocylinders

F_θ	= power between principal meridians
F_s	= power of the sphere
F_c	= power of the cylinder
L'_c	= emerging vergence toward circle of least confusion
L'_{p1}	= emerging vergence of first principal meridian
L'_{p2}	= emerging vergence of second principal meridian
ℓ'_c	= image position of circle of least confusion

Power Between Principal Meridians of a Spherocylindrical Lens

$$F_\alpha = F_s + F_c \sin^2\theta \tag{6-8}$$

Average Vergence Leaving a Spherocylindrical Lens - Circle of Least Confusion

$$L'_c = \frac{L'_{p1} + L'_{p2}}{2} \tag{6-9}$$

Location of the Circle of Least Confusion

$$\ell'_c = \frac{1}{L'_c} = \frac{2}{L'_{p1} + L'_{p2}} \tag{6-10}$$

NOTES AND YOUR OWN EQUATIONS

Thick Lenses and Lens Systems

n_1 = index to left of first surface or lens
n_2 = index of lens or medium between lenses
n_3 = index to right of last surface or lens
F_1 = power of first surface
F_2 = power of second surface
F = equivalent power of the lens in diopters
F_v = back vertex power
F_n = neutralizing power
ffl = f_n = front focal length
bfl = f'_v = back focal length
H = first principal plane
H' = second principal plane
N = first nodal point
N' = second nodal point
O = optical center

Back Vertex Power

$$F_v = \frac{F_1}{1 - \left(\frac{t}{n_2}\right)F_1} + F_2 \qquad (7\text{-}12)$$

Neutralizing Power

$$F_n = \frac{F_2}{1 - \left(\frac{t}{n_2}\right)F_2} + F_1 \qquad (7\text{-}13)$$

Equivalent Power

$$F = F_1 + F_2 - \left(\frac{t}{n_2}\right)F_1F_2 \qquad (7\text{-}16)$$

Primary and Secondary Focal Length

$$F = -\frac{n_1}{f} = -\frac{n_3}{f'} \qquad (7\text{-}17)$$

Back Focal Length

$$F_v = \frac{n_3}{bfl} = \frac{n_3}{f'_v} \tag{7-11}$$

Front Focal Length

$$F_n = -\frac{n_1}{ffl} = -\frac{n_1}{f_n} \tag{7-10}$$

Principal Plane Location

$$A_1H = n_1\left(\frac{t}{n_2}\right)\left(\frac{F_2}{F}\right) \qquad A_2H' = -n_3\left(\frac{t}{n_2}\right)\left(\frac{F_1}{F}\right) \qquad \text{(7-14)} \; \text{(7-15)}$$

Optical Center

$$A_1O = \frac{r_1 t}{r_1 - r_2} \tag{7-18}$$

Principal Plane - Nodal Point Location

$$HN = H'N' = f + f' \tag{7-19}$$

$$FN = f' \qquad F'N' = f \tag{7-20}$$

$$HH' = NN' \tag{7-21}$$

NOTES AND YOUR OWN EQUATIONS

Plane and Curved Mirrors

See Single Refracting Surfaces for symbols
δ = deviation
θ_i = incident angle
θ_r = reflected angle
α = angle between two inclined mirrors
F_{LM} = power of lens mirror
F_m = power of mirror

Deviation of Reflected Rays at a Plane Mirror

$$\delta = 180 - 2\theta_i = 180 - 2\theta_r \tag{8-1}$$

Two Mirror Deviation

$$\delta = 2(180^\circ - \alpha) \tag{8-2}$$

Power of a Curved Mirror

$$F = \frac{-2n}{r} \tag{8-4}$$

Reduced Object Vergence - Incident Vergence for a Curved Mirror

$$L = \frac{n}{\ell} \tag{8-6}$$

Reduced Image Vergence - Emergent Vergence for a Curved Mirror

$$L' = \frac{-n}{\ell'} \tag{8-7}$$

Primary and Secondary Focal Length of a Curved Mirror

$$\frac{r}{2} = f = f' \tag{8-9}$$

Lateral Magnification of a Curved Mirror

$$LM = \frac{h'}{h} = -\frac{\ell'}{\ell} = -\frac{f}{x} = -\frac{x'}{f'} \qquad (8\text{-}10)\ (8\text{-}11)$$

Power of a Lens Mirror

$$F_{lm} = 2F_l + F_m \tag{8-12}$$

NOTES AND YOUR OWN EQUATIONS

Index